Nanomaterials in the Wet Processing of Textiles

Scrivener Publishing
100 Cummings Center, Suite 541J
Beverly, MA 01915-6106

Advanced Materials Series

The Advanced Materials Series provides recent advancements of the fascinating field of advanced materials science and technology, particularly in the area of structure, synthesis and processing, characterization, advanced-state properties, and applications. The volumes will cover theoretical and experimental approaches of molecular device materials, biomimetic materials, hybrid-type composite materials, functionalized polymers, supramolecular systems, information- and energy-transfer materials, biobased and biodegradable or environmental friendly materials. Each volume will be devoted to one broad subject and the multidisciplinary aspects will be drawn out in full.

Series Editor: Ashutosh Tiwari
Institute of Advanced Materials
SE-58330 Linköping
Sweden
E-mail: director@iaam.se

Publishers at Scrivener
Martin Scrivener (martin@scrivenerpublishing.com)
Phillip Carmical (pcarmical@scrivenerpublishing.com)

Nanomaterials in the Wet Processing of Textiles

Edited by

Shahid-ul-Islam and B.S. Butola

Scrivener Publishing

WILEY

This edition first published 2018 by John Wiley & Sons, Inc., 111 River Street, Hoboken, NJ 07030, USA and Scrivener Publishing LLC, 100 Cummings Center, Suite 541J, Beverly, MA 01915, USA

For more information about Scrivener publications please visit www.scrivenerpublishing.com.

Wiley Global Headquarters
111 River Street, Hoboken, NJ 07030, USA

For details of our global editorial offices, customer services, and more information about Wiley products visit us at www.wiley.com.

Library of Congress Cataloging-in-Publication Data

ISBN 978-1-119-45984-2

Cover image: Pixabay.Com
Cover design by Russell Richardson

Set in size of 11pt and Minion Pro by Exeter Premedia Services Private Ltd., Chennai, India

10 9 8 7 6 5 4 3 2 1

Contents

Preface

Nanotechnology has been booming in many areas, including materials science, mechanics, electronics, optics, medicine, plastics, energy, electronics, and aerospace. Nowadays, it is playing an extraordinary role in the functional finishing of textiles and polymers. It has sought to improve existing material performances and develop fibers, composites, and novel finishing methods. Application of nanoparticles has imparted novel characteristics to textiles such as flame retardation, UV-blocking, water repellence, self-cleaning, and antimicrobial properties. This book on "Nanomaterials in the Wet Processing of Textiles" presents diverse aspects of nanomaterial application in the textile industry, approaches used for synthesis and application, as well as strategies to remove toxic effluents from wastewaters employing state-of-the-art nanoadsorbents. We have no doubt that the readers will benefit from this book which covers latest developmental work pertaining to functionalization of textiles using diverse nanomaterials. We would also like to thank all the authors who contributed the chapters and provided their valuable ideas and knowledge in this book. We are also very thankful to the editorial board, and in particular to Martin Scrivener for the generous cooperation at every stage of the book production.

Shahid-ul-Islam
B.S. Butola
Indian Institute of Technology Delhi
(IITD), Hauz Khas, New Delhi, India
December 15, 2017

1

Functional Finishing of Textiles via Nanomaterials

Azadeh Bashari*, Mina Shakeri, Anahita Rouhani Shirvan and Seyyed Abbas Noorian Najafabadi

Textile Engineering Department, Amirkabir University of Technology, Tehran, Iran

Abstract

The unique properties of nanomaterials have real commercial potential for the textile industry. In recent years, fine materials that are produced using nanotechnology have been used in the textile production process. Production of functional textiles is the main purpose of using nanomaterials or nanocomponents on natural fibers such as cotton, wool, silk and synthetic fibers such as polyester, nylon, and acrylic, as they possess various properties such as light resistance, antimicrobial, self-cleaning, fire retardant, etc.

Different kinds of nanostructures are used in textiles. For example, carbon and copper nanoparticles or polymeric nanostructures such as polypyrrol and polyaniline as electro conductive agents; aluminum, zinc oxides, and carbon nanotubes (CNTs) for increasing durability of fibers; antimicrobial agents such as silver, zinc oxide, and titanium dioxide (TiO_2); moisture absorbent agents such as TiO_2; self-cleaning nanostructures such as CNT, TiO_2, and fluoroacrylates; UV protection agents as TiO_2 and ZnO; nano porous structures such as silicon dioxide or carbon black in order to improve dye ability of fibers; and many advanced properties such as heat conducting or insulating or electromagnetic shielding via introducing CNT or vanadium dioxide and indium tin oxide to fibers, respectively.

In this chapter, the development of using nanostructures to improve the properties of textiles is discussed. For this reason, nanostructures used in finishing processes are presented, separately.

Keywords: Nanotechnology, functional finishing, antibacterial, anti-odor, deodorant, UV-protective, water repellent, self-cleaning, flame-retardant, wrinkle-resistance

**Corresponding author*: bashari@aut.ac.ir

Shahid-ul-Islam and B.S. Butola (eds.) Nanomaterials in the Wet Processing of Textiles, (1–70) © 2018 Scrivener Publishing LLC

1.1 Introduction

Textiles play a major role in the development and industrialization of countries. The increasing demand for modern functional textiles has led to the usage of new materials and technology. Therefore, high-tech materials and well thought-out fabric constructions can improve the wearing comfort and provide unique properties. Antimicrobial effects, UV-protection, flame retardancy, stain and water repellency features, and others are the most important requirements of textiles. Since textiles are now widely used in different application sectors such as clothing, pharmaceutical, medical, engineering, agricultural, and food industries, imparting these characteristics into textiles can increase their potential for different applications.

Nowadays, there is a new revolution in the textile industry with the apparition of new technologies, which could add special functions and prominent features to the fabrics. For example, there has been notable improvement in technologies for natural and synthetic textile finishing, smart fabrics, and high performance functional textiles. In this sense, nanomaterials play a vital role in technological evolution since they show interesting surface properties that allow increasing their effect in comparison with bulky traditional additives and materials. For instance, using conventional nanomaterials such as metal oxide agents, carbon-based materials, host-guest compounds, and so on are examples of nanostructured materials used in antimicrobial, deodorant, UV-protection, self-cleaning, and other common finishing methods. In addition, in view of the rising environmental awareness, using environment-friendly methods and materials is necessary in the finishing process. Therefore, using alternative materials with high environmental safety is preferred.

This chapter reviews the most relevant contributions of the use of nanoparticles for functionalize textile materials. In fact, in this section, the use of nanomaterials for providing new properties such as antibacterial activity, anti-odor properties, UV-protection, self-cleaning, crease resistance, and others is explained.

1.2 Antibacterial Textiles

With increasing population in recent years, the bacterial infection problems are becoming more and more serious in comparison to the past. Many microorganisms live in human's environment. The presence of these microorganisms on textiles can lead to unwanted consequences such as paling, staining, decrease in mechanical properties, and decaying of the textile. In addition, some environmental factors such as temperature and humidity and chemical materials that are used in textile finishing can provide an appropriate media for the microorganisms to grow and multiply [1].

Some species of bacteria have a covering capsule, which surround them and keep them from drying out and other external factors. Each bacterium is enclosed by a rigid cell wall composed of peptidoglycan, a protein-sugar (polysaccharide) molecule. The wall gives the cell its shape and surrounds the cytoplasmic membrane, protecting it from the environment. The most important role of antibacterial agents is to penetrate in the cell wall and inhibit the bacteria's living.

The best way to prevent growth and multiplication of bacteria is to destroy their appropriate conditions for living. Several factors provide an acceptable media for bacteria to live such as nutritional requirements, water, oxygen, and heat. These factors are easy to find in textiles due to their contact with the human body. That is why the human body can be a good place for the microorganisms to live and multiply [2].

Antibacterial modification of textiles prevents the growth of bacteria, fungi, alga, and other microorganisms on them. One or more chemical agents that can destroy the microorganism's structure are used in this process [3].

Antibacterial materials are divided into two main groups called bacteriostatic and bactericide.

- Bacteriostatic can link to the amino acids of the DNA in the bacterial structures and prevent the multiplication of the bacteria.
- On the other hand, bactericides can disrupt the routine metabolism of the bacteria and completely destroy them [4].

A practical antibacterial agent must be easy to use in textile finishing methods, stable and durable in different treatments, compatible with other finishing materials, and non-toxic for end users and environment [3].

The antibacterial finishing of textiles has recently become a very active research field and has a great significance among the other methods of modification of textiles. Several metal nanoparticles have been suggested for antibacterial finishing of textiles such as silver, zinc, and titanium. There are also some biopolymers like chitosan, alginate, and starch that are used extensively in the textile industry.

Metal nanoparticles are more effective in comparison to biopolymers due to their multi-targeted mechanism of action, high surface area-to-volume ratio, and unique properties of these nanoparticles. A large surface area of the nanoparticles increases the contact of the antibacterial agent with bacteria and fungi, which is an important advantage of nanoparticles. Among all the materials with antibacterial properties such as copper, zinc, silver, titanium, gold, chitosan, and alginate, silver have proved to be the most effective against bacteria in antibacterial textile finishing [5].

Metal nanoparticles can be induced into textile by sol–gel technique, magnetron sputter coating, plasma sputtering, layer-by-layer coating, and other methods. One of the widely used techniques for coating textile substrates is the combination of the sol–gel synthetic procedure with the "pad-dry-cure" method [1, 6].

Some of the nanostructures that can be used in antibacterial finishing of textiles are mentioned as follows.

1.2.1 Antibacterial Organic and Non-Organic Nanostructures

These nanostructures can be applied directly on the fabric or can be loaded on the textile via a chemical carrier. There are following two groups in this category.

- Non-organic and metal nanostructures and nano composites: As mentioned before, some nanoparticles such as metal oxides, copper nano crystals, carbon nanotubes (CNTs), and nano clay can be used for antibacterial finishing of textiles.
- Nanostructure-loaded carriers: In this method, several chemical materials such as nano spheres, nano/microcapsules, dendrimers, liposomes, and nano tubes can be used for loading of the antibacterial agent and delivering it to the surface of the fabric.

Some of the mentioned nanostructures are explained briefly as follows.

1.2.1.1 TiO_2 Nanoparticles

When the titanium dioxide (TiO_2) catalyst is irradiated with light of energy greater than or equal to its band gap energy, electron–hole pairs are generated that can induce redox reactions at the surface of TiO_2 [7]. The general scheme for the photocatalytic damage of microorganism cells by TiO_2 photocatalytic properties involves several steps:

1. The photo-excited TiO_2 catalyst produces electron–hole pairs that migrate to the TiO_2 surface.

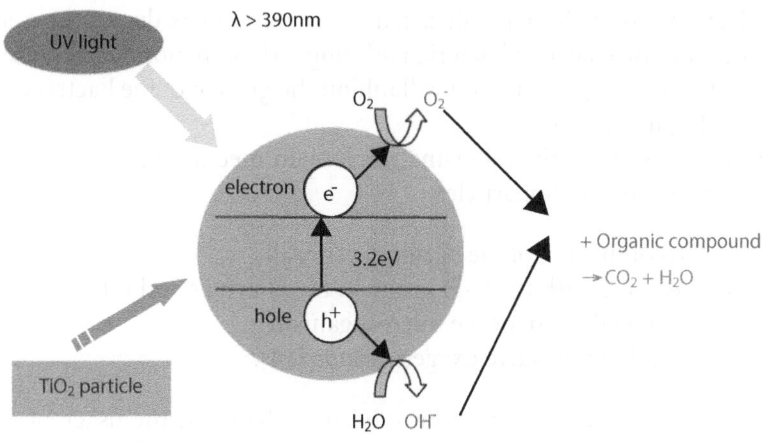

Figure 1.1 Photocatalysis mechanism of titanium dioxide [9].

2. Photo-generated holes in TiO_2 can react with adsorbed H_2O or OH- at the catalyst/water interface to produce highly reactive hydroxyl radicals and the electrons can react with oxygen vacancies to form superoxide ions.
3. Finally, the various highly active oxygen species generated can oxidize organic compounds/cells adsorbed on the TiO_2 surface, resulting in the death of the microorganisms [8].

By the combination of electron and electron hole with oxygen, carbon dioxide and water will form in an oxidation-reduction reaction (Figure 1.1).

Through this reaction, TiO_2 nanoparticles can destroy the cell wall of the bacteria and provide an effective protection against them [9, 10].

1.2.1.2 Silver Nanoparticles

Silver is a nontoxic inorganic metal, which is not harmful for the human body. Several studies have shown that silver has a broad antibacterial activity toward bacteria and fungi. It is believed that the antibacterial effect of silver nanoparticles comes from their strong affinity toward phosphorus or sulfur. Since the cell wall of the bacteria contains amounts of sulfur-containing proteins, silver nanoparticles can react with these groups and affect the bacteria performance [11, 12]. It was also reported that the released silver ions (particularly Ag^+) from silver nanoparticles can interact with phosphorus moieties in DNA, leading to inactivation of DNA replication, or react with sulfur-containing proteins, resulting in the inhibition of enzyme functions [13].

Silver nanoparticles have attracted many researchers due to their effective performance against bacteria and fungi. These nanostructures can be very reactive with proteins and will inhibit the growth of the bacteria completely (Figure 1.2) [9].

Scientists suggest the following three main mechanisms for antibacterial effect of silver nanoparticles:

- Decomposing of the bacterial cell wall
- Reacting with the S–H group in the enzymes and inhibiting the metabolism of the microorganism
- Creating the active oxygen groups [14].

In recent years, researchers have been studying on the usage of silver nanoparticles in nano composites with polypropylene fibers and other metal oxides [15].

1.2.1.3 ZnO Nanoparticles

Zinc oxide nanostructures are also photocatalyst materials, which has the same mechanism as TiO_2 nanoparticles. ZnO nanoparticles are widely

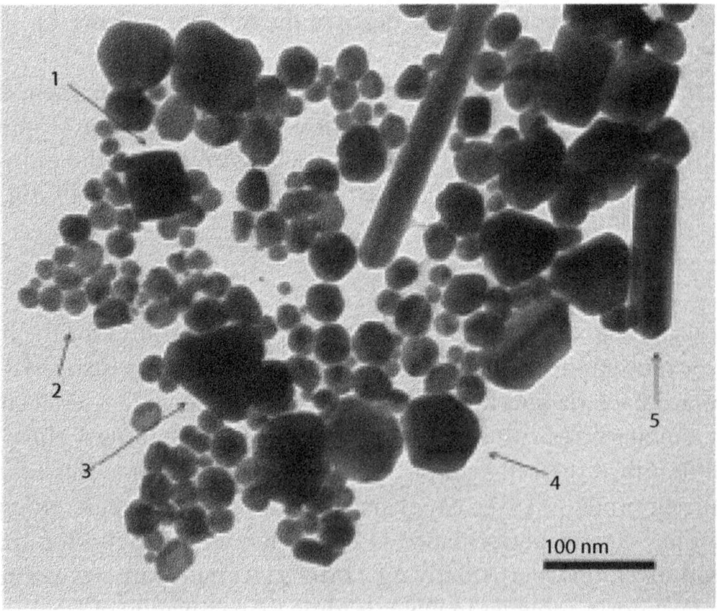

Figure 1.2 Close-up of the TEM image of silver nanoparticles in different shapes and sizes [5].

used now in antibacterial textiles due to their effective photocatalytic properties [9].

Recent studies indicate that the ZnO nanostructure show better antibacterial properties when doped with CuO molecules. It can be due to the narrower band gap of Zn structures when doped with copper and better photocatalysis activity in this condition [16].

In another study, antibacterial behavior of suspensions of zinc oxide nanoparticles has been investigated against *E. coli.* It was deducted that the effects of particle size, concentration, and the use of dispersants are large on the antibacterial behavior. The results show that the ZnO nano fluids have bacteriostatic activity against *E. coli.* It was also deducted that the antibacterial activity increases with increasing nanoparticle concentration and with decreasing particle size, while it turned out that particle concentration is more important than particle size. SEM analyses of the bacteria before and after treatment with ZnO nano fluids show that the presence of ZnO nanoparticles damages the membrane wall of the bacteria. Electrochemical measurements also show some direct interaction between ZnO nanoparticles and the bacterial membrane at high ZnO concentrations [17].

1.2.1.4 Chitosan

Chitosan is one of the natural biopolymers that has attracted researchers' attention due to the environmental and toxicological concerns about the use of heavy metals for the production of nanoparticles (Figure 1.3). It is assumed that the antibacterial effect of chitosan is because of the presence of amino groups, which gives chitosan a significant characteristic in decomposing the bacteria and fungi [5].

Chitosan's molecular weight is an important object in researches, which has a significant effect on the antibacterial activity of the textiles. Recent studies on the antibacterial activity of chitosan and chitosan oligomers have revealed that chitosan is more effective in inhibiting the growth of bacteria than chitosan oligomers. Furthermore, the antibacterial effect of

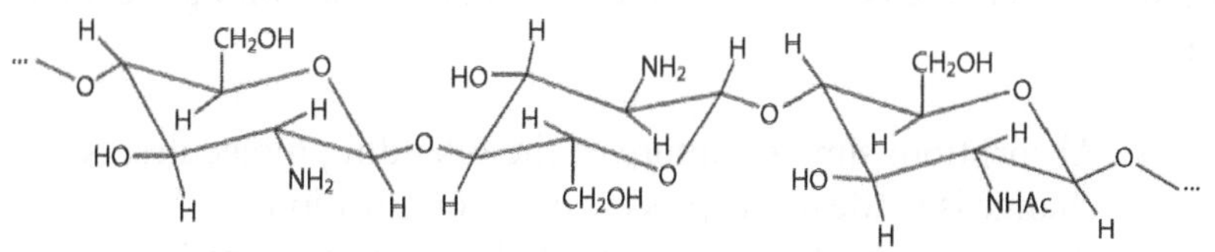

Figure 1.3 Chemical structure of a chitosan [6].

chitosan and chitosan oligomers is reported to be dependent on its molecular weight [18].

In a study, the molecular weight of chitosan and chitosan oligomers has been investigated. It was shown that chitosan has a better antibacterial activity in comparison to chitosan oligomers [19].

1.3 Anti-Odor Textiles

Anti-odor textiles are one of the main sectors of the textile industry, which have increased considerably over the last few years due to rising consumer health awareness. Especially in the main parts of the textile industry such as sportswear, underwear, socks, and shoes, using anti-odor textiles have received much attention and nanotechnology plays a vital role in this issue.

There are various types of unpleasant odors in our daily life, such as body odor caused by metabolism and aging, and odors of garbage and cigarettes. In terms of chemical structures, unpleasant odors are divided into the following three main groups:

- Fatty acids such as body odor, sweat odor, etc.
- Nitrogen compounds such as urine, the smell of fish, etc.
- Sulfur-containing compounds, such as excreta, etc.

Therefore, deodorant technology has attracted worldwide attention in different fields of applications. Among the various approaches, physical methods (absorption of odors by activated carbon), chemical methods (turning the odor into a common smell), and using aromatic compounds have been proposed to eliminate unpleasant odors [20]. Therefore, with the increasing number of people being sensitive to odors, using anti-odor technology is an obligation for textile producers.

1.3.1 Odor-Control Methods

In general, there are two different mechanisms for controlling the odors in textiles:

- Absorption: simple capture of the offending molecules. No change is made to the process of decomposition.
- Prevention: bacteria are prevented from multiplication; the offending molecules are not generated [21, 22].

1.3.1.1 Absorption Mechanism

According to the absorption mechanism, several types of nanostructured materials are used for deodorizing textiles, which is shown in Figure 1.4.

1.3.1.1.1 Cyclodextrins

Cyclodextrins (CDs) are a family of cyclic oligosaccharides with 6–8 glucopyranoside units which are known as α, β, and γ-CD [23]. Among different types of CDs, β-CD has a wide range of applications in the textile industry due to its low cost, ease of production, and ease of attachment to surfaces. Also, the size of pores makes it suitable for trapping a range of odor molecules [24]. One of the natural properties of CD is stacking the molecules on top of each other that increases the probability of absorption of odorous compounds. On the other hand, when the scent molecules are absorbed by CDs, the conical structure provides the effective possibility of maintaining the odor molecules. However, CDs as anti-odor agents have some problems. Since different odor compounds have various sizes and shapes, small molecules pass easily through the pores but large molecules cannot enter the inner structure of CDs. Therefore, CDs just maintain specific odor molecules with certain size [25]. CDs have been extensively used for the absorption of various odors in different applications such as controlling the odors caused by chronic wounds. For example, CDs can be incorporated

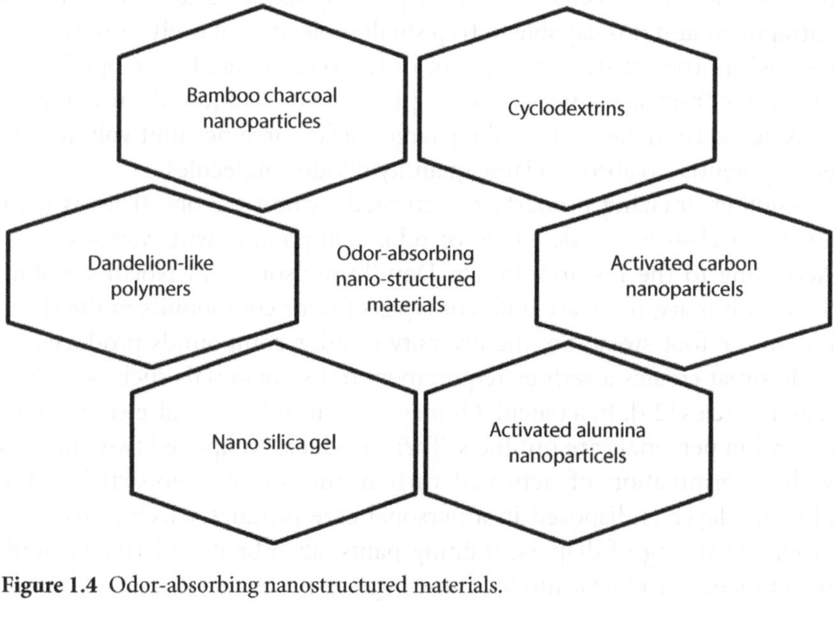

Figure 1.4 Odor-absorbing nanostructured materials.

into various materials used for wound healing such as hydrogels. They are used in applications outside of wound care for the control of foul odors. However, there are some limitations in the absorption process by CDs like slow diffusion rate and water deficiency. It seems that these problems can be solved by using the combination of CDs and conventional hydrocolloids, such as sodium carboxymethyl cellulose for effective elimination of wound malodor. This technology has been already introduced as "Exuderm Odorshield" brand (Medline Industries Inc, Mundelein, IL, USA) [26].

In a research, Narayanan *et al.* introduced an efficient wound odor removal by β-cyclodextrin-functionalized poly (ε-caprolactone) nanofibers. According to the results, the PCL/β-CD nanocomposites, by virtue of having their β-CD cavities free and unthreaded by PCL, could potentially be an ideal substrate for removing wound odors through the formation of inclusion compounds with odorants [27]. In the same study, Lipman *et al.* developed a new series of hydrocolloid adhesives based on CDs to provide an alternative technology for the adsorption of chronic wound odors. Their results showed that CD materials provide a new method of controlling the malodor associated with various types of wounds [28].

1.3.1.1.2 Activated Carbon Nanoparticles

Activated carbon is one of the most widely investigated absorbents in various fields such as water treatment, food industry, medicine, and the environment. It is considered as the most effective method for deodorizing in textiles. The mechanism of odor absorption by activated carbon is physical entrapment and it is capable to trap small molecules of body odors due to the high porous structure. Then, in order to eliminate the trapped scent molecules from a textile, a scouring treatment is required. In addition, activated carbon has an incredibly large surface area per unit volume and good potential to absorb a large quantity of odor molecules.

Another amazing property of activated carbon nanoparticles is their ability to absorb a wide range of odor compounds with various sizes. According to the research by Dr. Don Thompson from North Carolina State University, there are different types of odor compounds in the shoes due to the foot sweat. So, the diversity of odor compounds produced by body sweat creates a serious requirement to use materials such as carbon nanoparticles [29]. In a patent, Quincy *et al.* invented a dual-element odor control in personal care products. Their products comprised two portions with a formulation of activated carbon and silver nanoparticles. The obtained layer is disposed in a personal care product selected from the group consisting of diapers, training pants, absorbent underpants, adult incontinence products, and feminine hygiene products [30].

1.3.1.1.3 Bamboo Charcoal Nanoparticles

Bamboo is an Asian plant that is found in diverse climates from cold mountains to tropical or subtropical regions. Bamboo is one of the most notable natural resources because it is a fast-growing plant and used for various applications such as preparation of cellulose nanofibers and enhancing properties of synthetic fibers such as polyester.

Bamboo charcoal is prepared by five years of pyrolysis process of bamboo at high temperatures of about 800–1200 °C (Figure 1.5) and obtained from two different resources:

- Stems, leaves, and roots of the bamboo plant
- Bamboo wastes.

Therefore, the bamboo charcoal nanoparticles act like a sponge, catching any odor inside its fine pores. These nanoparticles are considered as a good candidate for odor absorption of textiles and the environment [31, 32]. Organic bamboo charcoal is 100% safe to use in fabrics, which are directly in touch to our skin. Therefore, bamboo charcoal nanoparticles have already been widely used with great success in different applications such as beauty products, water filters, foods, and medicines.

Bamboo charcoal powder's ultra-fine structure contains millions of tiny gaps, which allows 1 gram of bamboo charcoal powder to cover a surface area of 600 square meters. This super porous structure of bamboo charcoal powder absorbs bacteria, unwanted odor, and wicks away sweat up to 50% quicker than cotton.

According to the literatures, bamboo charcoal can absorb formaldehyde, benzene, toluene, ammonia, and chloroform at a rate of 16–19.39%, 8.69–10.08%, 5.65–8.42%, 22.73–30.65%, and 40.68%, respectively. Also, bamboo charcoal fiber effectively decomposes the microorganisms attached to its surface and in the air around it. Recently, Ettitude Company has introduced its new product, "Bacteria and Odor Control Bamboo Charcoal Bedding". In this product, bamboo charcoal's ultra-fine structure powder is added in the bamboo lyocell fabrication process to create the antimicrobial and anti-odor bedding fabric [33]. In addition, Greenyarn Company

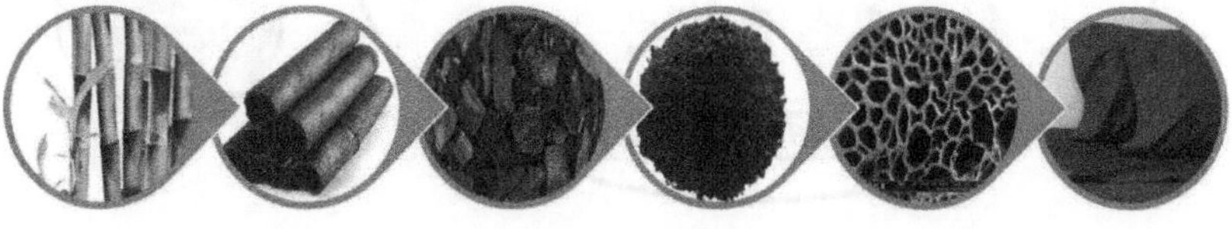

Figure 1.5 The production process of a bamboo nanoparticle.

has exposed their fabrics that contain bamboo charcoal nanoparticles for use in products such as socks and underwear. These bamboo charcoal nanoparticles have antibacterial and deodorizing properties [34].

1.3.1.1.4 Dandelion-like polymers

A novel method of producing an odor-controlling textile is using a polymer with odor-trapping ability (Figure 1.6). Milliken Company first developed this technology in 1865 as a housekeeping and cleanliness technology to control the unpleasant odor of carpets and furniture. This method was inspired by a dandelion structure as a polymer mass containing very small branches. When, a person wears such a textile with an odor-trapping technology, the small scent molecules released from the body are trapped by the dandelion structure. Following the odor-capturing process, the conventional scouring method is utilized to remove the odor compounds from the polymer. So, these kinds of textiles have reusability, which is the most important characteristic of dandelion-like polymers. However, short lifetime and absorption of a low range of odor compounds are the main drawbacks of these polymers [35].

1.3.1.1.5 Activated alumina nanoparticles

Activated alumina is prepared by de-hydroxylation of aluminum hydroxide. It is a highly porous material with a large surface area-to-volume ratio due to the many "tunnel like" pores. Activated alumina is used as an effective odor absorbent in a wide range of applications such as wastewater, anti-odor textile, and others via trapping the odor molecules in their pores [36]. In some researches, activated alumina nanoparticles with their tunnel-like pores have been used as an effective absorbent for different applications.

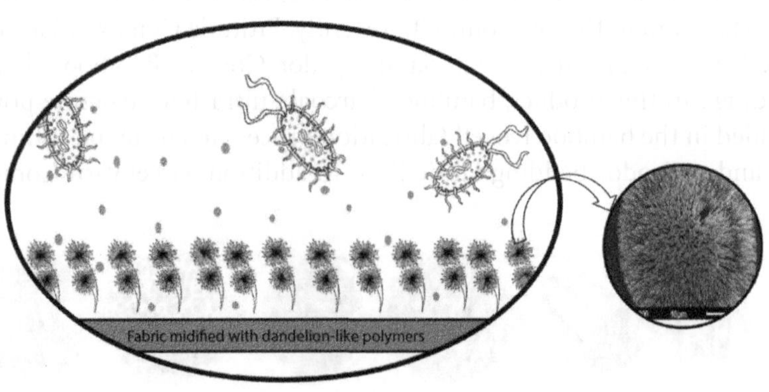

Figure 1.6 Odor-captured textiles with dandelion polymers.

For example, Jenkins *et al.* produced an absorbent composition with improved odor control suitable for use as an animal litter, comprising an absorbent material, activated alumina, and optional additives [37]. Furio *et al.* proposed an absorbent structure with odor-control material consisting of activated alumina, carbon, silica gel, zeolite, siliceous molecular sieve, and their mixtures for an adult incontinence garment, absorbent pad, or diaper [38].

1.3.1.1.6 Nano Silica Gel

Silica gel is a hard, granular, and porous substance made by precipitation from sodium silicate solutions treated with an acid. The main properties of nano silica gel such as porous structure and large specific surface area play a vital role in the odor absorption process [39]. However, nano silica gel is not commonly used as an odor absorbent in comparison with other odor absorption agents in different applications.

1.3.1.2 Prevention Mechanism

This approach is based on preventing proliferation of the odor-causing bacteria by different types of antimicrobial agents such as silver, titanium, zinc oxide nanoparticles, and so on, which are described in section 1.3.

1.4 Deodorant Textiles

Smelling a delightful aroma can be a very pleasurable experience and have physiological and emotional effects. Fragrance finishing of textiles has been greatly expanded in recent years and enhanced the benefit of the products.

The most natural aromatic substances are essential oils that are extracted from different parts especially aerial parts of plants such as leaves and flowers. These compounds are produced due to complex metabolic reactions in plant in order to protect the plant against various pathogens and insects and also reduce the tendency of some herbivorous animals [40]. Some of the biological activities of natural aromas such as antimicrobial, antiviral, anti-inflammatory, anti-mutation, anti-cancer, and antioxidant effects have been proven [41].

1.4.1 Aromatic Textiles with Nanocarriers

Although the aromatic textiles concept is not new, recently great progress has been made on the textile-finishing process in order to produce effective and long-lasting aromatic textiles. In the past, the aromatic extracts usually have been used via direct spraying on the surface of textiles. Today, encapsulation has been known as an effective method to increase the lifetime of odors on textiles. Micro/nano encapsulated fragrances are coated on textiles and released due to pressure or abrasion of textiles or garments [42].

For many years, researchers are seeking a new formulation for gradual release of drugs in the body. Today, drug delivery systems (DDS) such as lipid- or polymer-based nanostructures are characterized having great benefits in comparison with conventional methods of drug delivery [43]. In order to control the release rating, improve the efficiency, reduce the toxicity, and lower the side effects of aromatic compounds, the same DDS are used for essential oils [44].

Nanocarriers such as lipid carriers, nano emulsions, and biocompatible polymer nanostructures are able to protect aromatic compounds against oxidation or evaporation. Some types of nanocarriers are shown in Figure 1.7.

In the 1970s, researchers used nanoparticles for delivering and releasing vaccines and anticancer drugs for the first time. Recently, in addition to pharmaceutical compositions, the nanoparticles have been utilized to carry the aromatic compounds and protect them from optical and thermal decompositions and increase the lifetime and efficiency of these active materials.

1.4.1.1 Polymeric Nanocarriers

Polymeric nanocarriers include nanocapsules and nano sphere systems. Nanocapsules are nanoscale core/shell systems made from a polymer in which aroma is confined into the core surrounded by a polymer membrane, while nano spheres are matrix systems in which the aroma is physically dispersed.

Biocompatible natural or synthetic polymers are suggested for producing polymeric nanocarriers. The most common natural and man-made polymers in preparation of polymeric nanocarriers are chitosan, gelatin,

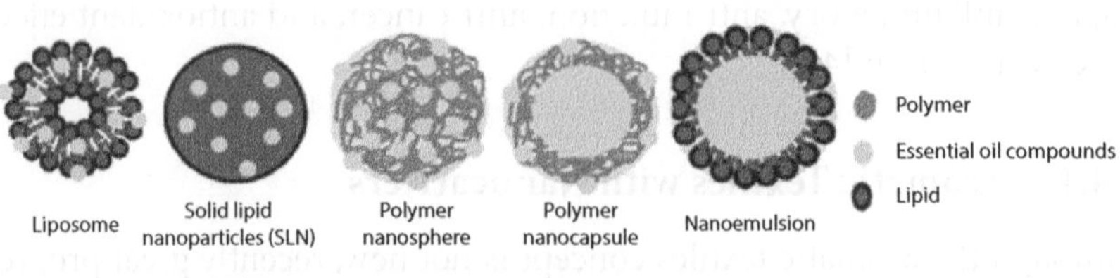

Figure 1.7 Aroma nanocarriers.

alginate, polylactides (PLA), poly (lactide co-glycolides) (PLGA), and poly acrylamide [45].

1.4.1.1.1 Methods of Producing Nanocapsules

This group of nanostructures can be prepared with different physical, chemical, physicochemical, and mechanical techniques, which are effective on size, particle-size distribution, thickness of shell, morphology, and other characteristics of the nanocapsules [46]. Different types of such methods are summarized in Figure 1.8.

Many methods have been developed for preparing polymeric nanocarriers containing aromatic compounds. The usage of a non-solvent compound to fabricate nanostructured deposition and solvent exchange are the most widely used methods. In this technique, polymer and aroma are dissolved in a suitable organic solvent or a mixture of several solvents, and then this organic solution, in the presence or absence of a surfactant, is introduced to water with applied shearing stress. The organic solvents are separated from the polymer via evaporation and a polymeric film or powder is made [47].

Corzani prepared a fragrance delivery system consiting of a mixture of 50% poly(tetramethylene glycol) having an a number average molecular weight (M_n) of approximately 2000 Da with 25 wt% of a rosin ester plasticizer and 25 wt% benzyl acetate with a selcted perfume.

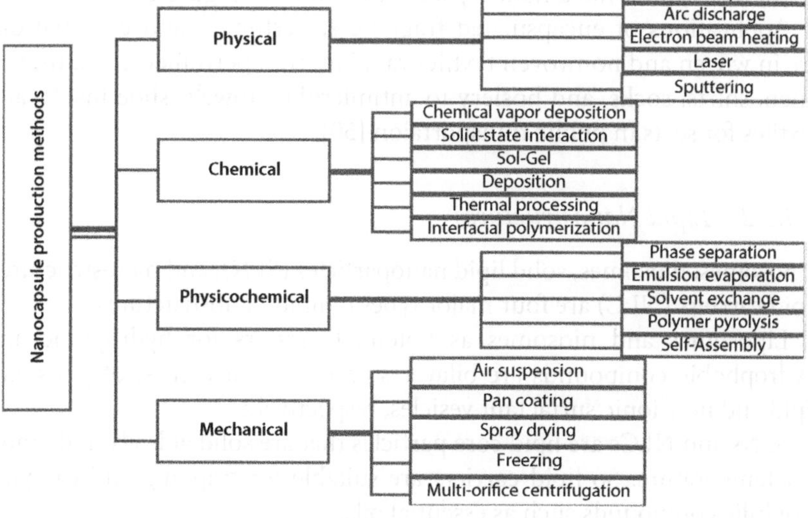

Figure 1.8 Nanocapsule production methods [46].

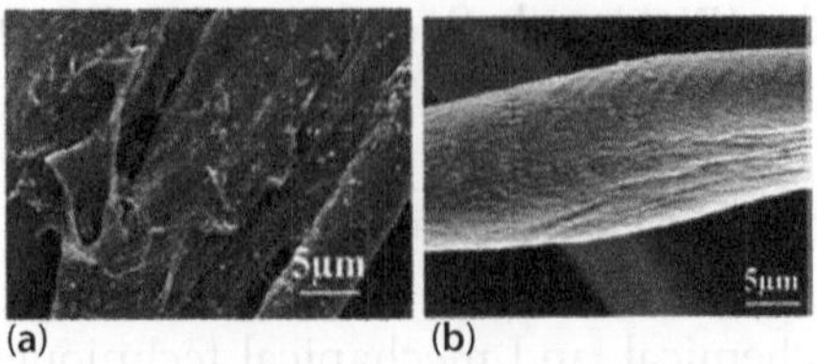

Figure 1.9 SEM images of the finished cotton fabrics with aroma (a) and untreated cotton fabric [49].

Yang prepared a fragrance delivery system consiting of 20% limonene and 3% γ-terpinene compounded with an ionic stabilizer, dihydropropyltrimonium chloride, fatty esters, and a silicob fluid.

A fragrance delivery system consisting of a nanostructure microcapsule shell of urea-formaldehyde or melamine-formaldehyde and containing an olfactive component was prepeared by Lie and had at least a 6-month shelf life when stored at 37 °C [48].

Encapsulating rose fragrance in nanoparticles in order to reduce the evaporation of volatile compounds and prepare the long-term fragrance-releasing textile has been proposed by Hu *et al.* Chitosan nanocapsules loaded with fragrance were prepared via the ionic gelification in acetic acid, Tween 80 as a surfactant, and sodium tripolyphosphate as a gelation compound under ultrasonication. Aroma-loaded nanocapsules were incorporated into the cotton fibers via immersion in nanoemulsion under vacuum [49]. The SEM micrographs of untreated cotton and treated cotton fibers with aroma-filled nanocapsules are shown in Figure 1.9.

Applications of encapsulated fragrances, perfumes, and essential oils are in woven and nonwoven textiles ranging from perfumed curtains, bed linen, shirts, socks, and hosiery to antimicrobial towels, shoe insols, and textiles for seats in public transportaion [50].

1.4.1.2 *Lipid Nanostructures*

Liposomes, niosomes, solid lipid nanoparticles (SLN), and nanostructured lipid carriers (NLC) are four major types of lipid nanostructures.

Liposomes and niosomes as potential carriers for hydrophilic and hydrophobic compounds are bilayer structures that consist of phospholipid and non-ionic surfactant vesicles, respectively.

SLNs and NLCs are lipid core particles that are solid at body and ambient temperature. So lipid carriers are suitable for trapping and carrying lyophilic compounds such as essential oils.

Lipid nanostructures are summarized in Table 1.1.

Table 1.1 Lipid nanostructures [51].

Properties	Structure	Type
Niosomes are non-ionic surfactant-based vesicles that are mostly used as drug, gene, and protein delivery systems.	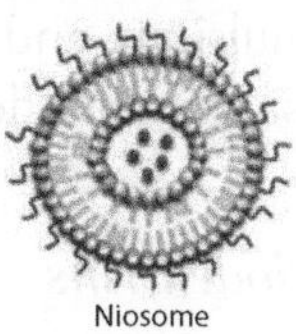	Niosome
Liposomes are tiny vesicles made out of at least one phospholipid bilayer that surrounds the aqueous space. Due to their amphipathic characteristics, liposomes are used as delivery systems for hydrophilic and hydrophobic compounds.	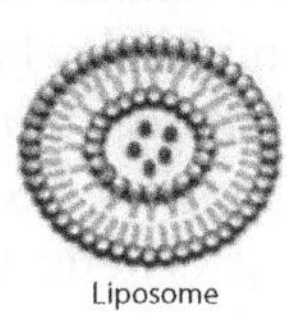	Liposome
SLNs are lipid vesicles with an average diameter of 10 nm to 1 μm with a solid lipid core matrix that is formed due to the presence of a surfactant (emulsifier) and can solubilize lipophilic molecules.	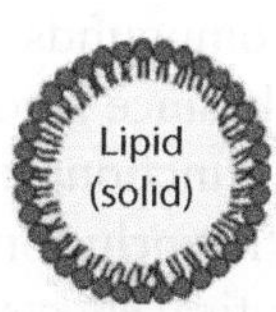	SLN
NLCs as the second generation of SLNs are composed of a solid lipids matrix (long chains) and liquid lipids (short chains) with an average diameter of 10 to 500 nm.	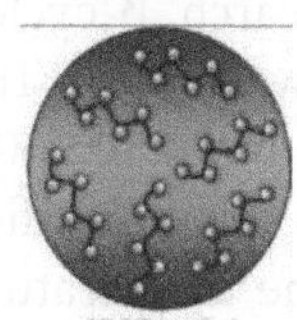	NLC

Increasing the stability of the drug delivery system and dissolving oil-based aromas in aqueous phase are the main factors for using lipid carriers in aroma finishing of textiles. Liposomes are the most reported nanocarriers for essential oils.

Dahms *et al.* invented a SLN dispersion for the controlled release of fragrances and/or aromas. In this formulation, SLNs with varying melting points are dispersed in an aqueous medium with emulsifiers that formed not only monolayers around lipid particles but also lyotropic liquid-crystalline membranes within the aqueous phase. The fragrance may be included both in the nanoparticles and/or in the emulsifier monolayers or the liquid-crystalline membrane layers. Oil-soluble and/or amphiphilic fragrance oils or aromas are preferably stored in the lipid particles while hydrophilic, water-slouble fragrances are incorporated into the aqueous phase [52].

On the other hand, nanostructured lipid carriers (NLCs) can be used in fragrance products to prolong the release of the fragrance. Fragrance evaporates fast from ethanolic solutions, delayed from the more viscous liquid lipid (oil) of o/w emulsions, and more delayed from the solid lipid of NLC due to reduced diffusion. Other delayed release products are insect repellents [53].

1.4.1.3 Cyclodextrins

Considering the volatile nature of fragrance compounds and essential oils, the fragrance molecules have to be formed as inclusion compounds with CD molecules in order to retain fragrances for a long time. To prepare aromatherapeutic textiles, β-cyclodextrin is the first choice as the host molecule, because β-cyclodextrin molecules are capable of forming inclusion compounds with organic compounds that fit into their own cone-shaped hydrophobic cavities. As a result of the inclusion, the physicochemical properties of the compounds are changed; for example, the vapor pressure of the volatile substance is reduced and the stabilities against light and air are enhanced. The functional textile with the "guest–host" effect may be achieved by anchoring the inclusion compound. The sedative effects for emotion and the pharmaceutical effects of essential oils were shown by Wang *et al.*

In this research, β-cyclodextrins were resolute in a mixture of alcohol and distilled water (1:3). The solution was emulsified in a high-speed mixer at a speed of about 10,000 rpm for 5 min. The lavender–alcohol solution was added to the emulsified solution within 30 min, with the aid of stirring, while the temperature was maintained at 40 °C for 2 h. The inclusion compound was fixed onto cotton with a low temperature binder by a conventional pad-thermofixing method. As a result, the rate of fragrance release of inclusion compounds decreases with time [54].

If CD is used as a carrier for aromatic compounds, this carrier should be bound to fibers in order to enhance the washing fastness of fragrance textiles. Since CD is not able to bind directly to the textile substrate, the usage of CD derivates or a cross-linking agent is recommended [55]. β-cyclodextrin modified with monochlorotriazine was synthesized by Sricharussin *et al.* and applied on cotton fabric. The pad–dry–cure method at high temperature (150 °C) was the best fixation way for this CD derivate on cotton fabrics. Sandalwood oil was sprayed onto the modified fabric to prepare aroma inclusion in CD [56].

In another research, Bndyopadyay *et al.* employed maleic anhydride as the modifying agent for finishing cotton fabric with β-cyclodextrin in the presence of ammonium peroxodisulphate, as a free radical polymerization catalyst and tetra sodium pyrophosphate as an esterification catalyst. The modified fabric had been treated with Thyme essential oil and the retention of fragrance and sustained release of essential oil had been assessed after repeated post-wash procedures. The results showed that fragrances held by CD can withstand the action of detergent wash for a significant number of wash cyles [57].

1.4.1.4 Dendrimers

Dendrimers are repetitively branched molecules that have attracted considerable attention in the textile industry (Figure 1.10). A dendritic structure has potential to act as a host molecule via abundant functional groups and hollow interior branches. They are able to include non-polar or charged guest molecules into their pockets by hydrophobic/hydrogen-bond interactions, or on the branches by electrostatic interactions and forming different types of inclusions or ion-pairs in aqueous solutions. In this regard, Akbari *et al.* used the second generation of PPI dendrimer to produce long-lasting fragrant semi-worsted fabric. They investigated the release behavior of ginseng and rosewater as fragrance candidates using an E-nose and reported excellent sustained-release property of treated fabrics for a long time [58].

1.4.2 Application of Aroma Textiles

- Cosmetic textiles: micro/nanocapsules containing moisturizing agents, essential oils, skin-cooling agents, vitamins, and pro-vitamins have been used as finishing agents for textiles.

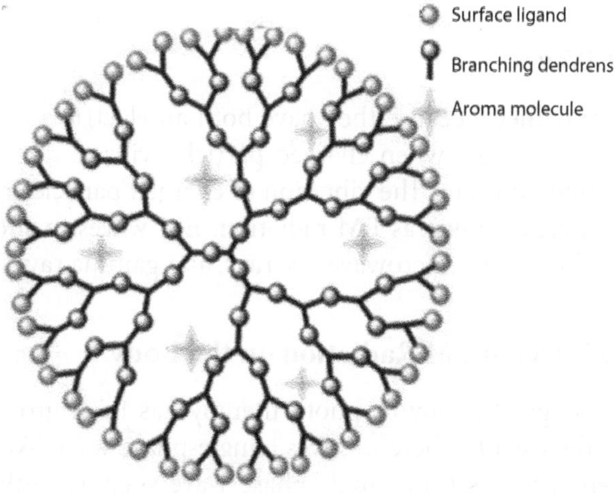

Figure 1.10 Basic dendrimer components.

- Aromatherapy via textiles: modified textiles such as bed linen, curtains, carpets and bedspreads with lavender, cinnamon, and citrus oils are very appropriate for a better sleep and lowering fatigue. On the other hand, the patients suffering from high blood pressure feel more comfortable after using relaxing aromatic herbal pillow with lavender, basil, or lemon oils.
- Sportswear: antibacterial effects and preserving the freshness and energy can be obtained in sports apparel via using essential oils such as lemon or orange.

1.5 Protective Textile Against Electromagnetic Radiation

The development of the electronics industry and the increasing usage of electronic products in communications, computing, automation, biomedical, aerospace, and others can lead to electromagnetic (EM) interference. In other words, radiation-induced waves can cause problems for electronic devices and health issues. Hence, the demand for strict protection of people and electronic devices against the unpleasant effects of EM signals and free charge problems has increased. This section discusses different types and properties of EM waves, followed by the introduction of methods and materials that improve protective finishing textiles against EM radiation. These methods include protective finishing textiles by using nanostructures and high performance fibers.

1.5.1 EM Waves

EM waves are named because they have both an electric and a magnetic component. They create when charged particles vibrate due to the various forces acting on them. The vibration of charged particles results in an emission of energy known as EM radiation. EM waves include infrared, ultraviolet, radio waves, microwaves, X-rays, and gamma rays.

1.5.2 The Effect of EM Radiation on the Body

Although that light therapy or photo therapy has been growing rapidly recently and the use of coherent waves (single-phase wave like laser light) or non-coherent waves (non-single-phase wave such as sunlight) in the infrared can accelerate wound healing, reduce pain and inflammation, and prevent tissues' death [59], the exposure of EM radiation on the body may lead to physical damage, stimulating nerves and muscles, and changes in the nerve cell performances [60]. Due to the vibration and heat generation by EM radiation in the human's body, DNA and RNA in cells may halt production, which will lead to abnormal chemical activity of cells and cell cancer [61].

The simplest and most effective way to prevent EM radiation hazards is to stay away from the radiation source. One of the primary methods used in protection from EM radiation is "shielding". Shielding can be described as reducing the EM field in a space by blocking the field with barriers made of conductive or magnetic materials. EM shielding materials are usually used to reduce the intensity of EM fields [62].

1.5.3 Shielding Materials Against the EM Waves

When an EM field is passed through an object, there are three phenomena that determine how the field strength is lost as it interacts with the object: absorption attenuation, attenuation due to reflection, and attenuation due to successive internal reflections. Thus, EM shields function based on the above-mentioned two major EM mechanisms: reflection from a conducting surface and absorption in a conductive volume. An EM wave striking a metallic surface encounters both types of losses. Part of the wave is reflected, while the remainder is transmitted and attenuated as it passes through the media. The combined effect of these losses (reflection and absorption) determines the effectiveness of the shield [63].

Absorption in an EM shield transforms EM energy into thermal energy. In general, the materials used for absorbing EM waves can be classified into two groups:

1. materials with high dielectric constant, such as $BaTiO_3$ and carbon particles,
2. materials with high permeability, such as Fe_3O_4 and ferrite materials [64].

Materials with a high dielectric constant absorb the electric energy and convert it into thermal energy, while materials with high permeability convert the magnetic energy into thermal energy [65].

In the past, the traditional approach for EMI shielding relied on the use of metallic materials, which supply excellent shielding effectiveness. In recent years, in addition to the metal layers, conductive polymers are used to protect the body against EM waves. Conventional textile fibers have sufficient isolation properties with resistance at around 10^{15} Ω/cm^2, which is more than the resistance required for applications of protection against EM waves (10^9–10^{13} Ω/cm^2). Therefore, conductive textiles are produced for improving protection against EM waves [61]. The basic mechanism of protection against EM radiation is improving charge transfer (electron or hole). In other words, the protective materials must contain electrical/magnetic dipoles, to have the ability to interact with the EM field.

Electric dipoles such as carbon particles and $BaTiO_3$ on textiles have high dielectric constants, whereas magnetic dipole on ferromagnetic substrates such as ferrite material and Fe_3O_4 cause high magnetic permeability. In other words, materials with a high dielectric constant attract electric power and convert it into thermal energy, while high magnetic permeable materials convert absorbed magnetic energy into heat energy [66].

Metals are the best protective materials against the EM waves, due to their conductivity and stability. However, these materials are often expensive and heavy and some problems such as expansion and oxidation or corrosion may occur during their use. In contrast, most of the fabrics made of synthetic fibers are electrically insulated and transfer EM field [66].

In recent years, flexibility and lightweight of conductive fabrics have attracted the attention of many researchers in order to produce protective textiles against EM waves. Conductive fabrics are prepared by conductive fibers or special finishing methods using a slight amount of conductive materials such as carbon. The following methods can be used to achieve the conductive textiles:

1. Lamination of conducting layers onto the fabric surface, conductive coating, zinc arc spraying, ionic plating, vacuum metallization, and sputtering
2. Adding conductive fillers, such as conductive carbon black, carbon fiber, CNTs, metal powder, and flake (Al, Cu, Ag, Ni) to the spinning dope of synthetic fibers
3. Incorporation of conductive fibers or yarns into the fabric [67].

1.5.3.1 Conductive Polymer

Conducting polymers are a new class of materials with an extensive delocalization of p-electron conjugated network. Among conducting polymers, polyaniline and polypyrrole have received widespread attention because of their environmental stability and moderate conductivity. Conducting polymers are unprocessible due to the extensive delocalization of p-electrons.

In order to solve this problem, different groups have prepared conducting composites by blending conducting polymers with insulating polymers so that the conducting composites retain the mechanical properties of the conventional polymers and electrical conductivity of the conducting polymers [68].

The polymerization of aniline and pyrrole on insulating fabrics such as polyester cloth, glass fabrics, and high silica cloth was carried out by *in situ* chemical oxidative polymerization. The chemical polymerization of pyrrole in the presence of textile substrates results in the formation of electrically conducting, polypyrrole-coated fabrics [69].

Engin *et al.* developed conductive fabrics by polymerizing aniline onto the polyamide (PA)-knitted fabrics. Hydrochloric acid as an acidic medium and ammonium persulfate as an oxidant were employed during the polymerization process. The results showed that the fabric treated with 0.5 M aniline concentration had the highest shielding effectiveness [70].

1.5.3.2 Metal Nanocoating

Despite the problems associated with the use of metals and conductive polymers for improving the protective properties of textiles, the use of nanoparticles as a very thin layer on textiles is much more favorable. Applying hydroxyl apatite and nanoparticles (TiO_2, ZnO, and Fe_2O_3) with other organic/inorganic materials on the textile surface can increase the protection effect against ultraviolet, infrared, and EM radiations, as well as improving abrasion stability, water repellency, and anti-microbial properties. Clinoptilolite nanoparticles, a silicate mineral microporous crystalline structure with tetrahydral configuration, have been applied on cotton fabric. This treated fabric was used as protective clothing against radioactive radiation [66].

Among the metal particles, applications of magnetic nanoparticle are the best option for improving the protection property against EM waves on textile fabrics. For this purpose, various methods for the synthesis and use of magnetic nanoparticles are used such as lumen loading (direct wet-end addition), nano-coated fibers, or *in situ* nanoparticle synthesis on the textile fabrics [71].

In situ synthesis of nanoparticles on textiles is the easy-to-do method and attracted researchers' attention for the production of multifunctional textiles. Polyester fabric with multiple protection properties against magnetic fields, antimicrobial, and catalytic activity was prepared by *in situ* synthesis of magnetic nanoparticles of iron (magnetite and hematite) using iron chloride, iron sulfate, and sodium hydroxide. The process was in two steps at different temperatures, 100 °C and 130 °C, and by usage of α-Fe_2O_3 and Fe_3O_4 nanoparticles, respectively. The treated fabric with magnetite and hematite has great potential in the protection against EM waves with high stability, excellent antimicrobial activity, and noticeable catalytic activity. Although this method improves textile protection properties, the color of magnetic nanoparticles restricted their application on light-colored textiles [72].

Classical methods of metallic layer deposition such as vacuum evaporation, electron beam evaporation, or chemical deposition of metalorganic compounds are not suitable for some fabrics made of polymeric fibers, such as polypropylene (PP) or polytetrafluoroethylene (PTFE), because of the properties of their surfaces. The deposited layer has very poor adhesion and is subject to wear. The only possibility is plasma activation of the surface followed by the deposition of a metal layer by vacuum evaporation. It is, however, a two-stage process. Ziaja *et al.* found that the metallization of PP fabric by pulse magnetron sputtering gives the possibility to obtain metallic layers with very good adhesion, which is impossible to get by other methods. The layer formed is very fine, which is measured in nanometers [73].

1.5.3.3 Carbon Nanostructures

In recent years, CNT/graphene-based polymer composites are commonly used as EM shielding materials, as they are lightweight, have ease of processing, and good electrical conductivity [74].

Bonaldi *et al.* produced flexible, lightweight, and porous conductive fabrics as EM shielding by knife-over-roll coating and using combinations of CNT, conductive polymer, and metal nanoparticles. The coating thickness was 100–200 μm, and the EM range tested was 200–1000 MHz. EM shielding fabrics of 95–99.99% (15–40 dB) were obtained, and CNT was found to be the most effective material [75].

Kowsari *et al.* created a magnetic functional graphene oxide (MFGO)/epoxy resin composite. The test results showed that a high thickness resulted in excellent microwave absorption [76].

Lin *et al.* utilized the yarn coating process to ensure that multiwall carbon nanotubes (MWCNT) can be arranged along the axial direction via a drawing force. In this method, PET yarns were used as the core and fed into the yarn coating machine. Then, PP/MWCNT composite pellets were coated over the PET yarns via melt extrusion. The conductive yarns were then made into conductive knitted fabrics and woven fabrics. Changing the structure of conductive fabrics and the number of lamination layers also significantly improved the EM shielding effect [77].

1.6 UV-Protective Textiles

Sunlight spectrum ranges from ultraviolet to infrared EM radiation. UV wavelength is divided to three parts as UV-A from 320 to 400 nm, UV-B from 280 to 320 nm, and UV-C from 200 to 280 nm, in which the energy per part is significantly increased respectively. While, the UV-C region is completely absorbed by the ozone layer, the UV-A and UV-B regions of wavelength can reach the earth's surface (Table 1.2) [78].

Too much exposure to short-wavelength EM radiation can lead to damage substance surface. In that case, exposure to UV radiation cause damaging and decaying of nonwoven car supplies, fading of colored clothes, and skin burning for lightweight summer garment wearer. This could result in a series of damaging health effects such as acceleration of skin aging, sunburn, erythema (skin reddening), promotion of photo-dermatitis, and even severe skin cancer (Figure 1.11) [80].

Table 1.2 Sunlight wavelength properties that reach earth [79].

Category	Wavelength (nm)	Relative intensity	Mean of photon energy (kj/mol)
UV-B ray	280–320	0.5	400
UV-A ray	320–400	2.4	350
Visible ray	400–800	51.8	200
Infrared ray	800–3000	42.1	63

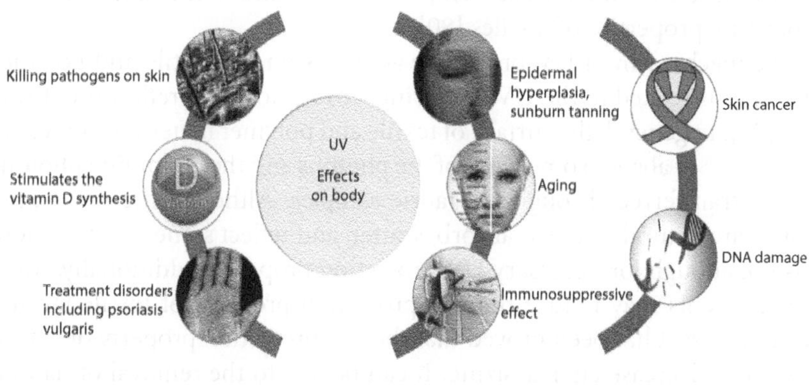

Figure 1.11 The effect of ultraviolet radiation on human skin (positive effects on the left and negative effects on the right).

1.6.1 The Necessity of Using UV-Protective Textiles

Nowadays, transformation of UV sun light has increased due to the destruction of ozone layer, and as a result, it has led to more skin contact with UV radiation. Therefore, modification of fabric properties as a layer to protect the skin from harmful radiation has been investigated more by researchers. Surface modification of textiles to protect against the UV radiation in children's, athletes, outdoor workwears, and leisure wear is more important [78].

UV protection properties often make use of a transparent layer of UV-absorbent materials on the surface of the fabric. In other words, chemical agents or nano inorganic materials such as UV absorber and UV blocker can be applied on textile fabrics to improve UV protection factor (UPF) and sun protection factor (SPF). UPF and SPF are described as the UV protection properties of a modified substance. Some factors such as additive agents, material and weaving method of the fabric, and shade of colored fabrics are able to play an important role in the protection of skin against UV radiation. It can consequently reduce the possibility of skin burning and cancers by increasing SPF of the fabrics. The UV protection property of fabrics and garments holds a great significance to end users all over the world and researchers are still studying on each of the effective factors that can influence the SPF of fabrics [45].

1.6.2 UV Protection Effect of Textile

The UV protection property of a textile fabric depends on fiber types, weaving structures, finishing modification, and color of fabric [81]. Azeem *et al.* critically discusses and modeled the effect of fabric structure parameters such as ends/inch, picks/inch, warp count, and weft count on the UV protection properties of textiles [80].

The mechanism of UV protective activities on the textile and polymer materials is related to the UV scattering, absorption, and reflection of surface [82]. In general, the surface of textile and polymer materials can reflect and scatter or absorb some parts of the photons and the remaining photons can be transferred through the fabric samples. Although the raw textile fabrics are naturally able to absorb, scatter, and reflect some UV radiation, it is not enough for necessary UV protection property. Additionally, some pretreatments may reduce the UV protection property of textile fabrics. For example, it has been proved that the UV protection property of cotton fibers will decrease after desizing. It can be due to the removal of natural pigments and lignin of cotton fabric after desizing [83].

1.6.2.1 UV-Protective Textiles with Nanomaterials

Modification of textile and polymer surfaces by chemical agents and inorganic nanoparticles is significantly investigated. Inorganic nanoparticles such as TiO_2, ZnO, SiO_2, Cu_2O, CuO, Al_2O_3, and reduced graphene oxide are more abundantly utilized owing to the thermal and chemical stability at high temperatures, permanent stability under UV rays, and non-toxicity compared to the organic ones [84–86].

Among the nano inorganic materials, nano TiO_2 and ZnO have some important advantages such as low cost [87], white appearance, photo catalytic oxidation capability [88], self-cleaning and UV-blocking properties [89, 90], and absorption of a larger fraction of the solar spectrum [87, 91]. Modification of textile and polymer substances by metal oxide as UV protection agents can improve the absorbency percentage in the UV region. However, by modifying textile and polymers via increasing the total reflectance and absorbance, the amount of transferred UV light can be reduced. Therefore, the UV protection of modified textile and polymers will increase.

The mechanism of the UV absorption characteristic is certified to the electronic structure of nano metal oxide. Indeed, modification of textile and polymers by the metal oxide nanostructures can absorb UV radiation with the energy of hν that is equal to or surpasses their band gap energy (Eg). The Eg of nanostructures lies in the UV region of the solar range, and therefore has excessive potential to absorb UV rays due to a great specific surface area [81, 92]. In other words, by increasing the surface-area-to-volume ratio of materials in nano scale, more absorption of UV rays can be enhanced [91]. For example, TiO_2 in nano scale shows superior UV-protection properties than that of the pigment TiO_2 in micro scale possessing its stronger UV absorption properties. Also, enormous weak scattering intensity visible for nano TiO_2 showed slight effect on the matrix color. Moreover, nano TiO_2 has higher UV-A and UV-B absorption capacity than most of the other metal oxides and it has the highest refractive index among metal oxides. Because of great specific surface area and high surface energy and activity, the TiO_2 nanoparticles have good affinity to textiles and polymers. Alternatively, when nano-TiO_2 is used as a UV protecting agent, the negative effect of blue shift on the absorption edge will decrease the absorption of UV-A close to visible light [92]. However, using TiO_2, nanoparticles have proved superior UV protection property for textiles and polymers. Some researchers used bi-component metal oxide to improve UV protection of textile and polymers. Ibrahim *et al.* used Ag/TiO_2 nanoparticles on linen/cotton (50/50%) fabrics for antibacterial activity, UV-protection property, self-cleaning ability, and durability to washing process. They claimed that these nanoparticles increase the UV-blocking and or scattering capability of UV-rays by changing the fabric color from white to brownish [93]. In another research, Cu_2O/ZnO nanoparticles were *in situ* synthesized to the cotton fabric surface in order to improve the UV protection property. The results revealed that the improving UV protection property is much more in the case of using Cu_2O/ZnO nanoparticles than using ZnO and Cu_2O nanoparticles alone on cotton fabric. They explained that ZnO nanoparticles have greater UV absorption and reflectance than Cu_2O nanoparticles, which guards next-to-treated cotton fabric and both are capable to shield fibers together by *in situ* formation of a thin layer of particles on the surface of the fabric. Consequently, when ZnO and Cu_2O nanoparticles were positioned on the cotton fabric together, more UV rays can be absorbed due to the narrower band gap of this bi-component and synergistic photo absorption activity can be obtained. By studying the effect of *in situ* parameters such as curing time on the UV protection property, the obtained results showed that by increasing the curing time, the UV protection property improved [94–97].

They revealed that the formation of Cu_2O/ZnO caused by the conversion of the residual of metal hydroxides to metal oxides could be increased under long exposure to heat of the curing process. Consequently, *in situ* synthesis of Cu_2O/ZnO nanoparticles on the cotton fabric showed a synergistic effect on the UV protection property [83]. Prasad *et al. in situ* synthesized ZnO nanoparticles by the spraying and dipping process in order to modify antibacterial and UV protection properties on treated cotton fabrics. They claimed that producing durable antibacterial and UV protection properties on cotton fabric is a result of negative charge of cellulosic chains under alkaline condition that provide abundant sites for the nucleation and formation of ZnO nanoparticles [98, 99].

UV or even visible light radiation causes some negative effects on textiles and polymers such as photo shading of dyed fabric, reduction of mechanical properties, and flaking and photo yellowing in wool fabrics. Therefore, using inorganic nanostructures can postpone these effects on textile fabric. For example, wool textiles have the lowest photo stability compared to other types of fibers, such as cotton or synthetic fibers. Therefore, light irradiation can cause photo yellowing on wool fabrics [100, 101]. Inorganic metal oxide nanoparticles such as TiO_2, ZnO, and Al_2O_3 delay of photo yellowing in wool textiles [102, 103]. It has been demonstrated that UV protection and flame retardancy properties on textile fabrics result in applying Mg–Al nano layered double hydroxide. In other words, Barik *et al.* claimed that good UV protection can be achieved together with both higher mechanical strength and flame retardant property at higher stiffness due to release of bound water and metal oxide residue through igniting [104].

A novel approach for making UV-protective textile is the use of nanofibers as a protective mat on fabric. Embedding of UV protective nanoparticles in the nanofibers can have a major synergistic effect on UV protection properties of a fibrous structure. Several studies have been carried on for the protection against UVR by nanofibers containing nanoparticles.

Lee has reported that an electrospun PU/ZnO thin layer significantly increases the UV blocking and this property enhances with increasing zinc oxide concentrations of the nanocomposite fiber web [105].

Dadvar *et al.* showed that the TiO_2 nanofilm coated on the PAN nanofibrous mats has a considerable effect on its UV protection. They also showed that the UV protection factor (UPF) of the mats increases with decreasing nanofiber diameter [106].

In a recent research, Merati *et al.* produced ZnO and MWCNT nanoparticle-embedded electrospun polyacrylonitrile (PAN) nanofibrous mats. The results showed that the UV protection properties of the PAN/MWCNT composite nanofiber mats of 3 g/m^2 were better than that of PAN/ZnO at nanoparticle concentrations of 2.5% and higher, while the PAN/ZnO mats of ≥5% ZnO nanoparticles provided enough UV protection properties. Consequently, the composite nanofiber mats with 5% and higher ZnO nanoparticles had enough UV protection properties to cover the requirements of the Australian/New Zealand standard [107].

In addition to metal nanoparticles, carbon nanostructures can also be used to create protection against ultraviolet radiation. For example, graphene as one of the carbon materials possesses high electrical/thermal conductivity, UV blocking, and hydrophobicity [108]. However, the use of graphene in textile application because of few polar groups and weak bonding force with fibers has been limited, but graphene oxide (GO), as a functionalized derivative of graphene with a large number of polar groups that can easily bind with fibers and disperse in water, is a good candidate as a textile finishing agent [109].

Cai *et al.* prepared UV protective cotton fabric via the functionalization of cotton by *in situ* thermal reduction of GO absorbed on cotton. Conventional dip and dry approach was utilized. Both transmittance values in UVA and UVB regions decreased visibly after the cotton fabrics were coated with GO nanosheets. Compared with untreated cotton (UPF: 14.9) and GO/cotton (UPF: 20.4), reduced-GO/cotton fabrics (UPF: 35.8) possessed higher UV blocking ability [110].

1.7 Water-Repellent Textiles

Surface wettability of substrate is influenced by chemical composition and geometry of the surface. Although wettability is satisfactory in many textiles, in some applications such as clothing, furniture, protective textiles, sportswear, and automotive textiles, water repellency or in other words low wettability of the surface is desired. In this section, various physical and chemical nano finishing methods for producing water and stain-repellent textiles have been reviewed.

1.7.1 Are Water-Repellent and Waterproof Finishing the Same?

A waterproof textile is one in which the pores, the open spaces between the yarns and fibers, are filled with substances that make integrated surface and very low air permeability. In simpler terms, water cannot pass through the waterproof fabrics; however, the surface of this type of fabric may get wet during raining.

A water-repellent textile is one in which the surface tension of the fabric has been decreased via a hydrophobic compound. The pores are not filled in this finishing process, so this type of textile is permeable to air and water vapor. Water tends to form droplets on a water-repellent fabric. The tendency of textiles to resist wetting is revealed by measuring the contact angle between water and substrate.

The size of angle reveals the repellency of the textile. When the contact angle is less than 90 degrees, the fabric is wettable. When the angle is above 90 degrees, the fabric is repellent or non-wettable (Figure 1.12) [111]. Various hydrophobic compounds such as waxes, petroleum-like molecules, soaps of polyvalent metals containing long-chain hydrocarbons, and compounds with low surface tension like fluorocarbons are used in order to produce water-repellent fabrics. Surface tensions and contact angles of polymers, which are used in textile finishing, are mentioned in Table 1.3.

In addition to the chemical finishing, water repellency is created on the fabric due to the surface roughness. An uneven surface compared to a flat

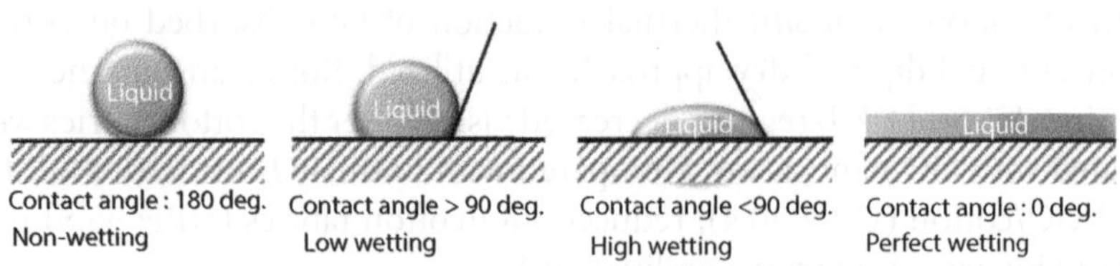

Figure 1.12 Different shapes of drops on a textile substrate.

surface with the same chemical structure usually shows a higher contact angle. The presence of roughness in micro/nano scale can lead to the super hydrophobic textiles with contact angle about 160 degrees.

Both surface-free energy and surface roughness are critical factors for wettability. As a natural sample with superhydrophobic performance, lotus leaves and insect wings are made up of micro/nano-scale structure. In many researches, nanoparticles were added to the surface of a substrate to change the surface tension or surface roughness by chemical and physical methods such as *in situ* synthesis, anodic oxidation, electrodeposition, chemical etching, plasma etching, laser treatment, electrospinning, chemical vapor deposition, sol–gel processing, spray-coating, etc. [113, 114].

1.7.2 Plasma Treatment

Plasma is ionized gas consisting of a mixture of free electrons, ions, and neutral atoms (depending on the degree of ionization) and is formed when sufficient energy is put into a gas. Under laboratory conditions, ionized gas is generated by using high electric fields, known as cold plasma or electric discharge. Ionized molecules with high energy can be bombarded on the surface of a given substrate. Most of the chemistry occurs on the surfaces within the reactor due to the high-energy ion bombardment. High-energy ions cause relatively unselective fragmentation of the surface-adsorbed species and a deposition or etching process results, depending on the power and molecules involved [115].

Low pressure plasma processes have been widely utilized for the modification of surface properties of natural and synthetic textiles. The surface

Table 1.3 Surface tensions and contact angles of conventional polymers in textile [112].

Polymer	Contact angel against water	Surface tension (mN/m)
PTFE (Polytetrafluoroethylene)	108	18.5
PVLF (Polyvinylidene fluoride)	82	25
PVF (Polyvinyl fluoride)	80	28
PE (Polyethylene)	94	31
PVC (Polyvinyl chloride)	87	39
PET (Polyethylene terephthalate)	81	53

properties, which depend on the feeding gas, input power, pressure, electrode geometry, and other operating conditions, allow to obtain etching, grafting, cros–s-linking and deposition with the same experimental apparatus [116]. Several research groups investigated the hydrophobicity of polymers and textiles using plasmas containing fluorocompounds such as tetrafluromethane [117], hexafluoroethane [118], hexaflouropropane [119], etc. Small molecules such as CF_4 or CF_6 as plasma additives can form short polymer chains hanging on the treated fabric surface. So despite their good initial hydrophobicity (contact angle > 120 degrees), they do not have acceptable washing or dry-cleaning fastness [120].

In order to improve water repellency and anti-stain characteristic of PET textiles, De Vietro *et al.* used plasma enhanced-chemical vapor deposition (PE-CVD) with a large fluorinated molecule for depositing a 100 nm thick layer with F/C ratio higher than 1.4 on textile and obtained a water contact angle around 150 degrees and oil contact angle around 120 degrees. The obtained fabrics were wear resistant, particularly if they were modified at high value of input power [121].

In another research by the surface of polypropylene fabric, samples were treated with hexamethyle disiloxane (HMDSO) to produce inorganic protective layers to reduce wettability. All HMDSO plasma-treated fabrics showed a significant decrease in demand wettability (water uptake) and a significant increase in contact angle values [122].

Hegemann used plasma treatment as a method to produce strain-repellent polyester fabric. The plasma activation in a low-pressure RF discharge with Ar/O_2 was found to support a subsequent wet chemical fluorocarbon impregnation of a densely woven PET fabric and a PET knit showing a multi-layer, open structure. The plasma process that first leads to a more wettable surface, thus enabled enhanced oil repellency at a reduced film thickness in the sub-micron range leaving the touch and the optic of textile fabrics unaffected [123].

Acrylic fabrics were processed with an RF atmospheric pressure plasma generator and afterward a fluorocarbon finish was applied through a traditional pad-dry-cure method. Two gas mixtures were tested (helium and helium/oxygen) with different plasma treatment times. The aging of the finishing was simulated through repeated accelerated laundry cycles. The water and oil repellencies were measured through standard test methods. While the initial water and oil repellency did not change, the plasma treatment improved the durability of the finish after artificial aging. The results presented in this research confirmed the benefit of applying an atmospheric plasma treatment as a preparation step for standard water and oil repellency finishing of acrylic fabrics. Although the repellency grade was not affected immediately after the chemical application, the improvement in durability has been shown at long-term usage. After artificially aging, plasma pre-treated fabrics decreased their performance more slowly than the untreated and this gap has been quantitatively demonstrated [124].

The atomic layer deposition (ALD) technique, similar to chemical vapor deposition (CVD), is a chemical process to deposit a thin layer or film by sequential use of a gas. Al_2O_3 and SiO_2 coatings deposited on cellulose-based materials have been reported to function as gas and vapor barriers [125]. Cotton fibers were modified with ZnO to make them conductive [126]. In the ALD process, a metal organic precursor (e.g., bis (diethylamido) silane, trimethylaluminium, and diethyl zinc) and a reactant (e.g., water) are sequentially exposed to a surface desired to be modified, resulting in a complementary sequence of self-limiting reactions. ALD allows preparation of dense and pin-hole free inorganic films. ALD is expected to be a promising method to produce super-hydrophobic surfaces with topography control [127].

1.7.3 Electrospinning

Electrospinning, as an effective technique to produce nanofibers, provides a mat of nano fibers with effective nanostructured roughness. Electrospinning of nanofibers with N, N-dimethylformamide (DMF) as a solvent for the polyacrylonitrile (PAN) polymer, was performed in order to compare it with a well-known commercial protective multi-layered fabric, Gortex™. It could be concluded that using the nanofiber membranes, instead of coating material (such as PTFE), presented acceptable water-repellent properties and water vapor permeability behavior [128].

In order to simulate the "lotus effect" in nature, polyvinylidene fluoride (PVDF), polyvinylidene fluoride/polyethylene glycol (PVDF/PEG), and polyvinylidene fluoride/polyethylene glycol/silicon dioxide (PVDF/PEG/SiO_2) nanofiber fabrics are fabricated through electrospinning and weaving technology. Hydrolyzing PEG and doping with SiO_2 nanoparticles on a hierarchical nanofiber structure in the PVDF/PEG and PVDF/PEG/SiO_2 nanofiber were performed. The results showed that the fiber structure of the PVDF/PEG composite nanofiber doped with SiO_2 nanoparticles after water scrubbing was composed of balls, which are formed by embedding the SiO_2 nanoparticles in the polymer, similar to the "small thorn" on the "lotus hill" because of the embossment of SiO_2 nanoparticles on the "ball" surface. The PVDF/PEG/SiO_2 nanofiber fabric exhibits super hydrophobicity with a water contact angle of 173.2 degrees [129].

Cakir *et al.* used electrospinning for coating of cotton fabric surface with the polyurethane/silica hybrid solutions. In this study, PU/silica hybrid solutions were prepared by the sol–gel method. Coating of the solution onto the fabric surface as a thin layer was carried out using an electrospinning process combined with a UV curing step. This thin layer cannot be obtained by dip-coating or other textile finishing methods. The innovative method of this study is to choose the electrospinning process to obtain a very thin coating layer instead of nanofibers onto just one surface of the fabric with the combination of UV-curing. Thus, the other surface of the fabric remains uncoated without any clothing comfort loss. Furthermore, during the electrospinning process, with the help of a continuous UV radiation, a well and deeply crosslinked coating layer can be obtained on the fabric surface. The effect of fluorinated silane-terminated urethane on the water-repellency was investigated by contact angle measurements. Water-repellency was evaluated by contact angle measurements. A sample that contained 50% inorganic part showed a contact angle of 154.5 degrees. After the domestic laundering step, this contact angle value decreased to 137 degrees because of the surfactant residue of the detergent [130].

1.7.4 Pulsed Laser Deposition

Pulsed laser deposition (PLD) is a highly versatile tool to prepare a functional thin film on textile surfaces. PLD is a dry process, very versatile, and has several advantages in the formation of thin films of functional materials. A laser is utilized as an energy source to evaporate a target material that is placed on a carousel. The laser plume deposits onto the surface of the substrate that is placed on the sample stage (Figure 1.13). Through PLD, it is possible to control the film thickness digitally at the angstrom scale. It is also possible to deposit a hetero-structure or multi-layer film whose structure differs from layer to layer by changing the ablation target appropriately. Among available organic materials, polytetrafluoroethylene (PTFE) has been the main target in PLD because it exhibits poor solubility in all solvents and has low surface adhesion. Note that conventional wet processes such as spin-coating methods are not applicable for the formation of thin PTFE films [127].

Deposition of polytetrafluoroethylene (PTFE) thin films on a cotton fabric was performed via PLD at room temperature. The PTFE-coated fabric showed super hydrophobic effect by a water contact angle of 151 degrees. The presence of uniformly grown PTFE grains with a grain size of about 50–70 nm was confirmed by scanning electron microscopy micrographs [131].

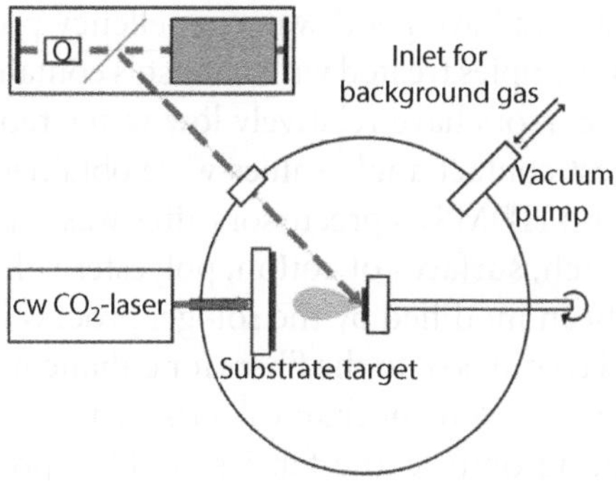

Figure 1.13 Scheme of the deposition chamber with Q-switched and substrate heating laser [132].

1.7.5 Sol–Gel Technique

The sol–gel technique is an effective and easy method for depositing coating onto the textile substrate in order to prepare water-repellent property. Large deposition area, uniform deposition on substrates with different shapes, and time saving are the main advantages of this finishing method. Yin and Wang prepared a water-repellent cotton fabric via coating hybrid SiO_2/hexadecyltrimethoxysilane (HTEOS)/γ-chloropropyltriethoxysilane (CPTS) sol. Despite the hydrophilic effect of cotton fabric because of massive hydroxyl groups on the surface of fibers, the contact angle of the treated fabric increased to 139.8 degrees. During this finishing process, the Si-O-Si chains via dehydrating and condensation reactions were deposited on the cotton surface and most of the OH groups on cellulose fabric were blocked by the hybrid coating. Silica nanoparticles caused a rough microsurface (mean roughness was about 16.4 nm and the root mean square was 21.1 nm) with a low surface energy and good hydrophobic water repellency [133].

Onar *et al.* obtained water- and oil-repellency properties for cotton fabrics by using the sol–gel process. To achieve this aim, they treated cotton fabrics by using the sol-gel process, with tetra ethylorthosilicate (TEOS), tridecafluorooctyl triethoxysilane (TDFOES), hexadecyltrimethoxysilane (HDMS), vinyltrimethoxysilane (VTEO), phenyltriethoxysilane (FES), 3-aminopropyl trimethoxysilane (AMMO), glycidyloxypropyltriethoxysilane (GLYEO), zirconium (IV) acetylacetonate (zrA), and titanium (IV) isopropoxide (TIPT) as precursors. It was found by using a spray test that fabric samples treated by nanosols have excellent water-repellency properties (90–100), while untreated fabric has hydrophilic properties. The fabric samples treated with nanosols containing TDFOES, HDMS, VTEO, ZrA, and TIPT as precursors have good water-repellency properties (90–100). However, the fabric samples treated with nanosols containing FES, AMMO, and GLYEO as precursors have relatively low water-repellency properties (70–80). The highest contact angle values were obtained for samples containing TDFOES and HDMS as precursors after washing [134].

In another research, surfaces of cotton, polyester (PES), and polyamide (PA) fabrics have been modified by the sol–gel process in order to achieve silica-based hybrid coatings onto the fibers for enhancing their water repellency, thermal stability, and mechanical properties. In order to achieve such a hybrid coating onto textile fabrics, a chloropropyltriethoxysilane (CPTS) precursor was utilized and hydrolyzed in the presence of hydrochloric acid (HCl) as a catalyst to form reactive silanol groups by replacing some ethoxy groups (-OC_2H_5) of CPTS by hydroxyl (–OH) groups. In the subsequent reaction of condensation, hydroxyl groups reacted between each other or with the textile fibers leading respectively to siloxane bonds (Si-O-Si) or covalent bonds (Si–O–C) in the case of cotton and hydrogen bond in cases of PES and PA. Regardless of the type of textile, the sol-gel-treated samples exhibit a hydrophobic surface in addition to efficient water-repellent properties. The best results were achieved with the sol containing the highest CPTS concentration whatever the impregnation time of the textiles [135].

1.7.6 Dendrimer

Dendrimers are repetitively branched macromolecules with a three-dimensional structure and their repellency effect depends on build-up crystal structures in a nanostructure (Figure 1.14). Good water repellency can be obtained via using dendrimers and combination of dendrimers with fluorocarbon polymers in order to force them into a more ordered structure leading to self-organization of the fluoro polymers on the outer layer of the textile [136, 137].

The repellency effect of dendrimers is based on their forming nano-sized crystalline structures that impart wash resistance through water repellency and high abrasion resistance. Sancaktaroğlu investigated changes in water and oil repellency values after treating cotton and cotton/polyester fabrics with products containing a dendrimer, a fluorocarbon, and a dendrimer–fluorocarbon combination. The dendrimer–fluorocarbon treatment gave the best oil and water repellency effect [138].

Namlıgöz *et al.* applied polymeric dendrimer-containing fluorocarbon, nano-sized fluorocarbon (FC) polymer, nano-silica acid, and conventional agents (e.g., a paraffin emulsion containing zirconium salt and a

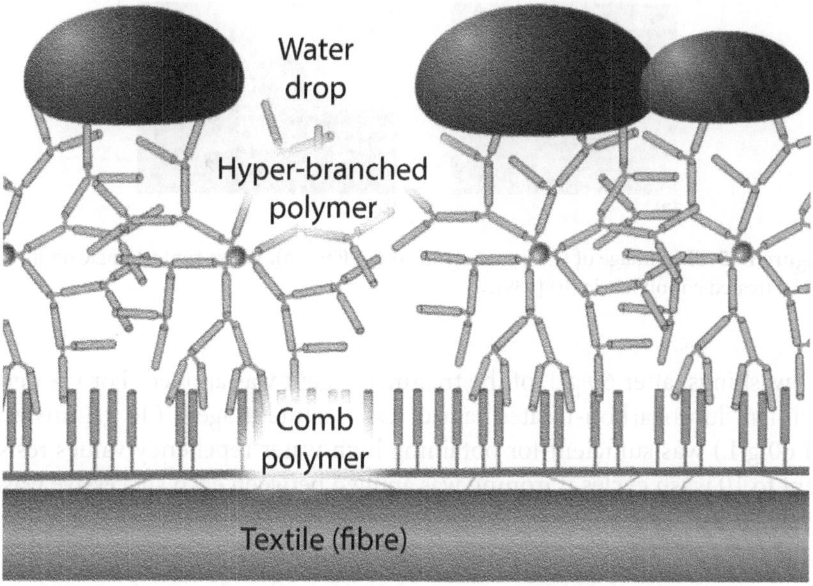

Figure 1.14 Water-repellent effect of ®RUCO-DRY ECO on textiles [139].

conventional fluorocarbon compound) to cotton fabrics. Water repellency results revealed new chemicals such as the polymeric dendrimer containing FC and the FC nanomolecule had better results than conventional ones in terms of performance and washing resistance. Stain release results for dirty motor oil and olive oil soiling clearly showed that especially the polymeric dendrimer containing FC, the FC nanomolecule, and the conventional FC compound had better results than other samples. When comparing the stain repellency performances of all the chemicals for tea soiling, they had similar and good results for coffee soiling without paraffin emulsion containing zirconium salt and nano-silica. The performances of all the chemicals were dependent on the concentration [140].

Atav *et al.* applied dendrimer, fluorocarbon, and dendrimer–fluorocarbon to undyed woven cotton fabrics using the pad-dry-cure method. The water repellency of the treated cotton fabrics tested in this study gave the following results: dendrimer-fluorocarbon treated > fluorocarbon treated > dendrimer treated. The optimal application conditions of the fluorocarbon and dendrimer–fluorocarbon-treated fabrics were 60 g/L of the treatment agent at 160 °C for 2 min and 60 g/L of treatment agent at 140 °C for 3 min, respectively. These treated fabrics gave water repellency values of 100 that could be maintained even after 10 successive wash cycles. The dendrimer-treated fabric gave a water repellency value of 50 after

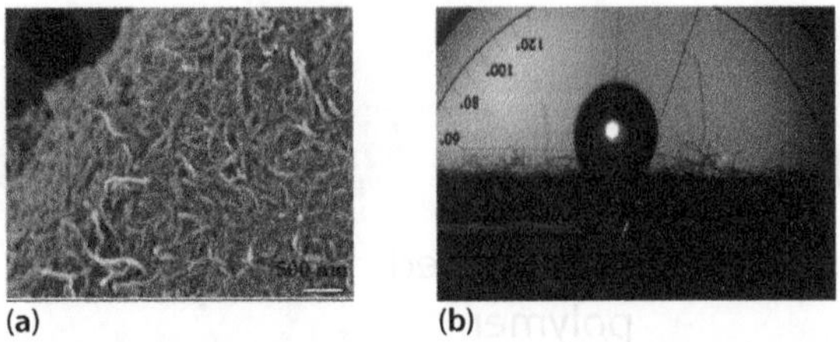

Figure 1.15 SEM image of CNT coating on cotton fiber (a). Water contact angle on the CNT-treated cotton fabric (b) [139].

10 washings, after 60 g/L of the treatment agent was applied. For the dendrimer–fluorocarbon-treated fabric, less treatment agent (40 g/L instead of 60 g/L) was sufficient for obtaining high water repellency values resistant to 10 wash cycles if ironing was applied between each successive wash cycle [138].

1.7.7 Carbon Nanotube

Another possibility to prepare water-repellent textile is employing CNTs with a long structure shape. However, it is difficult to grow CNT arrays on common textiles due to van der waals forces of CNTs bundles. The assembly of CNTs on the fabric surface can lead to create a lotus-like leaf surface at the nanoscale. Liu *et al.* showed that cotton fabric with perfect water absorbability, after surface grafting with CNTs-poly-butylacrylate composite via dip-dry-cure finishing method, showed super hydrophobic effect with water contact angle larger than 150 degrees (Figure 1.15) [139].

1.8 Self-Cleaning Textiles

One of the new intelligent textile products, which attract many researchers, is self-cleaning textiles. The development of self-cleaning textiles can extend the applications of the routine textiles, which have been used for a long time. One of the benefits of these products is their ability to do the self-cleaning process without any need to common laundering methods. In addition, the self-cleaning properties of textiles can increase the durability of the fabric [141].

In recent years, several scientists studied the modification of innovative textiles with new properties. Some of these properties such as the self-cleaning characteristic of textiles are inspired from nature.

There are two approaches for self-cleaning finishing of textiles.

- In the first approach, a super hydrophobic surface is provided for the textile, which has the self-cleaning properties.
- In the second approach, some nanoparticles such as titanium oxide are used to cover the textile fabric in nano scale. These nanoparticles can decompose the stains and dirt on the textile fabric in the presence of water, oxygen, and sunlight [142].

1.8.1 Self-Cleaning and Superhydrophobic Surfaces

For the preparation of superhydrophobic surfaces or films, the combination of surface roughness and low surface energy is required. From a viewpoint of surface roughness, hydrophobicity is competitive with its transparency because surface roughness becomes a scattering source of light. Thus, the precise control of roughness is required to satisfy both hydrophobicity and transparency. We have recently succeeded in developing a new process to prepare transparent superhydrophobic boehmite (AlOOH) or silica (SiO_2) films by using aluminum acetylacetonate (Al $(C_5H_7O_2)_3$, sublimation temperature 193 °C) as a sublimation material. Either the boehmite or the silica film is roughened by the sublimation of aluminum acetylacetonate after calcination. Then, transparent superhydrophobic films are obtained by subsequent coating of (fluoroalkyl)silane [143].

The self-cleaning properties of the textiles can be obtained via chemical and geometric modification of the surface. As a result, adhesion of dirt particles can be dramatically reduced and a real self-cleaning effect can be achieved. The effective surface contact area for dirt particles is extremely minimized by the surface structure and thus the adhesion will be very low.

In a rough surface, the adhesion energy of the particle to the solid surface is very low which is because of the roughness of the surface and small contact area. Consequently, dirt particles from a superhydrophobic surface are completely removed, in contrast to a smooth hydrophobic surface where the energy is higher allowing only for the relocation of particles. For a smooth surface, the adhesion energy between the particle and the solid surface is relatively higher than between the particle and the water drop.

1.8.1.1 Natural Superhydrophobic Surfaces

There are almost 200 types of different natural super hydrophobic surfaces, which have self-cleaning properties due to either having specific chemicals on their surface or a special morphology [144].

As mentioned earlier, scientists are trying to make use of the natural phenomenon to overcome the limitations of textiles. As an example, the use of textiles is limited due to their soiling and wetting behavior in many applications. That is why textiles are improved with a variety of treatments to meet the needs of the end users. Recently, several products have been invented that make use of the Lotus Effect and implement self-cleaning properties to a textile surface. This achievement can lead to a significant reduction in the cleaning requirement of such surfaces.

Using an efficient method, nature can achieve the self-cleaning properties, which has been perfectly realized on the leaves of the lotus plant. Besides this species, self-cleaning properties can be found on a variety of other biological surfaces, such as cabbage, reed, and nasturtium. The main function of nanostructured superhydrophobic surfaces in nature is probably the protection against pathogenic organic contamination such as bacteria or spores [145].

Scientists have fabricated superhydrophobic textiles by creating surface roughness by using low surface energy materials inspired by the self-cleaning properties of lotus leaves. These materials include organic silanes, fluorinated silanes, alkyl amines, and silicates [146]. Popular surface modification methods include wet chemical reactions, self-assembly and sol–gel, layer-by-layer deposition, polymerization reactions, colloidal template techniques, chemical vapor deposition (CVD), plasma treatment, and electrospinning [147].

The wetting behavior of the surfaces is determined by the measurement of the static contact angle of the water droplet with the surface. The water contact angle with the surface can show the hydrophobicity or hydrophilicity of the surface, which plays an important role in achieving the self-cleaning properties of the surface [148].

Although some chemicals such as fluorocarbons can help scientists to achieve the contact angle of 120 degrees, these surfaces do not fit in the self-cleaning category of textiles [149].

1.8.2 Superhydrophobic Finishing of the Textiles

Hydrophobicity of the textiles had been an important issue since more than 50 years ago. However, with the improvement in science and technology especially nanotechnology, scientists could achieve super hydrophobic textiles with self-cleaning properties.

Super hydrophobicity of the textiles can be obtained via using hydrogels, nano composites, and other nano structures. These treatments can be applied via different methods onto textiles. These methods can include simple ways such as padding, spraying, and surface coating to the more complex methods such as using nano structures, nano particles, nano wires, CNTs, and oxides of Zn and Ag [150].

The surface of textiles is different from other materials such as glass or metal plates in terms of flexibility and roughness. The roughness of the surface of the textiles makes it possible to have a super hydrophobic fabric via creating nano scale roughness.

Scientist suggests two methods for obtaining super hydrophobic textiles:

- Making roughness on the surface of the fabric using low surface energy materials
- Modification of a rough surface using low surface energy materials [151].

The important issue in the self-cleaning finishing of textiles is not only the creation of a super hydrophobic surface, but also maintaining the induced properties for a long time [152].

In order to fabricate the cotton fabric with super hydrophobicity, scientists suggest using chemicals such as metals, silanes, and some methods such as coating with gas and layer-by-layer finishing. Using some chemicals such as silver and zinc oxides can provide some other properties such as being antibacterial and conductive as well as super hydrophobicity.

Some methods for the production of self-cleaning textiles are represented in Table 1.4.

1.8.3 Modification of Textiles Using Photoactive Coatings

Recent researches have shown that via optical decomposition of organic substance using semi-conductive metals, self-cleaning surfaces can be obtained [153].

TiO_2 is one of the best candidates for a photocatalytic application. Nano-sized TiO_2 particles show high photocatalytic activity because they have a relatively large surface area per unit mass and volume, which facilitates the diffusion of the surface-generated charge carriers under light irradiation. Coating of textiles with TiO_2 has drawn attention of researchers due to its advantages such as cost saving, chemically stable, non-toxicity, and biocompatibility.

Titanium dioxide is used in self-cleaning of textiles with two approaches:

- Photocatalysis oxidation
- Super hydrophilicity properties.

Table 1.4 Several methods of production of self-cleaning textiles.

Synthetic fibers	Cotton
Silicone coating on the surface of polyester microfibers.	Using CNTs for making nano scale roughness.
Coating of synthetic fibers with an electro spun mat containing lowering surface tension materials.	Using silica nanoparticles containing amine groups.
Coating of polyamide 66 fibers on polyester.	Coating of perfluorocarbons using PECVD method.
	Using metals for making surface roughness and coating the surface with hydrophobic materials.

When the surface coated with TiO_2 is exposed to solar radiation, stimulated radicals on the surface of the textile can decompose organic substances such as dirt, stains, bacteria, and other materials to carbon dioxide and water [153].

Another approach is using TiO_2 with SiO_2, which shows great hydrophilicity when exposed to sunlight. In this situation, some of the oxygen molecules leave the surface and water molecules fill the empty positions. As a result, the surface is coated with a water coating, which can clean it continuously. This effect can remain on the surface of the textile until a couple of days after the sun exposure [142]. The synergic effect of using TiO_2 via photocatalysis effect and super hydrophilicity makes it an appropriate material for using it in self-cleaning products.

In one of the studies, TiO_2 nano particles and their derivatives were used to produce a transparent superhydrophobic surface. The results demonstrate that microstructure and surface roughness of the films were changed with increasing TiO_2 concentration, which attributed the size difference of the starting materials. Also the contact angles of the different films are also increased and the contact angle of the film containing 2 wt % TiO_2 was higher than 140 degrees [154].

In spite of numerous studies in the field of self-cleaning and super hydrophobic textiles, there are still some challenges relating to their short lifetime and fragile structures. Another problem is the tendency of these surfaces absorbing some types of dirt such as oily substances. Using some metal oxides such as titanium in comparison to superhydrophobic surfaces can provide a more durable and stable finishing on the textile fabric. Some other properties such as antibacterial, UV protective, and anti-odor surfaces can be obtained via using TiO_2.

In several studies, fibers are usually changes to a rough surface with nanostructures, such as nanoparticles, which face the issues to strengthen the interfacial interaction between the nanostructures and the substrate for application-oriented superhydrophobic surfaces. Alternatively, surface roughness could also be created directly from the bulk materials at the surface, resulting in hierarchical structures monolithic with the bulk substrate. The main advantage of this method is that these roughening structures possess strong adherence and mechanical stability [155, 156].

In one of the studies, PET fibers were roughened through chemical etching at the surface and durable superhydrophobic textiles could be prepared by following diffusion of fluoroalkylsilane into the fibers. However, diffusion of hydrophobic alkylsilanes into fibers shows dependence on specific equipment and conditions. It has also been suggested that some fluorochemicals have potential risks to human health and the environment [157].

In another study, the combination of a photoactive agent such as TiO_2 and a superhydrophobic agent such as SiO_2 was used to obtain the photocatalytic self-cleaning property on fluorine-free superhydrophobic cotton fabrics. The cotton fabrics coated with TiO_2–SiO_2 composite particles exhibit superhydrophobicity with a water contact angle of 160.5 degrees [158].

1.9 Flame-Retardant Textiles

Along with the rising trend of stringent fire safety regulations, demands for reduction in the fire hazards caused by highly combustible materials such as textiles and polymers have become a matter of significant importance. Therefore, numerous attempts have been made to improve the flame-retardant properties of textiles for a variety of applications.

Natural fibers especially cellulosic structures can ignite fast and may cause a fire. In order to improve self-extinguishing of these fibers, researchers developed flame-retardant agents for finishing of fabric. Indeed, flame-retardant finishing improved the resistance of the fabric against igniting. Furthermore, the flame spread rate decreased as a result of using flame-retardant finishing.

Another method was developing synthetic fibers to produce fire resistance types of regular fibers besides developing inherently flame- and ignition-resistant fibers created by aromatic structured polymeric chains such as aramids and the poly(meta-aramid) fiber Nomex (Du Pont). This was followed by the production of poly (para-aramid) fiber.

Kevlar (Du Pont), which has great modulus and tenacity, in addition to its great heat and fire resistance [159]. Lastly, some modacrylic and flame-retardant viscose fibers were developed for inherently fire-retardant fabrics. These fibers were usually used for firefighting clothes and anti-flash arc work wears. Based on the type of flame-retardant agent, there are different flame-retardant mechanisms which contain:

- Suppression of thermal decomposition by endothermic reaction (hydrated metal compounds),
- Dilution of decomposition products by releasing inert gas (nitrogen-based compounds)
- Trapping of free radicals in oxidation field (halogen-based compounds)
- Formation of surface layer to block heat and suppress diffusion of decomposition products (phosphorus-based compounds) [160].

1.9.1 Flame-Retardant Finishing Agents

Although active halogen compounds were almost used before to obtain fabric flame retardancy, halogenated flame retardants have been found having significant disadvantages for their toxic and corrosive properties [161]. In fact, halogen-based compounds are prone to cause incomplete combustion due to the mechanism of trapping free radicals and thus to generate a large amount of smoke [160]. Furthermore, some combustion products such as furans and dioxins cause some serious health effects and environmental risks. Therefore, researchers have developed halogen-free flame-retardant agents for fabric finishing. Flame-retardant agents, containing phosphorus, boron, nitrogen, silicon, and other nanostructures, have been investigated to develop the flame resistance of fabrics and polymers [161]. Among different halogen-free flame retardants, phosphorus-based agents show high flame-retardant properties. However, these compounds usually have low water resistance to cause significant reductions in insulation properties and appearance in high temperatures and high humidity [160]. Therefore, using non-halogen and non-phosphorous based flame retardant agents in different applications is preferred.

There are various nanostructured materials with fire-retardant features. In fact, nanostructures have been recently used to improve flame-retardant property of fibers by melt spinning polymer/nanoparticles or a combination of phosphorous material/nanoparticles through finishing treatments.

In this section, some flame-retardant nanomaterials used in textile finishing are described.

1.9.1.1 *Flame-Retardant Nanostructures*

1.9.1.1.1 Nano Clay

Clay exists naturally in the form of nanoscale platelets, stacked up to form particles about few microns. Most of the beneficial properties of clay as a filler material are only realized when the particles are dispersed and exfoliated to separate the individual nanoscale layers. So, exfoliated nano clays can exhibit excellent flame-retardant properties [162]. The most commonly used layered silicate in polymer nanocomposites is MMT, which has encouraged much improvement in the mechanical, thermal, flammability, and barrier properties of polymers owing to its high aspect ratio, large surface area, extraordinary modulus, and nano-scale dispersion [163].

Nano clay particles play a vital role in flame-retardant textile coating. However, the exact mechanism of the combustion inhibition caused by nanoclays is not yet known but it seems that when the polymer matrix is burned and gasified during combustion, the incorporated nanoclays accumulate at the surface and act as a barrier to oxygen diffusion. So, the burning process is slowed down. Also, the nano clays act as a barrier to flame propagation and promote the formation of chars but they have no effect on the after-flaming properties [164].

As mentioned earlier, nano clays containing polar functional groups such as OH, -NH- or NH_2 will lead to nano dispersion in polar and hydrogen-bonded polymers, for example, polyamides 6 and 6.6 and polyvinyl and polyacrylic and can be used for flame retardancy of fiber and fabric back coating finishing. Although cotton fibers have desirable characteristics such as comfort, good mechanical properties, biodegradability, and excellent wearing properties, it has some disadvantages such as high flammability and hydrophilicity. These properties are the reason for the limitations of the application of natural textiles in some branches such as workwear garments [165]. So, flame-retardant finishing on cotton fibers in some applications is necessary.

There are many researches about using clay nanoparticles as a flame retardant. For example, in order to fabricate flame-retardant finishing with different add-on values on cotton fabrics, the organosilicon-functionalized cyclotriphosphazene N_3P_3 [NH $(CH_2)_3$Si $(OC_2H_5)_3]_6$) was synthesized and applied to the cotton fabrics through a pad-dry-cure process. The LOI value increased from 20.5 to 24.5 and excellent flame retardancy was found durable after 30 cycles of the washing process [166]. Uddin *et al.* used the two types of local clay samples to impart the flame-retardant effects on bleached cotton fabric. It was found that the two types of clays used can enhance the flame retardancy of finished fabric and significant amount of char was obtainable with increasing clay concentration [167]. Keshin *et al.* used nanoclay on 100% greige and bleached pure conventional cotton fabrics by finishing process. They showed that by utilizing nanoclay with a proper application process, it is possible to impart flame-retardant characteristics to conventional cotton fabric samples [168]. A cotton/MMT nanocomposite was prepared by Delhom *et al.* by the addition of pretreated nano clay and cotton to 4-methylmorpholine-N-oxide as a cotton solvent. The nanocomposites composed of higher than 7 wt% nano clay showed char yield about 30 wt% [163].

The improvement of flame-retardant properties in polyamide/nano clay textile nanocomposites was also reported in the literature. Wu et al. proposed flame-retardant polyamide 6/nano clay/intumescent nanocomposite fibers through the electrospinning method. Based on the transmission electron microscopy results, there was good dispersion and exfoliation of nano clay platelets within individual nanofibers. The degradation temperature of nanocomposite samples was lower than pristine nylon 6 samples. Moreover, the difference in residual char weight after decomposition was significant [169].

1.9.1.1.2 Layered Double Hydroxide

Layered double hydroxide (LDH) known as brucite-like compound is anextensive class of anionic clays with divalent and trivalent metallic cations andanions. Since the range of metals and anions which can be usedin the LDH structure is quite extensive, the LDH properties are controllable [163].

In Pan *et al.*'s research, a layer-by-layer flame-retardant coating, assembled from MgAl-layered double hydroxides (MgAl-LDH) and alginate, was fabricated onto the surface of cotton fabric for reducing its flammability. The high flame-retardant ability of coated cotton fabrics is attributed to the fact that the MgAl-LDH-filled coating can form an inorganic protective layer, which can act as a barrier to heat and oxygen [170]. Also, Barik *et al.* applied the nano-MgAl-layered double hydroxide on the cotton fabric for enhancing flame retardancy, mechanical, and UV protection properties. It was found that the limiting oxygen index increased from 16.5 to 20.8 [104].

1.9.1.1.3 Nano Silica (SiO_2)

Silica nanoparticles are considered as a potential flame-retardant agent to improve flame-retardant properties of textile fabrics. Nano silica is an environment-friendly material compared to other nanomaterials. Also, they are cheap and their chemical behavior could be easily controlled by surface modification [171]. These materials are commonly used as a fire retardant agent in different researches.

Gashti *et al.* prepared the cellulose fibers using a polycarboxylic acid/hydrophobic silica nanocomposite coating with water-repellent and flame-retardant properties. The results indicated that the embedded silica nanoparticles modify the surface of fibers, increasing the thermal stability of substrate. This enhancement of thermal properties is attributed to the presence of the silica nanoparticles, which have high heat resistance, heat insulation effect, and mass transport barrier toward cotton molecular chains [172]. Also, Carosio *et al.* introduced a novel method to improve flame-retardant properties of textile fabric using multilayered thin films. In this work, PET fabrics were coated with silica nanoparticles using the layer-by-layer assembly method. Five bilayers of positively and negatively charged colloidal silica nanoparticles increased time to ignition and decreased heat release rate peak of PET fabric by 45% and 20%, respectively. This study demonstrates the ability to impart flame-retardant behavior using a water-based, environment-friendly protective coating [173].

1.9.1.1.4 Carbon Nanotubes

Carbon nanotubes (CNTs) have also been investigated as a potential flame-retardant filler material. They can reduce heat release in burning materials very effectively. Despite the relative high cost of manufacture for CNTs, the high performance of CNT-based flame retardants and the low concentration required could make them useful for various applications.

It is reported that combining a low concentration of multi-wall CNTs with the nanoclay filler produces a synergistic effect that is much more effective than either component alone.

In a recent study, CNT was stabilized on a cotton fabric using a vinyl phosphonic acid monomer as a crosslinking agent and benzophenone as a catalyst through UV irradiation. The results showed that the polyvinylphosphonic/CNT nanocomposite improved the thermal stability and decreased the flammability of the substrate [174]. In a research performed by Yin *et al.*, polyamide 6 (PA6) was melt-blended with an intumescent flame retardant (FR), multi-wall carbon nanotubes (MWNTs), and nano clay particles to produce multi-component FR-PA6 nanocomposites. The formulation combining both MWNT and clay showed the largest quantity of char residue and a synergistic effect in flame-retardant properties [175]. Dittrich et al. investigated the influence of various nano-carbon additives (5 wt%) on the flame-retardant properties of PP. Thermally reduced graphite oxide (TRGO) and multi-layer graphene (MLG250) were compared with spherical, tubular, and platelet-like carbon fillers including carbon black (CB), MWCNT, and expanded graphite (EG). According to their results, TRGO and MLG250 indicated higher flame retardation in the nanocomposites, which can be attributed to the well-exfoliated graphenes compared to the less exfoliated, micro-sized, expanded graphite or conventional spherical and tubular carbon nanoparticles [176].

1.9.1.1.5 Metal-based Nanoparticles

Metal-based nanoparticles are used in textile finishing in order to improve their flame retardation. Various mechanisms have been suggested about flame-retardant properties of metallic nanoparticles. Metallic nanoparticles can act as a barrier to heat and mass transfer, restrict the mobility of the polymer chains, absorb active species such as free radicals, and modify the degradation pathway of the polymer. Following, we present some reports of the metal-based nanoparticle application [163].

El-shafei *et al.* performed the flame-retardant and antibacterial finishing on cellulose fabrics using TiO_2 nanoparticles and chitosan phosphate. TiO_2 nanoparticles were prepared by the sol–gel method using titanium tetraisopropoxide. The obtained results indicated that cotton fabric gained flame retardancy and antibacterial properties [177]. Also in a same study, Lessan *et al.* proposed a novel flame-retardant woven cotton fabric using sodium hypophosphite (SHP), maleic acid (MA), triethanol amine (TEA), and nano TiO_2 through the pad-dry-cure method [178]. In another research based on using TiO_2, Hashemikia and Montazer immobilized TiO_2 nanoparticles on the surface of cotton/polyester-knitted fabrics using citric acid as a cross-linking agent and SHP as a catalyst to produce a multi-functional textile with durable flame retardation, anti-pilling, hydrophilicity, self-cleaning, antibacterial, and UV protection properties. The presence of TiO_2 nanoparticles and SHP as a phosphor source increased the char residue of the fabrics by 21% [179]. It is reported in Lam *et al.*'s research that nano titanium dioxide was used as a co-catalyst in the flame-retardant formulation of N-methylol dimethyl-phosphonopropionamide, melamine resin, and phosphoric acid for cotton fabrics to improve the treatment effectiveness and minimize the side effects. It was found that the nano-TiO_2 co-catalyst had a significant effect on decreasing the flame-spread rate [180].

In a same study, Lam *et al.* used zinc oxide (ZnO) nanoparticles as a co-catalyst for flame-retardant finishing on cotton fabrics. Their results showed that the flame ignited on the treated fabrics extinguished right after the removal of ignition source and by using 0.2% and 0.4% of nano-ZnO as a co-catalyst, the flame spread rate of samples decreased [181].

Also, Samanta *et al.* applied nano-ZnO on bleached jute fabric to impart a fire-retardant finish. Nano zinc oxide was prepared from aqueous Zinc acetate by the co-precipitation method in the presence of NaOH followed by calcination at 600 °C. Then, it was applied on the jute fabric by the pad-dry method. The coating had reasonably good LOI values for as many as five wash cycles [182].

In another study, the LbL coating has been used to obtain the flame-retardant cotton fabric by using Bio-based and phosphorus-free coating. Indeed, the flame-retardant cotton fabric was prepared by negatively charged alginate and positively charged polyethylenimine together with subsequent crosslinking of barium, nickel, and cobalt ions. Metal ions such as Cu^{2+}, Mg^{2+}, Ni^{2+}, Co^{2+}, and Ba^{2+} were practically used as crosslinking agents for alginate to introduce the flame-retardant properties on fabric and films. The char residues at 700 °C revealed that the added extras of barium, nickel, and cobalt metal ions enhanced thermal degradation of fabric to fabricate firm char residues and reduced the degradation rate of the matrix. Moreover, the char residue showed a superior washing durability of the LbL cotton fabric when treated by the metal ions as crosslinking agents than untreated metal ions of the LbL cotton fabric. In other words, the flame spread rate of the LbL cotton fabric treated by barium ions demonstrated 28% reduction than untreated cotton [183].

The *in situ* synthesis of silver and diphosphate malonate nanocomposites was used for antibacterial activity against *Staphylococcus aureus* bacteria and flame-retardancy properties on cotton blend and polyester fabrics. Indeed, silver nitrate was reduced to silver nanoparticles by a reducing agent and in the presence of diphosphate malonate, which acted as a capping agent. The results showed that by increasing diphosphate malonate in the nanocomposite, the size of silver nanoparticles will increase due to the reduction properties of diphosphate malonate. Moreover, comparing the effect of diphosphate malonate and silver nanoparticles on the flame retardancy property of treated textile fabrics, by increasing the diphosphate malonate concentration in the nanocomposite, more relative flame-retardant activity was resulted than increasing the silver nanoparticle concentration in the nanocomposite. The flame-retardancy properties of textile were considerably enhanced and char layer was solid enough by increasing the diphosphate malonate concentration in the nanocomposite. Indeed, the concentration ratio between the organic phosphate and silver nanoparticles showed a vital role in the nanocomposite effect on textile fabrics. The efficient effect of diphosphate malonate on the flame-retardant property leaded to the formation of a solid char layer upon the release of phosphoric acid which catalyzed degradation of cotton fabric, following char [184].

In general, the results demonstrated that most of the nanoparticles can improve the thermal stability and flame-retardant properties of the textile fabrics. The main mechanisms involved are the formation of a protective barrier layer, char promotion, and free radical trapping. Nevertheless, the extent of enhancement depends on several factors such as morphology and composition of the nanoparticle, compatibility between nanoparticle and polymer, dispersion of the nanoparticles in the polymer matrix, and migration speed of nanoparticles to the surface.

1.10 Wrinkle-Resistant Fabrics

Cellulose fibers are the most commonly used natural fibers in the textile industry. However, one of the major disadvantages of using cellulose fiber is low dimensional stability and wrinkle resistance. Cellulose fibers such as cotton have a high water absorbency and potential to make hydrogen bonds to water molecules. Therefore, water molecules penetrate into amorphous regions of cellulose and cause local swelling of the fibers. Subsequently, the cellulose chains displace from their original positions and hydrogen bonds re-form at new locations. As a result, fabrics become wrinkled in appearance [185]. In fact, there are two theories for the mechanism of anti-wrinkle finishing of cotton fabrics.

- In the deposition theory, small particles of finishing agents are dispersed in the amorphous region of the fibers and form hydrogen bonds with the fiber molecules. This may restrict the movement of the molecular chains and decrease deformation of the fabrics.
- In the cross-linking theory, the finishing agents form covalent bonds with fiber molecules, and join the adjacent molecular chains within the fibers. This gives the recoverability improvement of the deformed fibers and increases the crease resistance [186].

According to the researches, wool and silk fabrics have good ability to resist the formation of wrinkles. However, some studies have been recorded about wrinkle-free finishing on silk fabrics. Generally, there are two different physical and chemical methods in order to apply anti-wrinkle finishing on fabrics that are shown in Figure 1.16. Physical wrinkle-reduction techniques, such as heat and pressure, are effective ways of flattening fabrics. However, their effect is not permanent and wrinkles reappear due

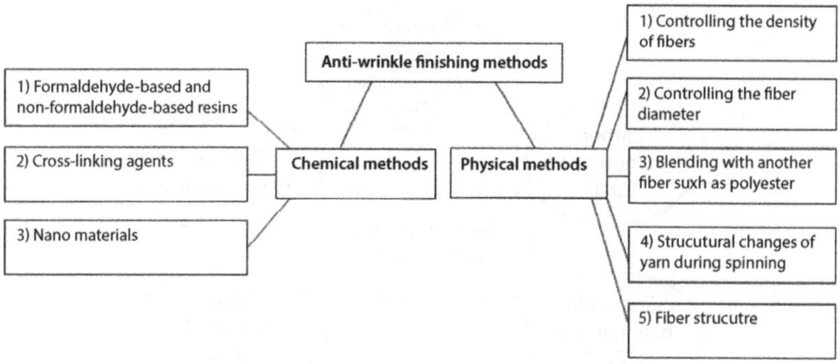

Figure 1.16 Physical and chemical methods of anti-wrinkle finishing.

to a range of shear, torsion, and compressive deformation forces applied in wear. Therefore, physical methods just reduce the wrinkle formation and do not have enough potential to improve crease resistance. However, the body's heat and humidity can relax the fabric and enhance the wrinkling of these deformational forces [187, 188].

Formaldehyde-based resins such as phenol formaldehyde, urea formaldehyde, melamine formaldehyde, and cross-linking agents such as glutaraldehyde and poly-carboxylic acid cause a significant increase in the crease-resistance properties of the textile. However, formaldehyde-based resins come with potentially dangerous side-effects and it is the main obstacle for efficient application of these types of materials [189]. Recently, using various nano materials in order to enhance the wrinkle resistance textiles has increased. In this section, different nano materials as anti-wrinkle agents have been studied.

1.10.1 Nanostructured Materials as Anti-Wrinkle Agents

In the conventional methods, in order to impart wrinkle-resistant properties to fabrics, resin was used previously. The most important drawbacks of conventional resin applications include decreasing of the strength of fiber, abrasion resistance, water absorbency, dye-ability, and breathability. To overcome the limitations of using resins, some researchers have employed nano structured materials with wrinkle-resistant features [190].

Nano materials play a vital role as a co-catalyst in anti-wrinkle finishing to promote the cross-linking between the fabric structure and the cross-linking agent. In fact, the presence of nano materials can enhance crease-resistant properties [186]. Different types of nano materials employed in the anti-wrinkle finishing process are represented in Figure 1.17.

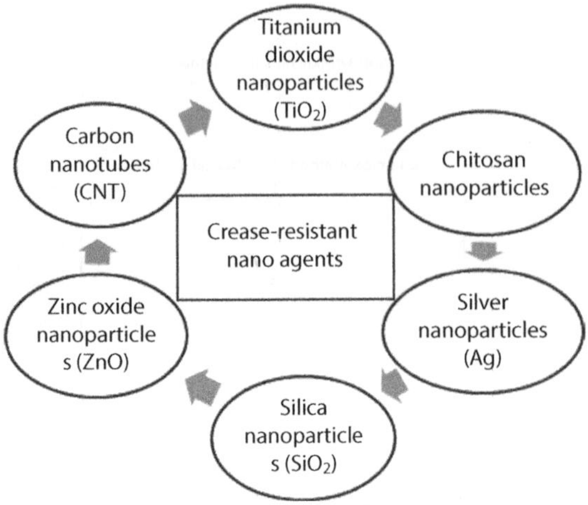

Figure 1.17 Some crease-resistant nano-agents for textile finishing.

1.10.1.1 Titanium Dioxide Nanoparticles (TiO_2)

Sodium hypophosphite (SHP) is considered as a catalyst in anti-wrinkle finishing of cotton via cross-linking compounds such as poly-carboxylic acid. However, high cost, producing shade changes, and serious environmental problems are the main drawbacks of using SHP. Recently using various co-catalysts such as titanium dioxide nanoparticles (TiO_2) have been investigated in order to enhance the crease recovery finishing process. In fact, nano-TiO_2 as a co-catalyst improves the finishing performance and decreases the side effects. In addition, nano-TiO_2 is very stable toward high temperature and pressure. Due to its large surface area, nano-TiO_2 provides sufficient number of active sites for the achievement of acceptable reaction rates [191]. On the other hand, nano TiO_2 can fill the amorphous region of the cellulose, and hence, the presence of nano TiO_2 inside the fiber would probably restrict the molecular movement of cellulose [192].

Recent studies have discussed the application of nano-sized TiO_2 in textile finishing such as wrinkle-freeness applications. For example, Lam *et al.* evaluated the wrinkle-resistant property of cotton specimens treated by butane tetra-carboxylic acid (BTCA) and catalyzed by SHP in the presence of nano-TiO_2. It was found that the addition of nano-TiO_2 could further enhance the wrinkle resistance of BTCA/SHP-treated cotton fabrics [191]. In a same study, Yuen *et al.* used nano-TiO_2 as a co-catalyst for improving wrinkle-resistance of cotton fabric in the presence of BTCA and SHP and the obtained results were successful in crease resistance [193]. Hezavehi *et al.* evaluated the wrinkle-resistant properties of cotton fabrics dyed by

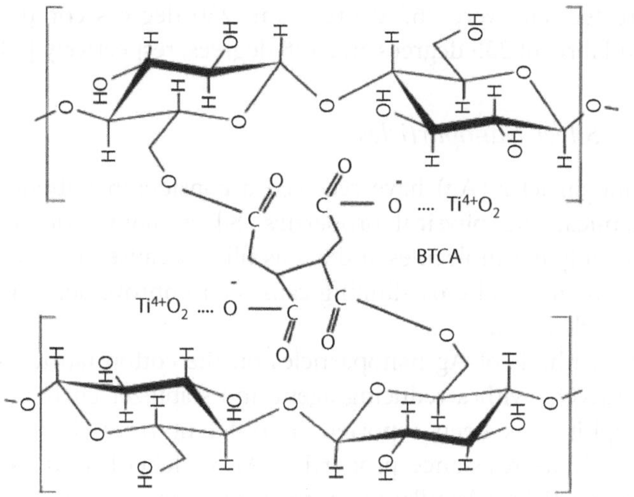

Figure 1.18 Formation of linkages between BCTA/cellulose chains and BCTA/nano-TiO_2 [195].

Direct Blue 2B in the presence and absence of TiO_2 nanoparticles. Also, the addition of BTCA can significantly enhance the crease resistance of fabrics, especially in the presence of SHP and nano-TiO_2 as a co-catalyst. In fact, TiO_2 nanoparticles restrict the molecular movement of cellulose chains, and leading to the enhancement of crease resistance of the samples. However, it seems that adding TiO_2 in the absence of BTCA causes better wrinkle-resistant property, but is not considerable [186]. Also, Karthik *et al.* performed the non-formaldehyde wrinkle-resistant finishing of cotton fabrics using citric acid as a cross-linking agent and SHP as a catalyst together with nano-TiO_2 as a co-catalyst compound. Based on the results, it was concluded that 10% CA, 10% SHP, and 0.1% nano-TiO_2 were the optimum concentrations required to enhance wrinkle-resistant properties without any negative effect on the other physical features of the treated cotton fabrics [194]. Nazari *et al.* optimized the resistance of bleached cotton fabric against creasing by using nano-TiO_2, sodium hypophosphite, and two different poly-carboxylic acids. Figure 1.18 represents the formation of linkages between BCTA/cellulose chains and BCTA/nano-TiO_2. The presence of nano-TiO_2 enhances the cross-linking between cellulose chain and cross linker. It has a positive effect on improving crease recovery angle, as well [195].

In another research, Bombyx mori silk was also treated with TiO_2 nanoparticles in chitosan by cross-linking reactions of citric acid and maleic anhydride. Based on the results, dry and wet crease recovery angles of the treated silk were 267 degrees and 250 degrees compared to the untreated fabric of 235 degrees and 178 degrees, respectively [185].

1.10.1.2 *Silver Nanoparticles*

Silver nanoparticles (Ag) have received a considerable attention due to their chemical and physical properties. Silver nanoparticles can enter inside the polymer molecules and act as filler because of their small size. This formed chemical cross-linking causes an improvement in the crease recovery angle [196].

In situ synthesis of Ag nanoparticles on the cotton fabric using silver nitrate ($AgNO_3$) with a reducing agent and stabilizer citric acid/sodium hypophosphite has been reported in different researches in order to improve wrinkle resistance properties. Also, Hebeish *et al.* synthesized silver nanoparticles (AgNPs) via a green method and treated the cotton fabric by AgNPs, 1,2,3,4-butane tetracarboxylic acid (BTCA) and sodium hypophosphite (SHP) as a catalyst. The obtained results show that the treated fabric has a significant crease recovery angle [197]. Montazer *et al.* applied the colloidal nano silver on the surface of cotton fabric and stabilized using BTCA. Their results showed that the winkle recovery angle of finished samples increased [198].

1.10.1.3 *Silica Nanoparticles*

Silica nanoparticles (SiO_2) penetrate easily into the interior of the cotton fiber and adhere more tightly to the fiber structure. Therefore, a covalent bond is formed between the hydroxyl group of cellulose and –Si–OH in silica nanoparticles. In fact, the thermodynamic affinity of fibers to the formation the non-covalent interactions such as dipolar–dipolar and hydrogen bond is the key factor for the adhesion of silica nanoparticles. The Si–O–cellulose covalent bonds in the treated fabrics make less contribution toward the adhesion of particles. Also, the cellulose in cotton fabric plays a decisive role in the intensity of the covalent bond between the hydroxyl group of cellulose and –Si–OH in silica nanoparticles. The bond formation can promote the wrinkle resistance property of the cotton fabrics. Also, it was observed that the increase of concentration of silica nanoparticles contributes more to the increase of the crease recovery angle of the fabric due to restriction of the molecular movement in the fiber structure [192].

Nallathambi *et al.* reported an efficient anti-wrinkle finishing method using silica nanoparticles, BTCA and SHP. Increasing the concentration of SiO_2 nanoparticles causes improved wrinkle recovery angle and strength in fabric [192]. In another research, SiO_2 nanoparticles and maleic anhydrate as a catalyst have been applied to silk fabric in order to improve wrinkle resistance [185]. Also, in a patent, Doneva *et al.* prepared a dispersion consisting of suitable inorganic nanoparticles like silica, titania, alumina, zinc oxide, and the mixed oxides class such as ITO (indium-tin oxide) in order to synthesis a silicone-based resin as an anti-wrinkle agent. The average particle size of dispersion was in the range of 10 to 50 nm and the amount of dispersion deposited on the fabric was between 0.25 and 0.5 wt%. As well as the enhancement of anti-wrinkling features, the nanoparticle depending on their structure can provide additional benefits such as anti-odor, water absorbency, and so on [188]. Patel *et al.* synthesized the silica nanoparticles and applied to cotton fabrics by the exhaust technique. There was an improvement in the crease recovery angle, which proves that the silica nanoparticles penetrated in between the polymer chain molecules. In addition, the results showed that the crease recovery properties of fabric improve with the increase in the concentration of silica nanoparticle and were found better in the exhaust method compared to the pad-dry-cure method. From the results, it seems that the incorporation of silica nanoparticles enhances the crease recovery angle without any negative effect on the fabric hand [199].

1.10.1.4 Zinc Oxide Nanoparticles

Based on the researches, ZnO nanoparticles, SHP as the catalyst, and BTCA or succinic acid as the cross linker have a positive effect on wrinkle resistance of the cotton fabric. Also, changing the parameters such as nanoparticle concentration, catalyzer concentration, and curing temperature can influence on the enhancement of the wrinkle recovery angle. For example, Ashraf *et al.* used ZnO nanoparticles in the curing process with the aim to replace the conventional catalyst and to enhance the crease recovery properties on the cotton fabric [200]. Shady *et al.* characterized the effects of zinc oxide nanoparticles on the performance characteristics of cotton, polyester, and their blends. Generally, the improvement greatly depends on the size and concentration of ZnO nanoparticles. As the size of ZnO nanoparticles increases, the crease recovery angles of the samples improve and this confirms the positive effect of ZnO nanoparticles on the anti-wrinkle properties [201]. Also, in a research performed by Farouk *et al.*, ZnO nanoparticles-chitosan composite was prepared for the finishing of textiles. The crease-resistant angle enhancement of the composite-finished fabric suggests that ZnO nanoparticles/chitosan composite form a network hybrid matrix on the fabric [202].

1.10.1.5 Carbon Nanotubes

Intermolecular cross-link between long-chain CNT and cellulose molecular chains via a crosslinking agent impart crease resistance properties to cotton fabric. In fact, cross-linkage between CNTs and the fabric creates a network and reduces the flexibility of molecular chains. Therefore, the crease recovery angle of fabrics is increased. For example, Amini *et al.* used multiwall CNTs in order to increase wrinkle-free properties of cotton fabric via succinic acid as the cross-linker and SHP. The obtained results confirm the notable effect of CNTs in reducing fiber flexibility [203]. In a patent, Alimohammadi *et al.* used CNTs for textile production. The CNT-embedded textiles are more antibacterial, conductive, crease resistant, flame retardant, thermally stable, and EM energy absorbent than raw textiles. They prepared seven cotton samples treated with a solution including the crosslinking agent, catalyst SHP, and CNT with different concentrations of 100, 250, 500, 1000, and 1500 ppm. According to their results, above a CNT concentration of 250 ppm, the crease-resistant properties of CNT-embedded cotton are better than raw cotton [204].

1.10.1.6 Chitosan Nanoparticles

Binding cellulose and chitosan nanoparticles as a polycation via intermolecular inclusion interaction decreases the possibility of molecular chains movement, hydrogen bonds breaking, and creasing the fibers. Also, the presence of a cross-linking agent is considered as a necessary factor to obtain desirable results. Although, chitosan nanoparticles have a positive effect on crease resistance, chitosan micro particles completely cover the surface and prevent the crosslinkage between the fabric and the cross-linking agent [205]. Lu *et al.* prepared hydroxypropyl chitosan (HCS) nanoparticles through ion-gel technology and applied it onto Antheraea pernyi (A. perny) silk fabric in the presence of 1,2,3,4-butane tetracarboxylic acid and sodium hypophosphite by a conventional pad-dry-cure process. According to their results, treatment of A. pernyi silk fabric with HCS nanoparticles strengthened the wrinkle-resistant property [206]. In another study, Cheng *et al.* synthesized chitosan nanoparticles with a size distribution of 15–45 nm and a good dispersity and stability. Then, they characterized the influences of nano-chitosan on antibacterial and wrinkle-resistant performances of Mulberry silk fabrics. The results indicated that Mulberry silk fabrics treated with appropriate mass fraction of nano-chitosan have an enhanced wrinkle-resistant performance [207]. Also, Zulfiqar *et al.* synthesized chitosan nanoparticles (CsNPs) by using ionic gelation method and then applied it on the cotton fabric utilizing the pad-dry-cure approach.

Based on the results, the higher crease resistance angle was due to the polycation nanoparticles which easily moved into the inter fiber region and bound the fibers to provide better crease recovery [208].

1.11 Future Trends and Challenges of Nano-Based Textiles

The present status of the global textile market is based on innovations in high-tech textile products, and nanotechnology is one of the prominent driving forces, which helps the technical textiles stay ahead of the competition. The textile industry is affected by two main factors, that is, market volatility and world-wide competition. Therefore, the industry needs to enhance its capability to produce and market high quality and added value products. Recently, with dramatic growth in nanotechnology, increased volumes of engineered nanoparticles have been produced which represent the potential functional benefits in a wide range of applications by integrating into textiles. Among different types of nano materials, polymeric nanostructures, metal oxides, CNTs, clay nanoparticles, carbon black, graphite nanofibers (GNF), and others provide an unprecedented level of textile performance such as being hydrophilic and hydrophobic, antimicrobial, anti-odor, stain-resistant, anti-static, wrinkle resistant, anti-UV, and so on. These materials are transparent due to their high surface area and nanoscale dimension, and do not blur the color and brightness of the textile substrates. They improve the properties of textiles by their own unique characteristic, as well. In fact, finishing with different nano materials can convert fabrics into versatile multifunctional materials.

Regardless of positive aspects of nano materials in textile finishing, there are still a lot of items to be taken into consideration. The first issue is cost, which in some cases limit the development of nano material coating on textiles and mass production. Development of products via nano structured materials remains an expensive affair and demands R&D spending in the textile industry. On the other hand, the durability of nano-finished textiles is notable. The challenge is to ensure the fastness of nano-structured finishing agents on fabric during usage and against different factors such as washing, rubbing, light, temperature, humidity, etc. However, regardless of its prosperous future, the main challenges that still need to be resolved in today's nano-finished textiles are as follows:

- Increase the durability and maintainability of nano-based properties on textiles after repeated using cycles.

- Due to the very small size of nanomaterials, they can potentially expose into the human body.
- Also, aggregation of nanoparticles and their uncontrollable release are other serious issues which threat the environment and human health.

It is expected that the future of nano-finished textiles be surely related to bio–nano textiles. New generation of the finishing agents integrate the nanoscale structure with eco-friendly properties, which produce dramatic characteristics without harmful effects on environment and the human health.

References

1. Perelshtein, I., Perkas, N., Gedanken, A., Ultrasonic Coating of Textiles by Antibacterial and Antibiofilm Nanoparticles. *Handbook of Ultrasonics and Sonochemistry*, 967, 2016.
2. Madigan M. T, Martinko J.M., *Brock Biology of Microorganisms*, Pearson Prentice Hall, Edition 11, 2006.
3. Mucha, H., Hofer, D., Assfalg, S., Swerev, M., Antimicrobial finishes and modifications. *Melliand Textil Int.*, 83, 238, 2002.
4. Edwards, J.V., Vigo, T.L., *Bioactive fibers and polymers*. 2001: ACS Publications.
5. Zille, A., Almeida, L., Amorim, T., Carneiro, N., Esteves, M.F., Silva, C.J., Souto, A.P., Application of nanotechnology in antimicrobial finishing of biomedical textiles. *Mater. Res. Express*, 1, 032003, 2014.
6. Fouda, M.M., *Antibacterial Modification of Textiles Using Nanotechnology*, in *A Search for Antibacterial Agents*, pp. 47–73, InTech, 2012.
7. Kamat, P.V., Photochemistry on nonreactive and reactive (semiconductor) surfaces. *Chem. Rev.*, 93, 267, 1993.
8. Linsebigler, A.L., Lu, G., Yates, J.T. Jr, Photocatalysis on TiO_2 surfaces: principles, mechanisms, and selected results. *Chem. Rev.*, 95, 735, 1995.
9. Wong, Y., Yuen, C., Leung, M., Ku, S., Lam, H., Selected applications of nanotechnology in textiles. *Autex. Res. J.*, 6, 1, 2006.
10. Montazer, M., Seifollahzadeh, S., Enhanced self-cleaning, antibacterial and UV protection properties of nano TiO_2 treated textile through enzymatic pretreatment. *J. Photochem. Photobiol.*,, 87, 877, 2011.
11. Matsumura, Y., Yoshikata, K., Kunisaki, S.i., Tsuchido, T., Mode of bactericidal action of silver zeolite and its comparison with that of silver nitrate. *Appl. Environ. Microbiol.*, 69, 4278, 2003.
12. Fong, J., Wood, F., Fowler, B., A silver coated dressing reduces the incidence of early burn wound cellulitis and associated costs of inpatient treatment: comparative patient care audits. *Burns*, 31, 562, 2005.

13. Selvam, S., Gandhi, R.R., Suresh, J., Gowri, S., Ravikumar, S., Sundrarajan, M., Antibacterial effect of novel synthesized sulfated β-cyclodextrin crosslinked cotton fabric and its improved antibacterial activities with ZnO, TiO_2 and Ag nanoparticles coating. *Int. J. Pharm.*, 434, 366, 2012.
14. Dastjerdi, R., Montazer, M., A review on the application of inorganic nanostructured materials in the modification of textiles: focus on anti-microbial properties. *Colloids Surf., B*, 79, 5, 2010.
15. Stevanato, R., Tedesco, R., New antibacterial acrylic fiber. *Chem. Fibers Int.*, 48, 480, 1998.
16. Nair, M.G., Nirmala, M., Rekha, K., Anukaliani, A., Structural, optical, photo catalytic and antibacterial activity of ZnO and Co doped ZnO nanoparticles. *Mater. Lett.*, 65, 1797, 2011.
17. Padmavathy, N., Vijayaraghavan, R., Enhanced bioactivity of ZnO nanoparticles—an antimicrobial study. *Sci. Tech. Adv. Mater.*, 9, 035004, 2008.
18. Jeon, Y.-J., Park, P.-J., Kim, S.-K., Antimicrobial effect of chitooligosaccharides produced by bioreactor. *Carbohydr. Polym.*, 44, 71, 2001.
19. No, H.K., Young Park, N., Ho Lee, S., Meyers, S.P., Antibacterial activity of chitosans and chitosan oligomers with different molecular weights. *Int. J. Food. Microbiol.*, 74, 65, 2002.
20. Interest, C. The Chemistry of Body Odours – Sweat, Halitosis, Flatulence & Cheesy Feet. http://www.compoundchem.com/2014/04/07/the-chemistry-of-body-odours-sweat-halitosis-flatulence-cheesy-feet, 2014.
21. Nissenken, S, Functionality assessment for textile: Deodorization testing. http://nissenken.or.jp/english/service/pdf/functionality_shoshu.pdf, 2017.
22. Sarıışık, M., Okur, S., Asma, Ş., Odor adsorption kinetics on modified textile materials using quartz microbalance technique, Proceedings of the International Congress on Advances in Applied Physics and Materials Science, Antalya, 121, 1, 2011.
23. Voncina, B., Vivod, V., *Cyclodextrins in textile finishing*, in *Eco-friendly textile dyeing and finishing*, pp. InTech, 2013.
24. Bhaskara-Amrit, U.R., Agrawal, P.B., Warmoeskerken, M.M., Applications of β-cyclodextrins in textiles. *Autex. Res. J.*, 11, 94, 2011.
25. Buschmann, H.-J., Knittel, D., Schollmeyer, E., New textile applications of cyclodextrins. *J. Incl. Phenom. Macrocycl. Chem.*, 40, 169, 2001.
26. Akhmetova, A., Saliev, T., Allan, I.U., Illsley, M.J., Nurgozhin, T., Mikhalovsky, S., A comprehensive review of topical odor-controlling treatment options for chronic wounds. *J. Wound. Ostomy. Continence. Nurs.*, 43, 598, 2016.
27. Narayanan, G., Ormond, B.R., Gupta, B.S., Tonelli, A.E., Efficient wound odor removal by β-cyclodextrin functionalized poly (ε-caprolactone) nanofibers. *J. Appl. Polym. Sci.*, 132, 42783, 2015.
28. Lipman, R.D., van Bavel, D., Odor absorbing hydrocolloid dressings for direct wound contact. *Wounds*, 19, 138, 2007.
29. Turk, A., Haring, R.C., Okey, R.W., Odor control technology. *Environ. Sci. Technol*, 6, 602, 1972.

30. Quincy III, R.B., Karandikar, B.M., MacDonald, J.G., Elshani, T., *Dual Element Odor Control In Personal Care Products*. 2011, Google Patents.
31. Bamboo Charcoal Nanoparticles in Fabrics for Antibacterial and Deodorizing Applications. http://www.azonano.com/nanotechnology-video-details.aspx?VidID=205, 2017.
32. Yu, X., Qin, A., Liao, L., Du, R., Tian, N., Huang, S., Wei, C., Removal of organic dyes by nanostructure ZnO-bamboo charcoal composites with photocatalysis function. *Adv. Mater. Sci. Eng.*, 2015, 2015, 2015.
33. Bacteria and Odour Control Bamboo Charcoal Bedding. https://www.indiegogo.com/projects/bacteria-and-odour-control-bamboo-charcoal-bedding#/, 2016.
34. Eco-Fabric Guide 2005, http://www.greenyarn.com/Ecofabric.pdf, 2005.
35. Zhang, D., Liu, Y., Fan, Y., Yu, C., Zheng, Y., Jin, H., Fu, L., Zhou, Y., Yan, D., Hierarchical self-assembly of a dandelion-like supramolecular polymer into nanotubes for use as highly efficient aqueous light-harvesting systems. *Adv. Funct. Mater.*, 26, 7652, 2016.
36. George, S., Gupta, A., Mondal, P., Overview of activated alumina defluoridation process, 2015.
37. Jenkins, D., Wheeler, D., Fritter, C., Shenoy, A., Deleeuw, D., *Absorbent composition with improved odor control*. 2004, Google Patents.
38. Furio, D.L., Johnson, T.L., Benson, D.H., *Absorbent structures with odor control material*. 1990, European Patent, EP 0389015 A2.
39. Zaki, T., Samy, M., Temraz, M., Synthesis and characterization of nano-silica gel aged under microwave radiation from white sandstone. *J. Sol-Gel Sci. Technol.*, 64, 224, 2012.
40. Bakkali, F., Averbeck, S., Averbeck, D., Idaomar, M., Biological effects of essential oils – A review. *Food Chem. Toxicol.*, 46, 446, 2008.
41. Shaaban, H.A.E., El-Ghorab, A.H., Shibamoto, T., Bioactivity of essential oils and their volatile aroma components: Review. *J. Essent. Oil Res.*, 24, 203, 2012.
42. Wang C. X, Chen Sh. L., Aromachology and its application in the textile field. *Fibres Text East Eur.*, 13, 41, 2005.
43. Maderuelo, C., Zarzuelo, A., Lanao, J.M., Critical factors in the release of drugs from sustained release hydrophilic matrices. *J. Control Release*, 154, 2, 2011.
44. Kumar, M.N.V.R., Nano and microparticles as controlled drug delivery devices. *J. Pharm. Pharm. Sci*, 3, 234, 2000.
45. Reis CP, Neufeld RJ, Ribeiro AJ, Veiga F. Methods for preparation of drug-loaded polymeric nanoparticles. *Nanomedicine*, 2, 8, 2006.
46. Zhang, Z.D., *in Encyclopedia of Nanoscience and Nanotechnology*. 2004, American Scientific Publishers. p. 77–160.
47. Hu, J., Xiao, Z.-B., Zhou, R.-J., Ma, S.-S., Li, Z., Wang, M.-X., Comparison of compounded fragrance and chitosan nanoparticles loaded with fragrance applied in cotton fabrics. *Text. Res. J.*, 81, 2056, 2011.

48. DeRosa, T.F, Next Generation of International Chemical Additives, 1st Edition, 578, Elsevier, 28th, August 2012, UK.
49. Bangham, A.D., Review of Lasic, Liposomes. *Biophys. J.*, 67, 1358, 1994.
50. Giamberini, M., Tylkowski, B., Prieto, S.F., *Microencapsulation: Innovative Applications*, 219, Germany: De Gruyter, 2015.
51. Zorzi, G.K., Carvalho, E.L.S., von Poser, G.L., Teixeira, H.F., On the use of nanotechnology-based strategies for association of complex matrices from plant extracts. *Rev. Bras. Farmacogn.*, 25, 426, 2015.
52. Dahms, G., Jung, A., aSeidel, H., *Compositions for the targetted release of fragrances and aromas*. 2014, Google Patents.
53. Müller R.H., Staufenbiel S., Keck C. M., Lipid nanoparticles (SLN, NLC) for innovative consumer care & household products. *Transdermal Deliv.*, 9, 18, 2014.
54. Wang, C.X., Chen, S.L., Fragrance-release property of β-cyclodextrin inclusion compounds and their application in aromatherapy. *J. Ind. Text.*, 34, 157, 2005.
55. Lindner, K., Using cyclodextrin aroma complexes in the catering. *Food/Nahrung*, 26, 675, 1982.
56. Sricharussin, W., Sopajaree, C., Maneerung, T., Sangsuriya, N., Modification of cotton fabrics with β-cyclodextrin derivative for aroma finishing. *J. Text. I.*, 100, 682, 2009.
57. Bandyopadyay, S., Das, D., Retention and sustained release of fragrance by Cyclodextrin functionalized cotton fabric modified using maleic anhydride. *Flavour Frag. J.*, 32, 207, 2017.
58. Somaye, A., Mehdi, A., Mohammad Haghighat, K., Firoz Mehr, M., Preparation of long-lasting fragrant worsted fabrics using polypropyleneimine (PPI) dendrimer. *Int. J. Cloth Sci. Tech.*, 28, 699, 2016.
59. Obendorf, S.K., *19 – Improving the functionality of clothing through novel pesticide protection, in Functional Textiles for Improved Performance, Protection and Health*, pp. 434–461, Woodhead Publishing, USA, 2011.
60. Vojtech L., Neruda M., Design of radiofrequency protective clothing containing silver nanoparticles. *Fibres Text East Eur*, 21, 6, 2013.
61. Scott, R.A., *21 – Military protection**, in *Textiles for Protection*, pp. 597–621, Woodhead Publishing, Boca Raton, Florida, 2005.
62. Chudasama, D., Protection against electromagnetic radiation with textile material. *IJEETC*, 3, 5, 2013.
63. Roh, J.-S., Chi, Y.-S., Kang, T.J., Nam, S.-w., Electromagnetic shielding effectiveness of multifunctional metal composite fabrics. *Text. Res. J.*, 78, 825, 2008.
64. Chung, D.D.L., Materials for electromagnetic interference shielding. *J. Mater. Eng. Perform.*, 9, 350, 2000.
65. Motojima, S., Noda, Y., Hoshiya, S., Hishikawa, Y. Electromagnetic wave absorption property of carbon microcoils in 12–110 GHz region. *J. Appl. Phys.* 94(4), pp 2325–2330, 2003.

66. Rosace, G., *16 – Radiation protection finishes for textiles A2 – Paul, Roshan*, in *Functional Finishes for Textiles*, pp. 487–512, Woodhead Publishing, 2015.
67. Chen, H.C., Lee, K.C., Lin, J.H., Koch, M., Fabrication of conductive woven fabric and analysis of electromagnetic shielding via measurement and empirical equation. *J. Mater. Process. Technol.*, 184, 124, 2007.
68. Erdoğan, M.K., Karakişla, M., Saçak, M., Preparation, characterization and electromagnetic shielding effectiveness of conductive polythiophene/poly(ethylene terephthalate) composite fibers. *J. Macromol. Sci. A*, 49, 473, 2012.
69. Kuhn, H.H., Child, A.D., Kimbrell, W.C., Toward real applications of conductive polymers. *Synth. Met.*, 71, 2139, 1995.
70. Engin, F.Z., Usta, I., Development and characterisation of polyaniline/polyamide (PANI/PA) fabrics for electromagnetic shielding. *J. Text. I.*, 106, 872, 2015.
71. Fragouli, D., Bayer, I.S., Di Corato, R., Brescia, R., Bertoni, G., Innocenti, C., Gatteschi, D., Pellegrino, T., Cingolani, R., Athanassiou, A., Superparamagnetic cellulose fiber networks via nanocomposite functionalization. *J. Mater. Chem.*, 22, 1662, 2012.
72. Harifi, T., Montazer, M., In situ synthesis of iron oxide nanoparticles on polyester fabric utilizing color, magnetic, antibacterial and sono-Fenton catalytic properties. *J. Mater. Chem. B*, 2, 272, 2014.
73. Koprowska, J., Ziaja, J., Janukiewicz, J.. *Plasma metallization textiles as shields for electromagnetic fields*. in *2008 International Symposium on Electromagnetic Compatibility – EMC Europe*. 2008.
74. Danlée, Y., Bailly, C., Huynen, I., Thin and flexible multilayer polymer composite structures for effective control of microwave electromagnetic absorption. *Compos. Sci. Technol.*, 100, 182, 2014.
75. Bonaldi, R.R., Siores, E., Shah, T., Characterization of electromagnetic shielding fabrics obtained from carbon nanotube composite coatings. *Synth. Met.*, 187, 1, 2014.
76. Kowsari, E., Mohammadi, M., Synthesis of reduced and functional graphene oxide with magnetic ionic liquid and its application as an electromagnetic-absorbing coating. *Compos. Sci. Technol.*, 126, 106, 2016.
77. Lin, Z.-I., Lou, C.-W., Pan, Y.-J., Hsieh, C.-T., Huang, C.-H., Huang, C.-L., Chen, Y.-S., Lin, J.-H., Conductive fabrics made of polypropylene/multi-walled carbon nanotube coated polyester yarns: Mechanical properties and electromagnetic interference shielding effectiveness. *Compos. Sci. Technol.*, 141, 74, 2017.
78. Chen-Yu, J.H., Wong, S.-W., Effects of repeated treatments of perspiration, weathering, and laundering on the ultraviolet protection of a naturally colored lightweight cotton fabric. *Cloth Text Res. J.*, 35, 128, 2017.
79. Çakır, B.A., Budama, L., Topel, Ö., Hoda, N., Synthesis of ZnO nanoparticles using PS-b-PAA reverse micelle cores for UV protective, self-cleaning and antibacterial textile applications. *Colloids Surf., A*, 414, 132, 2012.

80. Azeem, A., Nazir, A., Abid, S., Sarwar, Z., Munir, U., Modelling of fabric structure and shade for UV protection properties using full factorial design. *J. Text. I.*, 108, 1800, 2017.
81. Wang, R.H., Xin, J.H., Tao, X.M., UV-blocking property of dumbbell-shaped ZnO crystallites on cotton fabrics. *Inorg. Chem.*, 44, 3926, 2005.
82. Yang, H., Zhu, S., Pan, N., Studying the mechanisms of titanium dioxide as ultraviolet-blocking additive for films and fabrics by an improved scheme. *J. Appl. Polym. Sci.*, 92, 3201, 2004.
83. Noorian, S.A., Hemmatinejad, N., Bashari, A., One-pot synthesis of Cu_2O/ZnO nanoparticles at present of folic acid to improve uv-protective effect of cotton fabrics. *J. Photochem. Photobiol.*, 91, 510, 2015.
84. Chien, T.-W., Chu, H., Removal of SO_2 and NO from flue gas by wet scrubbing using an aqueous $NaClO_2$ solution. *J. Hazard. Mater.*, 80, 43, 2000.
85. Selishchev, D.S., Karaseva, I.P., Uvaev, V.V., Kozlov, D.V., Parmon, V.N., Effect of preparation method of functionalized textile materials on their photocatalytic activity and stability under UV irradiation. *Chem. Eng. J.*, 224, 114, 2013.
86. Hasani, M., Montazer, M., Electro-conductivity, bioactivity and UV protection of graphene oxide-treated cellulosic/polyamide fabric using inorganic and organic reducing agents. *J. Text. I.*, 108, 1777, 2017.
87. Saravanan, R., Karthikeyan, S., Gupta, V.K., Sekaran, G., Narayanan, V., Stephen, A., Enhanced photocatalytic activity of ZnO/CuO nanocomposite for the degradation of textile dye on visible light illumination. *Mater. Sci. Eng., C*, 33, 91, 2013.
88. Bazant, P., Kuritka, I., Munster, L., Kalina, L., Microwave solvothermal decoration of the cellulose surface by nanostructured hybrid Ag/ZnO particles: a joint XPS, XRD and SEM study. *Cellulose*, 22, 1275, 2015.
89. Becheri, A., Dürr, M., Lo Nostro, P., Baglioni, P., Synthesis and characterization of zinc oxide nanoparticles: application to textiles as UV-absorbers. *J. Nanopart. Res*, 10, 679, 2008.
90. El.Shafei, A., Abou-Okeil, A., ZnO/carboxymethyl chitosan bionano-composite to impart antibacterial and UV protection for cotton fabric. *Carbohydr. Polym.*, 83, 920, 2011.
91. Montazer, M., Amiri, M.M., Mohammad Ali Malek, R., In situ synthesis and characterization of nano zno on wool: influence of nano photo reactor on wool properties. *J. Photochem. Photobiol.*, 89, 1057, 2013.
92. Farouk, A., Textor, T., Schollmeyer, E., Trarbuk, A., Grancacic, A. M., Sol-gel-derived inorganic-organic hybrid polymers filled with ZnO nanoparticles as an ultraviolet protection finish for textiles. *Autex. Res. J.*, 10, 5, 2010.
93. Ibrahim, N.A., El-Zairy, E.M., Eid, B.M., Emam, E., Barkat, S.R., A new approach for imparting durable multifunctional properties to linen-containing fabrics. *Carbohydr. Polym.*, 157, 1085, 2017.
94. Hsu, C.-H., Chen, L.-C., Lin, Y.-F., Preparation and optoelectronic characteristics of ZnO/CuO-Cu_2O complex inverse heterostructure with GaP buffer for solar cell applications. *Materials*, 6, 4479, 2013.

95. Xu, C., Cao, L., Su, G., Liu, W., Liu, H., Yu, Y., Qu, X., Preparation of ZnO/ Cu_2O compound photocatalyst and application in treating organic dyes. *J. Hazard. Mater.*, 176, 807, 2010.
96. Jiang, T., Xie, T., Chen, L., Fu, Z., Wang, D., Carrier concentration-dependent electron transfer in Cu2O/ZnO nanorod arrays and their photocatalytic performance. *Nanoscale*, 5, 2938, 2013.
97. Liu, K., Zhang, J., Gao, H., Xie, T., Wang, D., Photocatalytic property of ZnO microrods modified by Cu2O nanocrystals. *J. Alloys Compd.*, 552, 299, 2013.
98. Prasad, V., Arputharaj, A., Bharimalla, A.K., Patil, P.G., Vigneshwaran, N., Durable multifunctional finishing of cotton fabrics by in situ synthesis of nano-ZnO. *Appl. Surf. Sci.*, 390, 936, 2016.
99. Shaheen, T.I., El-Naggar, M.E., Abdelgawad, A.M., Hebeish, A., Durable antibacterial and UV protections of in situ synthesized zinc oxide nanoparticles onto cotton fabrics. *Int. J. Biol. Macromol.*, 83, 426, 2016.
100. King, A.L., Millington, K.R., Trace metals can affect hydroxyl radical production and yellowing of photo-irradiated wool. *J. Text. I.*, 104, 648, 2013.
101. Montazer, M., Seifollahzadeh, S., Enhanced self-cleaning, antibacterial and UV protection properties of nano TiO_2 treated textile through enzymatic pretreatment. *J. Photochem. Photobiol.*, 87, 877, 2011.
102. Zhang, M., Tang, B., Sun, L., Wang, X., Reducing photoyellowing of wool fabrics with silica coated ZnO nanoparticles. *Text. Res. J.*, 84, 1840, 2014.
103. Montazer, M., Pakdel, E., Reducing photoyellowing of wool using nano TiO_2. *J. Photochem. Photobiol.*, 86, 255, 2010.
104. Barik, S., Khandual, A., Behera, L., Badamali, S.K., Luximon, A., Nano-Mg–Al-layered double hydroxide application to cotton for enhancing mechanical, UV protection and flame retardancy at low cytotoxicity level. *Cellulose*, 24, 1107, 2017.
105. Lee, S., Developing UV-protective textiles based on electrospun zinc oxide nanocomposite fibers. *Fiber. Polym.*, 10, 295, 2009.
106. Dadvar, S., Tavanai, H., Morshed, M., UV-protection properties of electrospun polyacrylonitrile nanofibrous mats embedded with MgO and $Al2O_3$ nanoparticles. *J. Nanopart. Res*, 13, 5163, 2011.
107. Merati, A.A., Mousavi Shoushtari, A., Mirzaei, J., A comparison between the UV protection of PAN/ZnO and PAN/MWNT composite nanofiber mats. *J. Text. I.*, 1, 2017.
108. Pu, X., Li, L., Liu, M., Jiang, C., Du, C., Zhao, Z., Hu, W., Wang, Z.L., Wearable self-charging power textile based on flexible yarn supercapacitors and fabric nanogenerators. *J. Adv. Mater.*, 28, 98, 2016.
109. Higginbotham, A.L., Lomeda, J.R., Morgan, A.B., Tour, J.M., Graphite oxide flame-retardant polymer nanocomposites. *ACS Appl. Mater. Interfaces*, 1, 2256, 2009.
110. Cai, G., Xu, Z., Yang, M., Tang, B., Wang, X., Functionalization of cotton fabrics through thermal reduction of graphene oxide. *Appl. Surf. Sci.*, 393, 441, 2017.

111. Rowan, J.R., Gagliardi, D., Properties of water-repellent fabrics *J. Res. Natl. Bur. Stand.*, 38, 15, 1947.
112. Mahltig, B., *13 - Hydrophobic and oleophobic finishes for textiles A2 - Paul, Roshan*, in *Functional Finishes for Textiles*, pp. 387–428, Woodhead Publishing, 2015.
113. Park, S., Kim, J., Park, C. H., Superhydrophobic textiles: Review of theoretical definitions, fabrication and functional evaluation. *J. Eng. Fiber. Fabr.*, 10, 18, 2015.
114. Yuyang, L., Xianqiong, C., Xin, J.H., Hydrophobic duck feathers and their simulation on textile substrates for water repellent treatment. *Bioinspir. Biomim*, 3, 046007, 2008.
115. Tendero, C., Tixier, C., Tristant, P., Desmaison, J., Leprince, P., Atmospheric pressure plasmas: A review. *Spectrochim. Acta, Part B*, 61, 2, 2006.
116. Vohrer, U., Müller, M., Oehr, C., Glow-discharge treatment for the modification of textiles. *Surf. Coat. Technol.*, 98, 1128, 1998.
117. Sigurdsson, S., Shishoo, R., Surface properties of polymers treated with tetrafluoromethane plasma. *J. Appl. Polym. Sci.*, 66, 1591, 1997.
118. Sun, D., Stylios, G.K., Fabric surface properties affected by low temperature plasma treatment. *J. Mater. Process. Technol.*, 173, 172, 2006.
119. Li, S., Jinjin, D., Improvement of hydrophobic properties of silk and cotton by hexafluoropropene plasma treatment. *Appl. Surf. Sci.*, 253, 5051, 2007.
120. Hodak, S.K., Supasai, T., Paosawatyanyong, B., Kamlangkla, K., Pavarajarn, V., Enhancement of the hydrophobicity of silk fabrics by SF6 plasma. *Appl. Surf. Sci.*, 254, 4744, 2008.
121. Nicoletta De Vietro, L.B., Lambertini, V., Placenza, B., Fracassi, F., Plasma treatment for preparing durable water repellent and anti-stain synthetic fabrics for automotive applications. *J. Surface Eng. Mater. Advanced Technol.*, 5, 103–109, 2015.
122. Sarmadi, A.M., Ying, T.H., Denes, F., HMDSO-plasma modification of polypropylene fabrics. *Eur. Polym. J.*, 31, 847, 1995.
123. Hegemann, D., Stain repellent finishing on fabrics. *Adv. Eng. Mater.*, 7, 401, 2005.
124. Ceria, A., Hauser, P.J., Atmospheric plasma treatment to improve durability of a water and oil repellent finishing for acrylic fabrics. *Surf. Coat. Technol.*, 204, 1535, 2010.
125. Hirvikorpi, T., Vähä-Nissi, M., Harlin, A., Marles, J., Miikkulainen, V., Karppinen, M., Effect of corona pre-treatment on the performance of gas barrier layers applied by atomic layer deposition onto polymer-coated paperboard. *Appl. Surf. Sci.*, 257, 736, 2010.
126. Jur, J.S., Sweet, W.J., Oldham, C.J., Parsons, G.N., Atomic layer deposition of conductive coatings on cotton, paper, and synthetic fibers: Conductivity analysis and functional chemical sensing using "All-Fiber" capacitors. *Adv. Funct. Mater.*, 21, 1993, 2011.

127. Song, J., Rojas, O. J., Approaching super-hydrophobicity from cellulosic materials: A Review *Paper Chemistry Nordic Pulp & Paper Research Journal V*, 28, 216, 2013.
128. Bagherzadeh, R., Latifi, M., Najar, S.S., Tehran, M.A., Gorji, M. Kong, L., Transport properties of multi-layer fabric based on electrospun nanofiber mats as a breathable barrier textile material. *Text. Res. J.*, 82, 70, 2012.
129. Zhou, Y., He, J., Wang, H., Nan, N., Qi, K., Cui, S., *Fabrication of superhydrophobic nanofiber fabric with hierarchical nanofiber structure*, in *E-polymers*. 2017. p. 249.
130. Cakir, M., Kartal, I., Yildiz, Z., The preparation of UV-cured superhydrophobic cotton fabric surfaces by electrospinning method. *Text. Res. J.*, 84, 1528, 2014.
131. Daoud, W.A., Xin, J.H., Zhang, Y.H., Mak, C.L., Pulsed laser deposition of superhydrophobic thin Teflon films on cellulosic fibers. *Thin Solid Films*, 515, 835, 2006.
132. Krämer, A., Kunz, C., Gräf, S., Müller, F.A., Pulsed laser deposition of anatase thin films on textile substrates. *Appl. Surf. Sci.*, 353, 1046, 2015.
133. Yin, Y., Wang, C., Water-repellent functional coatings through hybrid SiO_2/HTEOS/CPTS sol on the surfaces of cellulose fibers. *Colloids Surf., A*, 417, 120, 2013.
134. Onar, N., Mete, G., Aksit, A., Kutlu, B., Celik, E., Aysun Aksit, Bengi Kutlu, Erdal Celik, Water- and oil-repellency properties of cotton fabric treated with Silane, Zr, Ti based nanosols *IJTST*, 4, 84, 2015.
135. Boukhriss, A., Boyer, D., Hannache, H., Roblin, J.-P., Mahiou, R., Cherkaoui, O., Therias, S., Gmouh, S., Sol–gel based water repellent coatings for textiles. *Cellulose*, 22, 1415, 2015.
136. Froehling, P.E., Dendrimers and dyes – a review. *Dyes. Pigm.*, 48, 187, 2001.
137. Schindler, W.D., Hauser, P.J., *6 – Repellent finishes*, in *Chemical Finishing of Textiles*, pp. 74–86, Woodhead Publishing, Boca Raton, Florida, 2004.
138. Atav, R.B., Berkay dendrimer technology for water and oil repellent cotton textiles. *AATCC J. Res.*, 3, 16, 2016.
139. Liu, Y., Wang, X., Qi, K., Xin, J.H., Functionalization of cotton with carbon nanotubes. *J. Mater. Chem.*, 18, 3454, 2008.
140. Namligoz, E. S., Bahtiyari, M. I., Hosaf, E., Coban, S., Hosaf E., Coban S., Performance comparison of new (dendrimer, nanoproduct) and conventional water, oil and stain repellents. *Fibres Text East Eur*, 17, 76, 2009.
141. Yuranova, T., Mosteo, R., Bandara, J., Laub, D., Kiwi, J., Self-cleaning cotton textiles surfaces modified by photoactive SiO_2/TiO_2 coating. *J. Mol. Catal. A: Chem.*, 244, 160, 2006.
142. Tung, W.S., Daoud, W.A., Self-cleaning fibers via nanotechnology: a virtual reality. *J. Mater. Chem.*, 21, 7858, 2011.
143. Nakajima, A., Fujishima, A., Hashimoto, K., Watanabe, T., Preparation of transparent superhydrophobic boehmite and silica films by sublimation of aluminum acetylacetonate. *J. Adv. Mater.*, 11, 1365, 1999.

144. Parkin, I.P., Palgrave, R.G., Self-cleaning coatings. *J. Mater. Chem.*, 15, 1689, 2005.
145. Latthe, S.S., Terashima, C., Nakata, K., Fujishima, A., Superhydrophobic surfaces developed by mimicking hierarchical surface morphology of lotus leaf. *Molecules*, 19, 4256, 2014.
146. Daoud, W.A., Xin, J.H., Tao, X., Superhydrophobic silica nanocomposite coating by a low-temperature process. *J. Am. Ceram. Soc.*, 87, 1782, 2004.
147. Afzal, S., Daoud, W.A., Langford, S.J., Superhydrophobic and photocatalytic self-cleaning cotton. *J. Mater. Chem., A*, 2, 18005, 2014.
148. Roach, P., Shirtcliffe, N.J., Newton, M.I., Progress in superhydrophobic surface development. *Soft matter.*, 4, 224, 2008.
149. Li, X.-M., Reinhoudt, D., Crego-Calama, M., What do we need for a superhydrophobic surface? A review on the recent progress in the preparation of superhydrophobic surfaces. *Chem. Soc. Rev.*, 36, 1350, 2007.
150. Han, J.T., Zheng, Y., Cho, J.H., Xu, X., Cho, K., Stable superhydrophobic organic-inorganic hybrid films by electrostatic self-assembly. *J. Phys. Chem. B*, 109, 20773, 2005.
151. Ma, M., Hill, R.M., Superhydrophobic surfaces. *Curr. Opin. Colloid Int.*, 11, 193, 2006.
152. Zheng, J., Feng, J., Zhong, M., Fabricating polymer superhydrophilic/superhydrophobic surfaces by replica-molding method using Caco_3 particles as template [J]. *Acta Polymerica Sinica*, 10, 006, 2010.
153. Veronovski, N., Rudolf, A., Smole, M.S., Kreže, T., Geršak, J., Self-cleaning and handle properties of TiO_2-modified textiles. *Fiber. Polym.*, 10, 551, 2009.
154. Nakajima, A., Hashimoto, K., Watanabe, T., Takai, K., Yamauchi, G., Fujishima, A., Transparent superhydrophobic thin films with self-cleaning properties. *Langmuir*, 16, 7044, 2000.
155. Xue, C.-H., Zhang, P., Ma, J.-Z., Ji, P.-T., Li, Y.-R., Jia, S.-T., Long-lived superhydrophobic colorful surfaces. *Chem. Commun.*, 49, 3588, 2013.
156. Xue, C.-H., Ma, J.-Z., Long-lived superhydrophobic surfaces. *J. Mater. Chem., A*, 1, 4146, 2013.
157. Xue, C.-H., Li, Y.-R., Zhang, P., Ma, J.-Z., Jia, S.-T., Washable and wear-resistant superhydrophobic surfaces with self-cleaning property by chemical etching of fibers and hydrophobization. *ACS Appl. Mater. Interfaces*, 6, 10153, 2014.
158. Xu, B., Ding, J., Feng, L., Ding, Y., Ge, F., Cai, Z., Self-cleaning cotton fabrics via combination of photocatalytic TiO_2 and superhydrophobic SiO_2. *Surf. Coat. Technol.*, 262, 70, 2015.
159. Horrocks, A.R., Flame retardant challenges for textiles and fibres: new chemistry versus innovatory solutions. *Polym. Degrad. Stab.*, 96, 377, 2011.
160. Iwata, M., Fukushima, H., Watanabe, T., Non-halogen and non-phosphorous flame-retardant polymer. *Fujikura Technical Review*, 31, 2014.
161. Yang, J.-C., Liao, W., Deng, S.-B., Cao, Z.-J., Wang, Y.-Z., Flame retardation of cellulose-rich fabrics via a simplified layer-by-layer assembly. *Carbohydr. Polym.*, 151, 434, 2016.

162. Soutter, W. Nanomaterials for Environmentally Friendly Flame Retardants. S, https://www.azonano.com/article.aspx?ArticleID=3087, 2012.
163. Norouzi, M., Zare, Y., Kiany, P., Nanoparticles as effective flame retardants for natural and synthetic textile polymers: application, mechanism, and optimization. *Polymer Rev.*, 55, 531, 2015.
164. Ghosh, A., Nano-clay particles as textile coating, International Journals of Engineering and Science, 11(5), pp 34–36, 2011.
165. Przybylak, M., Maciejewski, H., Dutkiewicz, A., Wesołek, D., Władyka-Przybylak, M., Multifunctional, strongly hydrophobic and flame-retarded cotton fabrics modified with flame retardant agents and silicon compounds. *Polym. Degrad. Stab.*, 128, 55, 2016.
166. Wang, S., Sui, X., Li, Y., Li, J., Xu, H., Zhong, Y., Zhang, L., Mao, Z., Durable flame retardant finishing of cotton fabrics with organosilicon functionalized cyclotriphosphazene. *Polym. Degrad. Stab.*, 128, 22, 2016.
167. Uddin, F., Areeb, T., Hussain, A., Nadeem A., Local Clay Mineral in Flame Retardant Finishing of Cotton Fabric. Chamical and Materials Engineering, 3(1), pp. 6–15, 2015.
168. Keshin, R., Gocek, I., Arslan, M. H., *Improving Flame Retardant Characteristics of Cotton Fabric by Incorporating Montmorillonite Nanoclay*, in *CICEMA 2017: 19th International Conference on Engineering Materials and Applications*. 2017, International Journal of Materials and Metallurgical Engineering Copenhagen, Denmark, Istanbul Technical university: Turkey.
169. Wu, H., Krifa, M., Koo, J.H., Flame retardant polyamide 6/nanoclay/intumescent nanocomposite fibers through electrospinning. *Text. Res. J.*, 84, 1106, 2014.
170. Pan, H., Wang, W., Shen, Q., Pan, Y., Song, L., Hu, Y., Lu, Y., Fabrication of flame retardant coating on cotton fabric by alternate assembly of exfoliated layered double hydroxides and alginate. *RSC Adv.*, 6, 111950, 2016.
171. Youssif, M. I., El-Maghraby, R. M., Saleh, S. M., Elgibaly, A., Silica nanofluid flooding for enhanced oil recovery in sandstone rocks. Egypt. J. Pet, in press, 2017.
172. Gashti, M.P., Alimohammadi, F., Shamei, A., Preparation of water-repellent cellulose fibers using a polycarboxylic acid/hydrophobic silica nanocomposite coating. *Surf. Coat. Technol.*, 206, 3208, 2012.
173. Carosio, F., Laufer, G., Alongi, J., Camino, G., Grunlan, J.C., Layer-by-layer assembly of silica-based flame retardant thin film on PET fabric. *Polym. Degrad. Stab.*, 96, 745, 2011.
174. Gashti, M.P., Almasian, A., UV radiation induced flame retardant cellulose fiber by using polyvinylphosphonic acid/carbon nanotube composite coating. *Composites Part B: Engineering*, 45, 282, 2013.
175. Yin, X., Krifa, M., Koo, J.H., Flame-retardant polyamide 6/carbon nanotube nanofibers: Processing and characterization. *J. Eng. Fab. Fibers (JEFF)*, 10, 2015.
176. Dittrich, B., Wartig, K.-A., Hofmann, D., Mülhaupt, R., Schartel, B., Flame retardancy through carbon nanomaterials: carbon black, multiwall nanotubes,

expanded graphite, multi-layer graphene and graphene in polypropylene. *Polym. Degrad. Stab.*, 98, 1495, 2013.

177. El-Shafei, A., ElShemy, M., Abou-Okeil, A., Eco-friendly finishing agent for cotton fabrics to improve flame retardant and antibacterial properties. *Carbohydr. Polym.*, 118, 83, 2015.
178. Lessan, F., Montazer, M., Moghadam, M., A novel durable flame-retardant cotton fabric using sodium hypophosphite, nano TiO_2 and maleic acid. *Thermochim Acta*, 520, 48, 2011.
179. Hashemikia, S., Montazer, M., Sodium hypophosphite and nano TiO 2 inorganic catalysts along with citric acid on textile producing multi-functional properties. *Appl. Catal., A*, 417, 200, 2012.
180. Lam, Y., Kan, C., Yuen, C., Effect of titanium dioxide on the flame-retardant finishing of cotton fabric. *J. Appl. Polym. Sci.*, 121, 267, 2011.
181. Lam, Y., Kan, C., Yuen, C., Flame-retardant finishing in cotton fabrics using zinc oxide co-catalyst. *J. Appl. Polym. Sci.*, 121, 612, 2011.
182. Samanta, A.K., Bhattacharyya, R., Jose, S., Basu, G., Chowdhury, R., Fire retardant finish of jute fabric with nano zinc oxide. *Cellulose*, 24, 1143, 2017.
183. Pan, Y., Wang, W., Liu, L., Ge, H., Song, L., Hu, Y., Influences of metal ions crosslinked alginate based coatings on thermal stability and fire resistance of cotton fabrics. *Carbohydr. Polym.*, 170, 133, 2017.
184. Attia, N.F., Morsy, M.S., Facile synthesis of novel nanocomposite as antibacterial and flame retardant material for textile fabrics. *Mater. Chem. Phys.*, 180, 364, 2016.
185. Yetisen, A.K., Qu, H., Manbachi, A., Butt, H., Dokmeci, M.R, Hinestroza, J.P., Skorobogatiy, M., Khademhosseini, A., Yun, S.H., Nanotechnology in textiles. *ACS Nano*, 10, 3042, 2016.
186. Hezavehi, E., Shahidi, S., Zolgharnein, P., Effect of dyeing on wrinkle properties of cotton cross-linked by butane tetracarboxylic acid (BTCA) in presence of titanium dioxide (TiO_2) nanoparticles. *Autex. Res. J.*, 15, 104, 2015.
187. Gruntfest, I., Gagliardi, D., Wrinkle resistance of fabrics. *Ind. Eng. Chem.*, 41, 760, 1949.
188. Doneva, T.A., Griffiths, L.G., Mahammadi, M.S., Ormandy, K.A., Southey-Davis, H.M., *Fabric Care Compositions*. 2006, Google Patents.
189. Yang, C.Q., Xu, L., Li, S., Jiang, Y., Nonformaldehyde durable press finishing of cotton fabrics by combining citric acid with polymers of maleic acid. *Text. Res. J.*, 68, 457, 1998.
190. Morshed A. M. A., Iqbal, H., Hasan, K., impact of nanotechnology in the area of textile apparel finishing. Bangladesh Textile Today, 3(3), pp. 53–55, 2010.
191. Lam, Y., Kan, C., Yuen, C., Effect of concentration of titanium dioxide acting as catalyst or co-catalyst on the wrinkle-resistant finishing of cotton fabric. *Fiber. Polym.*, 11, 551, 2010.
192. Nallathambi, G., Ramachandran, T., Rajendran, V., Palanivelu, R., Effect of silica nanoparticles and BTCA on physical properties of cotton fabrics. *Mater. Res.*, 14, 552, 2011.

193. Yuen, C., Ku, S., Kan, C., Cheng, Y., Choi, P., Lam, Y., Using nano-tio 2 as co-catalyst for improving wrinkle-resistance of cotton fabric. *Surf. Rev. Lett.*, 14, 571, 2007.
194. Karthik, T., Rathinamoorthy, R., Murugan, R., Enhancement of wrinkle recovery angle of cotton fabric using citric acid cross-linking agent with nano-TiO_2 as a co-catalyst. *J. Ind Text*, 42, 99, 2012.
195. Nazari, A., Montazer, M., Rashidi, A., Yazdanshenas, M., Moghadam, M.B., Optimization of cotton crosslinking with polycarboxylic acids and nano TiO_2 using central composite design. *J. Appl. Polym. Sci.*, 117, 2740, 2010.
196. Chattopadhyay, D. P., Patel, B. H., improvement in physical and dyeing properties of natural fibres through pre-treatment with silver nanoparticles. IJFTR, 34(4), pp 368–373, 2009.
197. Hebeish, A., El-Bisi, M., El-Shafei, A., Green synthesis of silver nanoparticles and their application to cotton fabrics. *Int. J. Biol. Macromol.*, 72, 1384, 2015.
198. Montazer, M., Alimohammadi, F., Shamei, A., Rahimi, M.K., Durable anti-bacterial and cross-linking cotton with colloidal silver nanoparticles and butane tetracarboxylic acid without yellowing. *Colloids Surf., B*, 89, 196, 2012.
199. Patel, B., Chaudhari, S., Patel, P., Nano silica loaded cotton fabric: Characterization and mechanical testing. *Res. J. Eng. Sci.*, 2278, 9472, 2014.
200. Ashraf, M., Irshad, F., Umar, J., Farooq, A., Ashraf, M.A., Development of a novel curing system for low temperature curing of resins with the aid of nanotechnology and ultraviolet radiation. *RSC Adv.*, 6, 81069, 2016.
201. Shady, K., Michael, M., Shimaa, H.. *Effects of zinc oxide nanoparticles on the performance characteristics of cotton, polyester and their blends.* in *AIP Conf. Proc.* 2012: AIP.
202. Farouk, A., Moussa, S., Ulbricht, M., Textor, T., ZnO nanoparticles-chitosan composite as antibacterial finish for textiles. *Int. J. Carbohydrate Chem.*, 2012, 1–8, 2012.
203. Amini, A., Zohoori, S., Mirjalili, A., Karimi, L., Davodiroknabadi, A., Improvement in physical properties of paper fabric using multi-wall carbon nanotubes. *J. Nanostructure Chem.*, 4, 103, 2014.
204. Alimohammadi, F., Parvinzadeh, M., Shamei, A., *Carbon nanotube embedded textiles.* 2011, Google Patents.
205. Aly Sayed, A., Ali, H., Samaha sayed, H., Antibacterial finishing agent; chitosan; chitosan citrate; cotton; crease resistance. IJFTR, vol 29(2), 2004.
206. Gong, C., Qi, T., Wei, X., Qu, Y., Wu, Q., Luo, F., Qian, Z., Thermosensitive polymeric hydrogels as drug delivery systems. *Curr. Med. Chem.*, 20, 79, 2013.
207. Wang, C., Lin, H., Chen, Y., Lu, Y., Influences of nano-chitosan on anti-bacterial and wrinkle-resistant performances of mulberry silk fabrics. *Silk.*, 1, 004, 2013.
208. Raza, Z.A., Anwar, F., Fabrication of chitosan nanoparticles and multi-response optimization in their application on cotton fabric by using a Taguchi approach. *Nano. Struc. Object*, 10, 80, 2017.

2

Antimicrobial Textiles Based on Metal and Metal Oxide Nano-particles

Mangala Joshi* and Anasuya Roy

Department of Textile Technology, Indian Institute of Technology, New Delhi, India

Abstract

With the increasing surge in recent times toward better health and hygiene around the globe, considerable research attention has been focused on developing textile materials possessing antimicrobial functionality. In such times, when microbial resistance to synthetic organic compounds like drugs and antibiotics are rampantly increasing, metal and metal oxide-based inorganic nano-particles (NP) have taken a pivotal role in preparing such antimicrobial fibers and textile materials. Incorporation of antimicrobial NP, for example, NP of Ag, Cu, ZnO, Au, and titania, in fibers and fabrics via nano-coating or nano-finishing presented promising results and excellent bioactivity. These active nano-reinforced textile structures are potential candidates for application in various areas such as sportswear, upholstery, hospital curtains, surgery gowns, medical textiles, wound dressings, bandages, etc. However, research concerning its nano-toxicological impact to human health and environment is still at its infancy and has considerably been questioned its several applications. This chapter reviews the state-of-the-art NP-based antimicrobial textile fibers and fabrics, both man-made and natural with special emphasis on toxicological properties and its effect on commercial applications.

Keywords: Fibers, textiles, metal NP, antimicrobial, nano-toxicological studies

**Corresponding author*: mangala@textile.iitd.ac.in

Shahid-ul-Islam and B.S. Butola (eds.) Nanomaterials in the Wet Processing of Textiles, (71–112) © 2018 Scrivener Publishing LLC

2.1 Introduction

Textile materials provide ample ambient conditions of temperature, humidity, and nutrition for pathogenic micro-organisms to thrive and multiply on their surface and proliferate into the surrounding environment [1–3]. The exponential multiplication of micro-organisms is assisted by internal factors of the human body, such as accumulation of sweat, dead skin cells, oil secreted by sebaceous glands, as well as external parameters like dust, stains caused by spilled food and drinks, nature of finishing chemical, etc. As a consequence, imparting antimicrobial property to textile fabrics and fibers by different methods such as finishing or coating has become one of the utmost significant research field in the modern world [4–6]. The global antimicrobial finishing market is expected to rise by a compound annual growth rate of 20% to 25% as per recent market surveys and is estimated to reach 2.9 billion dollars by 2018 [7]. In addition to providing improvement in human living standards, it controls allergic reactions, lethal infections, cross-infections of microbe-transmitted diseases, and other harmful health effects [8–11]. Antimicrobial finishes on textiles prevent microbial attack in the material which is responsible for maintaining its aesthetic and functional value. Degradation of the fibers and dyes leading to loss of local color or undesirable staining is prevented along with unpleasant odor associated with bacterial colonization. Currently, antimicrobial fabrics and fibers are increasingly being used in various sectors such as protective, medical, technical and decorative, and sportswear textiles with vast application areas from hospital environment to daily household items [12–16].

Microbes are defined as a class of living micro-organisms, both unicellular and multicellular, that include bacteria, fungi, algae, virus, archaea, and protozoa [17–19]. The phyla of fungi includes yeasts (unicellular), rusts, smuts, mildews, molds, and mushrooms (multicellular). A large number of microbes co-exist in the human society with harmony, in the immediate surroundings and inside the bodies of living organisms [20–23]. Although it is conceived that majority of microbes are pathogenic and exert harmful ill-effects on the health, in reality, majority of them are completely harmless [24–26] and some even help us in bodily functions like digestion (*Escherichia coli*) and preparation of nutrient-containing food such as bread (*Saccharomyces cerevisiae*) [27], citric acid (*Aspergillus niger*) [28], riboflavin (*Eremothecium sp.*) and even has medicinal uses such as antibiotic isolation [29]. Some bacteria do not cause fatal or high-risk diseases but are associated with unpleasant smell and overall bad hygiene [30]. The human skin itself is a home to a microbiota of diverse species ranging from bacteria and viruses to yeasts, prokaryotes, and fungi

[31]. The environment humidity (moist or dry) and the composition of the clothing enclosing the skin have a strong effect on the type of residing bacteria [32–34]. An interesting study [35] found significantly different microbiomes on different clothing fabrics. Highest malodor was observed for polyester fabrics with infestations of *Micrococci*, *Enhydrobacter*, and *Propionibacterium sp.* owing to inherent low moisture uptake. Cotton fabrics showed a predominant strain of *Staphylococcus hominis* while both the fabrics showed growth of *Staphylococcus epidermidis.* Wool clothing promoted growth of almost all bacteria and nylon fabrics exhibited promoted growth of *Propionibacterium sp.*, *Staphylococcus sp.*, and *Enhydrobacter sp.* The *Propionibacterium sp.* are mainly responsible for unpleasant odor associated with nylon socks and thrives well in most areas such as foot [36].

The present decade has witnessed frequent outbreaks of infectious diseases with increase in microbial resistance to antibiotics and drugs based on organic structures [37]. Conventionally, textiles were treated with antimicrobial formulations based on the antimicrobial property of complex organic compounds such as quaternary ammonium salts, N-halamines, halogenated phenols like triclosan, polybiguanides, etc. [38–43]. Alternatively, antimicrobial textile substrates were developed using natural extracts, biopolymers, and bio-based NP [44–49]. Depending upon the mechanism of antimicrobial action of the active agent impregnated within the textile material, the textile may be termed as "bacteriostatic" when the growth of micro-organisms is inhibited or "bacteriocidal" when the active agents impart antimicrobial action by killing the bacteria corresponding to bound and leeching types of antimicrobial agents, respectively [38].

With the advent of nano-technology, considerable research studies have focused on a new class of antimicrobials which are inorganic in nature like metal and metal oxide NP based on Ag, Cu, CuO, ZnO, and titania NP [50–54] unlike the conventionally used organic antimicrobials. These biologically active nano-materials are highly potent even at very low concentrations while having zero microbial resistance with other advantages such as multi-targeted action, unique small size, large surface-to-volume ratio, and unique physiochemical properties. Several studies are oriented toward achieving antimicrobial textile materials with incorporation of these inorganic NP that provide fast and easy functionalization with excellent effectiveness and long durability [55–62]. However, owing to their metallic nature, the NP above the threshold limit can hamper normal cell division and proliferation even in humans due to the generation of reactive oxygen species (ROS) which can cause inflammation and allergy along with other minor complications [63–64]. This drawback has put a leash on the ever-expanding market of nano-scale-based antimicrobials. The toxicological data of these NP are elucidated primarily in water and cell media, which is a far cry from the complex *in vivo* environment of the human body [65–66]. Although Ag salt-based burn creams and gels are a commercial triumph, Ag NP having significantly higher affectivity face difficulty in penetrating the market due to unavailability of sufficient data confirming its non-toxic nature [67–68].

Various standard test methods, both qualitative and quantitative, are available for testing antimicrobial-finished textile fibers and fabrics. The AATCC Test Method 100: 1999 and AATCC Test Method 147: 1998 are most popular methods to test the efficacy of antimicrobial textile materials although AATCC Test Method 90: 1998 is also frequently employed. The AATCC Test Method 147: 1998, Parallel Steak method, is a fast and qualitative measurement of the antimicrobial activity in which a rectangular cut fabric sample (25 mm ′ 50 mm) is placed on a sterile nutrient agar plate parallel streaked with inoculations of Gram-positive or Gram-negative strain. The amount of bacterial growth on the petridish and on the textile after the incubation period of 18–24 h gives a qualitative method of antibacterial activity. In the AATCC Test Method 100: 1999, Colony Counting method, a swatch of fabric with a certain weight or dimension is inoculated with bacterial solution and incubated at ambient conditions. After the swatches are removed, the number of bacteria in the extract is determined using serial dilution and agar plating technique. This is an accurate quantitative test method and by far the most popular as the results are reproducible. The AATCC Test Method 90: 1998 Disk Diffusion Test is only suitable for antimicrobial fabrics that have leaching chemicals that can detach itself from the fabric and form a zone of inhibition. In this test procedure, the small swatch of fabric is placed on the center of surface-inoculated pertidish. The clear zone surrounding the sample is quantified after 18–24 h incubation period. Other test methods for antimicrobial textiles include ASTM test methods such as the ASTM E 2149 Test Method for determining the antimicrobial activity of antimicrobial agents under dynamic contact conditions, the ASTM E 2180 Test Method for determining the activity of incorporated antimicrobial agents in polymeric or hydrophobic materials, and the ASTM E 2722 Test Method for using Seeded-Agar for the screening assessment of antimicrobial activity in fabric and air filter media.

The antifungal properties of a textile material are generally evaluated using the AATCC Test Method 30: 1999. The standard provides four methods for evaluation of fungal growth on the textile substrate to be classified as "“no growth”, “microscopic growth”, or “macroscopic growth” accompanied with the determination of tensile strength loss due to fungus-induced deterioration in the fabric. The first method tests the extent of growth of fungi in the test sample of dimension 4 × 15 cm buried and incubated in fungi-inoculated soil for 2 to 16 weeks at 28 °C. The second and third method involves growth of fungi *Chaetomiumglobosum* and fungi *A. niger*, respectively, in an inoculated petridish having a circular (4 cm diameter) or rectangular specimen (4 × 15 cm) placed at the center. The last method involves a 2.5 × 7.6 cm specimen-introduced sealed jar seeded with three fungi mixtures at a high humidity environment for 14 to 28 days. The extent of fungal growth in the samples is taken as a direct measure of antifungal activity.

2.2 Antimicrobial NP Used in Functionalization of Textiles

2.2.1 Ag NP: Synthesis and Antimicrobial Activity

Ag has been used as a sign of prosperity since ancient times by royal families with Ag cutleries and crockeries considered epitome of high standard life [55]. The excellent biological activity of Ag was long known and has been used through ages as a medium to fight infectious ailments. Ag compounds and ions have been used for centuries to treat contaminated water to render it potable and life-sustaining [51, 55]. Among the different metallic NP used in antimicrobial finishing of textiles, Ag NP is by far the most prominently used due to highest effectiveness at lowest concentration. Ag NP inhibit almost 650 different types of pathogenic bacteria and five strains of yeast; however, its affectivity toward mildew, mold, and fungus is limited, probably because it falls into the category of non-essential elements of nutrition [69–71]. In the present decade, Ag-based finishes both in NP and bulk form have unprecedented market application in not only textile, but also plastic, chemical, pharmaceuticals, and hygiene industries [72–73]. Ag resin columns and Ag zeolites have been used since long in water filters and pharmaceutical applications [74–75]. Ag salts such as silver sulfadiazine have been extensively used in burn bandages and wound dressings [76–77]. The use of $AgNO_3$ as a finishing chemical to impart antimicrobial activity is often accompanied with a dark brown–black stain on exposure to air, heat, or light caused by uncontrolled reduction of Ag ions to form bigger metallic Ag particles [78–79].

The size, shape, and crystallinity of Ag NP have a profound effect on its antimicrobial behavior with the non-spherical shapes such as polygonal, triangular, and poly-hierarchical morphologies which have a higher inhibitory effect than spherical and disc-shaped NP [80–81]. However, the method of synthesis, that is, reduction route, type of chemicals, and process parameters, also affects the end properties [68]. The synthesis of Ag NP can be broadly categorized into top-down and bottom-up approaches. In the bottom-up approach, which is used predominantly for synthesis purpose, Ag NP are produced by both physical deposition and chemical precipitation methods. Chemical deposition routes often require harsh non-environment-friendly surfactants and reducing agents. The precursor salt is most often $AgNO_3$ as it offers advantages of comparatively low cost and chemical stability although silver acetate is also in use [82]. The reducing agents used are sodium borohydride, sodium citrate, and hydrazine hydrate as mostly used; however, ethanol, DMF, ascorbic acid, aldehydes, and isopropanol are also employed [83–86]. The capping agents used to cease particle growth and prevent agglomerated superstructures are mainly PVP and citrate salts with uses of amines, CTAB, sugars, and PVA also mentioned in the current literature [68, 84–86]. With growing concern about environmental hazards with the release of toxic byproducts and reagents into the immediate surroundings, green synthesis has found significant focus in recent times. Plant extracts, biological agents, and microbe-mediated metabolism byproducts that are capable of donating free electrons are increasingly reported as potential reducing agents to combat and replace hazardous reducing chemicals [87–90]. Physical deposition routes of synthesizing Ag NP include UV-irradiation, calcination at moderately elevated temperatures, and plasma treatment [91–93]. The top approach for synthesis which employs laser ablation techniques and nanolithography to reduce particle down is comparatively less reported due to difficulty in controlling the morphology and thereby leading to imperfectly shaped NP [68].

2.2.2 Titania NP: Synthesis and Antimicrobial Activity

TiO_2 or titania is the natural oxide form of titanium and has been used generously in a variety of applications in cosmetic, paints, paper, plastic, and textile industry for decades [95–96]. Titania is a semi-conducting material with one of the strongest photo-catalytic activities with high oxidizing power, long-term stability, and low cost [97]. The high refractive index and good light-scattering property of the material make it possible to impart considerable whiteness and opacity to its product even at very low concentrations. Titania is considered to be chemically and biologically inert, and hence accepted as a food and pharmaceutical additive for artificial food coloring, dental pastes, tablets, oral capsules, suspensions, and dermal ointments and creams [54, 98–99]. With breakthrough research on nano-science and nano-technology and development of nano-materials as an improved version of the parent materials with unique value-added properties, TiO_2 NP or nano-titania became increasingly important having unique photocatalytic, optical, UV-shielding, and antimicrobial properties making it suitable for a plethora of applications in various market sectors [95, 99, 100–101]. In the textile industry, nano-titania is popular in "self-cleaning" textiles for obtaining impurity-free surfaces and self-disinfecting nature by self-decomposition of dirt and organic pollutant molecules under accelerated photo-catalytic reaction [50, 102–104].

Crystalline TiO_2 exhibits three different polymorphs—anatase, rutile, and brookite—among which anatase exhibited the highest photocatalytic activity. However, the commercial nano-titania powder available for antibacterial studies contains 80:20 mixture of anatase and rutile [105]. Titania NP are generally obtained by the sol–gel precipitation method in which titanium isopropoxide is normally stirred with ammonium hydroxide, and the end-point of the reaction is confirmed when a change of color from purple to white precipitate is obtained [95,98]. When UV light is illuminated, a positive hole is photo-generated in the valence band when excitation causes electrons to jump from valence to conduction band. Reactive oxygen species (ROS) such as O_2^*, H_2O_2, and HO^* are generated via redox reactions between electron-hole and adsorbed oxygen and water. ROS are known to induce oxidative stress inside the bacterial cell and contribute toward bactericidal action [98–100]. Studies shown on *E. coli* bacteria demonstrates that OH radicals were major contributors toward oxidative stress; its high reactivity makes crevices on the outer wall that subsequently leads to drainage of cytoplasmic fluid [100]. Nano-titania with spherical crystallites <30 nm diameter exhibits strong photo-catalytic disinfection action against many strains of bacteria (*Streptococcus mutans, Streptococcus epidermis, E. coli, Bacillus sp.),* fungi (*Aspergillus sp.*), algae, and virus [99, 104–105], although their MIC values are pronouncedly less than Ag-based systems [106–107].

2.2.3 Cu NP: Synthesis and Antimicrobial Activity

Due to its inherent antimicrobial nature, Cu along with Ag has been employed by the human society since pre-historic ages with documented uses in coins, utensils, and inexpensive jewelry. Its hygienic and antibacterial properties were well-recognized and gained popularity in common masses, being less expensive as compared to Ag since the advent of human civilization [108–109]. In ancient times, pots or bottles were made of Cu where water was stored for hygienic purposes. In fact, old-age Hindu traditions believed throwing coins in water would fulfill wishes, and it was considered a good practice as Cu coins purified the water, making it potable. Cu, unlike Ag, is an essential trace element that is present in all living tissues and is crucial for growth and development of many micro-organisms including human [110–111]. Cu is a well-established element with antibacterial, antiviral, antifungal, insecticidal, and molluscicidal properties [112–113]. Cu salts such as $CuSO_4$ have been extensively used as a fungicide in agricultural applications [114]. However, its toxicity at higher doses, especially to marine and aquatic life, has restricted its widespread commercial applications in recent times [115]. With the advent of nano-technology, research on the synthesis of Cu and CuO NP was focused to bring antimicrobial effects at lower concentration and less toxicity [38, 55,116].

The current literature on the synthesis of Cu NP can be classified broadly into three categories by physical, chemical, and biological means. Like synthesis of Ag NP, most researchers focus on the bottom-up approach by wet chemical routes as the scope of controlling the size of NP is higher. In wet chemical routes, the precursor Cu salts such as $CuSO_4$, in aqueous or ethanol media using reducing agents such as sodium borohydride, hydrazine, ascorbic acid, etc., are reported [117–118]. Recently, extracts from plants, that is, extracts from the plant *Terminalia arjuna* and from the leaves of *Lawsonia inermis* (henna), are employed as reducing agents and/or capping agents [119–120]. These methods of reduction are environment friendly and avoid direct usage of harsh reducing agents. The reverse emulsion method containing surfactant-stabilized oil-in-water micro-emulsion acting as nano-reactors for the formation of Cu NP is another well-known procedure where shape and size are easily controllable [121–122]. The physical routes for synthesis include microwave irradiation, thermal treatment, electrochemical synthesis, ultra-sonication technology via acoustic cavitation, etc. [38, 123]. In an attempt to develop non-toxic, reliable, and green process routes, efforts are being put into modern studies to focus on biological synthesis [124–125]. The bacterial strain of *Pseudomonas stutzeri* isolated from soil was successful in producing cubic-shaped Cu NP of 50- to 150-nm diameter [125]. The synthesis of CuO NP is generally accomplished by hydrolysis of precursor Cu salt followed by dehydration. CuO NP relatively has more physical and chemical stability than pure Cu NP [126]. Cu and CuO NP are found effective against a wide range of both Gram-positive and Gram-negative bacteria, for example, *Staphylococcus aureus, Bacillus subtilis, E. coli, Micrococcus luteus,* and *Klebsiella pneumoniae* as well as different strains of fungi such as *Aspergillus flavus, A. niger*, and *Candida albicans* [55, 127]. The antifungal activities of Cu-based NP were higher than Ag NP of a similar diameter as Cu is relevant to the biochemical reactions of organisms and is readily taken up by fungi that consider Cu as a nutrient [128].

2.2.4 ZnO NP: Synthesis and Antimicrobial Activity

Zn belongs to the class of essential elements required for vital functioning, as it controls and catalyzes many cellular reactions responsible for structural activity [129–130]. It is responsible for growth and differentiation, utilized in redox processes, and they regulate osmotic pressure as well as maintain ionic balance. In lower organisms, as the Zn concentration reaches higher levels, depending on the type of micro-organism, it starts exerting its cytotoxic nature [131–132]. It has been postulated that it hampers the generation time of an organism by prolonging the lag phase of the cycle responsible for growth. Additionally, it alters the semi-permeable nature of the cell membrane preventing calcium uptake by controlling the enzymes responsible for membrane permeability [133–137].

Nano-ZnO represents a significant class of metal oxide NP possessing excellent antimicrobial action with lowest toxicity to human cells [132]. Its virtues make it highly recommendable in the food industry and in biomedical applications where an optimal balance has to be sustained between safety to human health and potent biocidal property [138–139]. Its UV shielding property has been rampantly used as active ingredients in skin lotions, creams, and sunscreens [138–140]. ZnO NP are also used in sensors, electronic devices, energy-harvesting materials, and in optical absorption due to its semi-conducting, photo-catalytic, and piezoelectric nature [140]. Different shapes of ZnO nano-structures, namely nano-rods, nano-spheres, nano-wires, nano-needles, and nano-rings, have been reported by controlling physical parameters and chemical parameters such as temperature, pH, solvent type, and precursor concentration [140–141]. Some of the most adopted technologies for synthesis of ZnO nano-structures include thermal evaporation, hydrothermal synthesis, sol–gel technique, simple thermal sublimation, self-combustion, polymerized complex method, vapor–liquid–solid technique, double-jet precipitation, and chemical solution synthesis [138–141]. Spherical shaped ZnO NP synthesized by the wet chemical process using a Zn precursor salt such as Zn acetate dihydrate and strong alkali like sodium hydroxide have currently been explored for antimicrobial activity and have been found to exhibit strong microbial resistance to gram-positive and gram-negative bacteria like *E. coli*, *S. aureus*, *P. aeruginosa*, *B. subtilis*, *S. acidaminiphila*, and *P. aeruginosa* [141–143]. The excellent antifungal behavior of ZnO NP is considerably higher than that of non-essential elements such as Ag and has been shown to be effective against white-rot fungi, mildew, and algae [128].

2.3 Application of NP onto Textile Substrates

2.3.1 Application of Ag NP on Textiles

Among different metal NP used for antimicrobial functionalization of textile materials, Ag NP are by far the most preferred choice over other NP, because of high potency at low concentration. Additionally, most research attention is focused on antimicrobial functionalization of cotton fabrics, mainly because it is most prone to microbial degradation and is used extensively in non-implantable medical textiles such as wound dressings, bandages, gauze fabrics, uniforms, and caps [4, 38, 57, 144]. Surface modification or finishing has been done using various application techniques such as dip-coating, pad-dry-cure method, layer-by-layer coating, plasma treatment, or *in situ* reduction [55]. According to the International Standard Organization, an NP is defined as a particle as a whole having a diameter <100 nm in all three dimensions. However, any particle between 100 nm and 1000 nm, which essentially contains clusters of individual NP, is also termed as NP in the textile world [55]. By this definition, some of the Ag salts impregnated into the textile fabric are also NP as they are clusters of Ag crystals in nano-meter range. With vast expanding areas of research conducted on antimicrobial Ag NP-based chemistry, excellent bioactive textile fabrics are readily available. However, although the binding mechanism of Ag NP onto textile substrates is not fully elucidated, effectiveness of antimicrobial activity is greatly reduced with strong chemical binding in the interface. This raises a huge concern for the detrimental drainage of nano-Ag into the immediate environment with successive washing cycles [145–146]. The toxicological shock represents a potential research area where future investigation must be directed to achieve commercial success of NP-based biochemistry onto textile substrates [67]. Unlike thermoplastic polymers, where incorporation of Ag NP can be carried out at various stages, impregnation in fibers and in fabrics can only be done at the final stage in which NP are applied as a finishing technique. Emam *et al.* [147] argue that functional household cellulosic textiles show a significant pick up of Ag NP and subsequent antibacterial activity through simple immersing in $AgNO_3$ aqueous solution and therefore the researches which articulate incorporation of reducing and stabilizing agents, whether physical or chemical or bio-based, only complicate the methodology without bringing out better functional attributes. However, the concentration of the Ag nano-sol has a significant control over its laundered antimicrobial activity. Low concentrations (less than 10 ppm) may not be suitable for long-term antimicrobial activity and high concentrations upto 50 ppm may be required to achieve a desirable effect.

The morphological shape of the Ag NP is a determining factor in deciding the extent of bio-activity. Different morphologies of synthesized Ag NP—spheres, flat nano-plates, nano-prisms, polygonal and hierarchical by chemical reduction route have been applied on different cotton fabrics using ultra-sonication technology at suitable temperature [148]. The fabrics treated with hierarchical Ag NP evaluated against *E. coli* and *S. aureus* showed the best activity of >91% after five washing cycles.

Several studies are tuned toward *in situ* reduction of Ag NP directly on the textile substrate. Perelshtein *et al.* [149] prepared antibacterial cotton, nylon, and polyester fabrics using the dip-coating method using a solution of $AgNO_3$ salt precursor and ethylene glycol in aqueous media. The fabric was further sono-chemically treated for immobilization of Ag NP effective against *E. coli* and *S. aureus*. Liu *et al.* [150] reported a simple yet effective one-pot synthesis method for fabricating a cotton bandage for biomedical applications with excellent activity toward *E. coli* and *S. aureus*. The nano-coating on cotton was carried out by dipping cotton fabric in ethanol-mediated $AgNO_3$ and butylamine solution. Lee *et al.* [151] developed bioactive cotton and a polyester-woven fabric by simply dipping in Ag nano-sol stabilized in ethanol. Rajavel *et al.* [152] demonstrated yet another simple method for producing cotton fabrics with durable antimicrobial finishing by dipping the fabric in $AgNO_3$ and subsequently sodium borohydride solution under static conditions. The authors optimized the concentrations of the solution to maximize loading and distribution and reported that low concentrations of both precursor (1 m M Ag NO_3) and reductant (0.5 m M borohydride) are beneficial for optimum results. The treated fabrics showed yellowish coloration and bioactivity against *B. subtilis* and *S. sonnei* for potential use in wound dressing. Antimicrobial cotton fabrics pre-activated with potassium hydroxide, dipped into $AgNO_3$ and ascorbic acid solution, and further treated with hexadecyltrimethoxysilane to coat with a low surface energy barrier layer were reported by Shateri-Khalilabad *et al.* [153]. The same research group reported pre-activation of cotton fabrics with alkali treatment to maximize loading [154]. The *in situ* NP-embedded hydrophobic fabric showed good laundering stability and antimicrobial action against *E. coli* and *S. aureus*. The carboxylate and the amine groups present in the polymeric chains of polyamide fibers were used as a reducing agent for obtaining Ag NP by Montazer *et al.* [155]. The

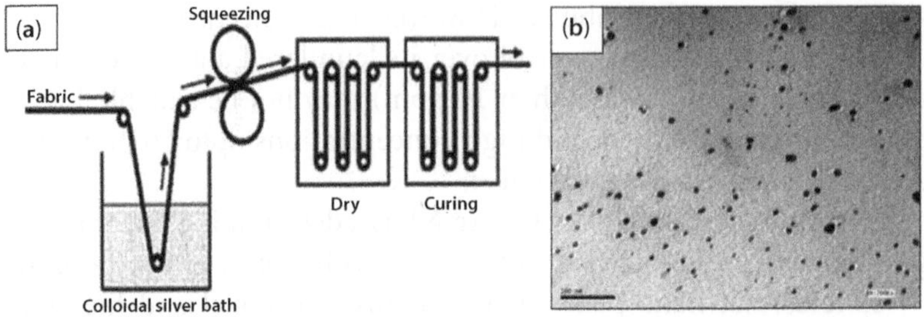

Figure 2.1 (a) Antibacterial finishing of cotton fabrics by pad-dry-cure. (b) TEM micrograph of silver nanoparticles with a concentration of 500 ppm. Adapted with permission from reference [86].

antibacterial fabric was obtained by treating it with Ag-ammonia complex at boil with 20–150 nm Ag NP forming *in situ* on its surface. The fabric exhibited excellent antibacterial activity against *S. aureus* and *E. coli* with >98% bacterial colony reduction after 20 washes. A schematic of the pad-dry-cure method and the corresponding TEM image of the Ag NPs as outlined by Hebeish *et al.* [86] has been outlined in Figure 2.1.

Apart from *in situ* reduction, considerable research activities were carried out to fabricate antimicrobial textiles using *ex situ* precipitated Ag NP. El-Rafie *et al.* [156] reported antibacterial cotton fabrics finished with Ag NP by dipping in a colloidal solution of Ag NP and an acrylate-based binder. The prepared fabrics were effective against *E. coli* and *S. aureus*. Hanh *et al.* [157] prepared polyester–cotton-blend fabric coated with Ag NP by the pad-dry-cure method. The Ag NP were previously prepared *ex situ* by γ-radiation of $AgNO_3$ chitosan solution, chitosan acting as a stabilizing as well as binding agent. The fabric showed acceptable performance as a wrapper in patient's bed in the hospital where chances of cross-infections are high. Alkali-hydrolyzed alginate solution has been used as a reducing and stabilizing agent for the synthesis of Ag NP and further used for finishing of cotton fabrics by the pad-dry-cure method. In another study by Hebeish *et al.* [158], carboxy methyl cellulose hyrdrogel-based Ag NP has been used as a finishing agent on cotton fabrics through the pad-dry-cure route. Multifunctional wool fabrics functionalized with different morphologies of *ex situ* prepared Ag NP by the dip coating method were developed by Tang *et al.* [159]. The prism and disk-shaped NP imparted brilliant and uniform coloration of blue, red, and yellow to wool fabrics due to localized surface plasmon resonance. Montazer *et al.* [160] reported antimicrobial nylon/Ag NP nano-composite fabric using colloidal Ag nano-sol and 1,2,3,4-butanetetracarboxylic acid (BTCA) as a binder as shown as

Figure 2.2 SEM images of coverless nylon: (a) 100× and (b) 15,000×, nylon fabric covered by silver nanoparticles/BTCA (c) 15,000× and (d) 30,000×. Adapted with permission from reference [160].

Figure 2.2. The multifunctional fabric showed antimicrobial activity and resistance to creasing without undesirable color formation. In another study by Yuranova *et al.* [161], polyester and polyamide fabrics were pretreated with both RF-plasma and Vacuum-UV radiation for surface modification to maximize Ag loading and further treated with $AgNO_3$. Zhang *et al.* [162] developed *ex situ* nano-colloidal Ag NP using an amino-terminated hyper-branched polymer as a reducing agent and a finishing agent for cotton fabric. Yeo *et al.* [163] reported a bicomponent PP sheath-core filament with Ag NP incorporated in the sheath part for enhanced bioactivity. Majumder *et al.* [164] presented a model-based study on the effect of knitting parameters on antimicrobial activity for knitted PET/Ag NP nano-composite fabrics. The authors reported a decrease in the count of yarn, and an increase in machine gauge increases antimicrobial activity and the fabrics showed ~1% of antimicrobial loss after 25 washes. A comparison study between *in situ* and *ex situ* approach for deposition of Ag NP on cotton fabric was presented by Perera *et al.* [165]. The authors concluded that *in situ* approach functioned better in terms of strength of antimicrobial activity and durability along with easier control of shape, size, and location of Ag NP on the fabric. Further, *in situ* approach is capable of providing antimicrobial action at low concentrations; thereby, leeching out of chemicals is also minimized.

The LBL nano-coating method for immobilization of Ag NP is generally used for nylon, silk, and cotton using alternate dipping into dilute solutions of poly(diallyldimethylammonium chloride) (PDADMAC) and Ag NP capped with poly(methacrylic acid) (PMA) [166]. Dubas *et al.* [167] reported better uniformity in nano-coating for silk fibers than nylon fibers, the antibacterial efficiency being 80% and 50% bacterial colonies reduction for silk and nylon, respectively, against *S. aureus*. Sataev *et al.* [168] invented a novel method of nano-coating in which fabrics are first treated with $CuSO_4$ solution by either dipping or spraying and further treated with phosphine to obtain copper phosphide and silver nitrate to obtain a silver phosphide coating of 40–600 nm thickness. The fabrics retained bactericidal efficacy after 10 washes. Ag NP (~20 nm) in another study reported by Budama *et al.* [169] were synthesized as reverse micelle cores using polystyrene-block-polyacrylic acid (PS-b-PAA) as a copolymer, polymerized *in situ* by atom transfer radical polymerization. Further, cotton fabrics were padded with the copolymer-Ag solution enabling esterification reaction between hydroxyl groups of cellulose and corona (PAA). The fabrics were effective against *E. coli* for up to 5 washes and *S. aureus* for up to 20 washes. Cotton gauze fabrics for wound healing applications finished with Ag oxide NP were reported by Gouda *et al.* [170]. The NP synthesis was carried out *in situ* by dipping the fabrics in successive baths of precursor salt containing $AgNO_3$ and reducing bath containing hydrogen peroxide and sodium hydroxide at pH 9.5. The fabric was further batched and dried and showed effective reduction of colonies against *S. typhimurium*, *S. aureus*, *C. albicans*, and *A. flavus*. It has been opined by Dastjerdi *et al.* [171] that there is binding force playing between the textile substrate and the inorganic NP due to difference in surface energies between the two phases and hence the authors devised a functional coating based on polysiloxane modified with nano-Ag which is applied to PET fabric by the pad-dry-cure route. The polysiloxane polymeric layer on the fabric makes it resistant to heat, light, and microbial attack in addition to providing it with anti-static property, eliminating the use of spin finish step.

Since Ag NP have extreme tendency to agglomerate and form clusters of metallic Ag, thereby decreasing its antimicrobial potency manifold, considerable research has been oriented toward achieving stable NP with supporting materials targeted toward maximum achievable functional augmentation. Silica-Ag core shell NP represent a significantly important class, with silica core offering advantages of sufficient compatibility, inert nature, and excellent heat resistance to support the peripherally located Ag NP [172]. Extremely lower sizes of Ag NP can be developed up to 1–2 nm in diameter deposited on the silica core having 270 nm diameter using one-pot chemical synthesis route as reported by Nischala *et al.* [172]. Cotton fabrics with silica-Ag NP immobilized by the pad-dry-cure method exhibited excellent activity against *E. coli* and retained its bioactivity up to acceptable standards after 10 washing cycles. Silica-Ag core shell particles have also been explored on natural fibers like wool and silk and man-made fibers like polyester and polyamide. Ag NP capped with dodecanethiol 1–5 nm in size was synthesized by the two-phase arrested micro-emulsion technique, and subsequently applied to the cotton fabric by padding it in nano-Ag sol, as reported by Tarimala *et al.* [173]. Cotton dressing for burn wounds with good bioactivity has been reported by Parikh *et al.* [174] using Ag-modified sodium carboxy methyl cellulose.

Titania doped with Ag NP by the sol–gel synthesis route using titanium isopropoxide and $AgNO_3$ as precursor salts has been another significant research area addressing toward the stabilization issue stemming from nano-scale-sized Ag NP. Degoutin *et al.* [175] reported woven cotton fabric, spun-bonded PP fabric, and a cotton/non-woven PET blend fabric with durable bioactivity against *E. coli* and *S. aureus*. The fabrics except the woven cotton fabric have been surface-modified in the first step to make it hydrophilic using RF plasma and acrylic acid as a monomer. All the fabrics are subsequently impregnated with titania-doped Ag NP using pad-dry-cure and electrospray process. Anatase titania coated with 1% Ag NP has been used to prepare isotactic polypropylene filaments by the melt compounding route in a twin screw extruder in concentration of 0.2–0.5 wt% as reported by Dastjerdi *et al.* [176]. The addition of a TiO_2–Ag nano-filler did not hamper processability and showed comparable mechanical activity with a neat filament. Montazer *et al.* [177] developed Ag-loaded titania NP through the photo-reduction route for functionalization of wool fabrics. The immobilization in wool fabrics effective against *E. coli* and *S. aureus* was carried out via padding using citric acid as a cross-linker.

The last decade witnessed tremendous growth in Ag NP-incorporated textiles fabricated through physical deposition methods to minimize toxicity associated with chemicals required for wet reduction for Ag NP. This includes techniques such as ultrasonic frequency treatment, UV irradiation, and plasma treatment. Lu *et al.* [178] developed an antimicrobial silk fabric with crystalline Ag NP immobilized on the surface by UV irradiation. UV irradiation was also employed by Montazer *et al.* [179] to produce a colloidal Ag NP solution using Tollen's reagent as a precursor and PVP as a stabilizing agent. The SEM images of the fibres are shown in Figure 2.3.

Figure 2.3 SEM images of cotton fabrics: (a) untreated and (b–d) treated with 35 ppm of Ag_2O. Adapted with permission from reference [179].

The *ex situ* prepared Ag NP were padded onto the polyamide fabric, and the fabric showed >99% reduction of bacterial colonies against *E. coli* after 20 washing cycles.

Plasma pre-treatment promoting surface activation of hydrophobic fibers to make it hydrophilic and assessable to functional groups is greatly beneficial for polyester and polyamide fibers. Deng *et al.* [180] developed an antimicrobial non-woven PET fabric where Ag NP were introduced by the dip-drying process between two layers of the organosilicon thin film (10–50 nm) and attached to PET by the atmospheric pressure plasma system. Radetić et *al.* [181] reported a series of antimicrobial textiles based on durable and uniform Ag NP nano-coating by varying the plasma sources such as low-temperature air radio frequency (RF), dielectric barrier discharge (DBD), and corona discharge. Their reports indicate that the strength of antimicrobial activity and deposition uniformity is better tunable with RF plasma, although handling the system is difficult than DBD and corona discharge. Nano-Ag coating assisted by high power impulse magnetron sputtering on PET fabrics has been developed by Chen *et al.* [182]. This method is advantageous as high density plasma is capable of producing adhesion between film and PET at comparatively lower temperature of substrates. Surface activation of PET fabrics using corona air plasma following immobilization of nano-Ag has also been reported by Gorenšek *et al.* [183].

With increasing environmental biohazards created by harsh reductants and solvents required for NP synthesis, green nano-technology focusing on sustainability and negative externalities has emerged as the most promising field for further research in this area. In an attempt to reduce environmental toxicity which is a matter of great concern of date, natural resources such as starch, natural gums, cyclodextrin, and chitosan used as a reducing and stabilizing agent have been reported. Chitosan, a natural polysaccharide and a derivative of glucose family formed from chitin found in the exoskeleton of arthropods, is a non-toxic, biodegradable, and polycationic biocompatible antibacterial agent [184–185]. Ag cations are known to form chelate complexes with amino and hydroxyl groups of chitosan enabling chemical modification and functionalization. The emulsion containing a synergistic complex of chitosan-Ag is applied to fabrics by the conventional pad-dry-cure procedure, or the finishing technique has an excellent activity toward *E. coli* and *S. aureus* and shows resistance against alkaline laundering [186]. Such finished fabrics have the advantages of low add-on percentage preventing undesirable color generation usually associated with Ag NP-finished fabrics. Cotton fabrics bound with naturally synthesized Ag NP through the pad-dry-cure route with an excellent antibacterial activity have been demonstrated by Hebeish *et al.* [86]. The NP have been synthesized using hydroxypropyl starch as a reducing and stabilizing agent. The authors further showed that the presence of binder was essential to retain antimicrobial activity after laundering cycles. Mahltig *et al.* [187] developed Ag NP supported on dendrimers as a finishing agent for obtaining antimicrobial textiles, the thermal fixation, and the generation of which controlled the extent of the antimicrobial activity. Green synthesis of Ag NP was also attempted using *Erigeron annuus* (L.) pers flower extract used as a reducing and capping agent by Velmurugan *et al.* [188]. The Ag NP were further attached to cotton fabrics under ultrasonic treatment and were found to be effective against *S. epidermis* and *B. linens.* Ravindra *et al.* [189] reported Ag NP (~20 nm) formed on cotton fibers using *Eucalyptus citriodora* and *Ficus bengalensis* plant extracts. Balakumaran *et al.* [190] reported Ag NP synthesis using a fungal mat of *Aspergillus terrus.* The Ag NP, 8–20 nm in size, was further padded to cotton fabric to obtain antibacterial and antifungal cotton fabric effective against *Bacillus subtilis, S.* aureus, *Methicillin-resistant Staphylococcus aureus (MRSA), E. coli, Klebsiella pneumoniae, Pseudomonas aeruginosa, Penicillium sp., Aspergillus fumigatus, A. niger*, and *Rhizoctonia oryzae* with excellent laundering durability. Balamurugan *et al.* [191] used *P. pterocarpum* flower extract to synthesize Ag NP to be coated on cotton fabrics.

2.3.2 Application of TiO_2 NP on Textiles

Titania NP are chiefly noteworthy because of its photo-catalytic activity for the application in self-cleaning textiles [192]. However, literatures mentioning its antimicrobial properties are also available. *In situ* generation of titania NP in both phases—anatase and rutile—onto cotton textile substrate has been demonstrated in a simple one-step synthesis method using ultrasonic irradiation by Perelshtein *et al.* [193]. Crystalline titania NP obtained by acoustic cavitation and deposited *in situ* without necessitating further heating of the fabric to elevated temperatures (300 °C for anatase and 600 °C for rutile) have been first reported. Through photo-catalysis invisible-light, the fabrics showed activity against *E.coli, S. aureus*, and *C. albicans. In situ* sono-chemical synthesis of anatase titania NP on cotton fabrics at low temperatures was reported by Sadr *et al.* [194]. In another method outlined by Daoud *et al.* [195], titania NP are prepared *ex situ* by the usual sol–gel synthesis and then applied to cellulosic fibers by the pad-dry-cure method followed by hydrothermal treatment at comparatively low temperature (97 °C) to obtain anatase phase in the nano-coated fibers. The cellulosic fibers demonstrated antibacterial behavior under UV and dark conditions making the process suitable for preparing antimicrobial low heat resistance materials such as wood, biomaterials, plastics, and paper. Arain *et al.* [196] developed cotton fabrics impregnated with commercially procured AgCl–titinia colloid and chitosan through the pad-dry-cure process with a reduction of 100% against *S. aureus* and 98% against *E. coli.* Kangwansupamonkon *et al.* [197] used *ex situ* precipitated apatite-coated titania NP for dip-coating of cotton fabrics. The porous apatite coating over NP functions as adsorbents for pollutants and contaminants and the NP showed resistance against *E. coli, S. aureus*, MRSA, and *Micrococcus luteus* in dark and visible and black light conditions. The nano-coated cotton fabrics were obtained by padding through a solution containing coated titiania NP, xanthan gum as a thickening agent, and an acrylic-based binder, and subsequently dried and cured. The fabrics showed higher antimicrobial activity in black light irradiation than in visible light or under dark conditions.

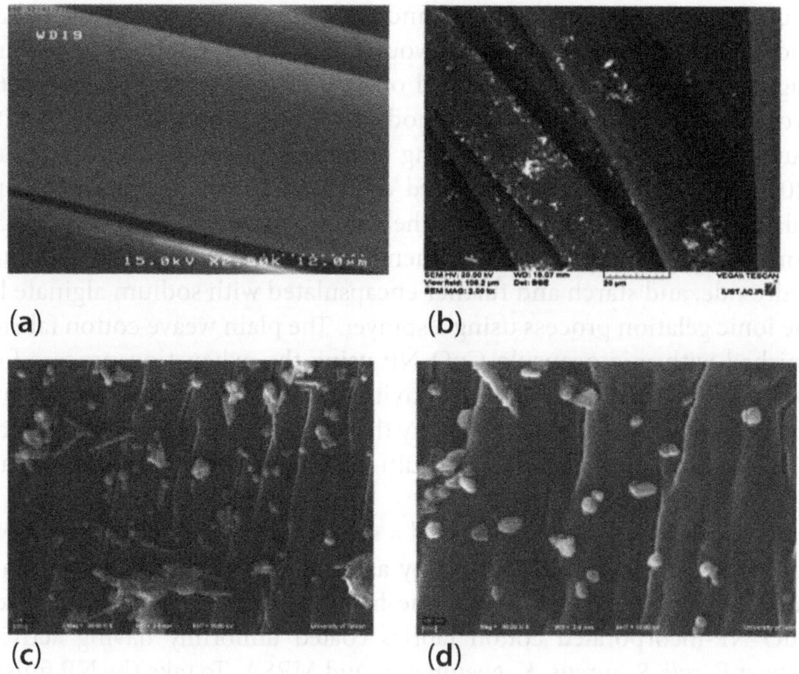

Figure 2.4 SEM images of the nylon fabric: untreated (a) 15000 × and treated with copper nano-particles (b) 2000 ×, (c) 20000 ×, (d) 40000 ×. Adapted with permission from reference [198].

2.3.3 Application of Cu NP and CuO NP on Textiles

The literatures on finishing of textiles where Cu and CuO NP are employed were much less compared to Ag-based finished agents. Komeily-Nia *et al.* [198] reported nylon/Cu NP composite fabrics impregnated with *in situ* reduced Cu NP (~85 nm size) by dipping in consecutive solutions containing ascorbic acid and cetyl trimethyl ammonium bromide (CTAB). The SEM images of the Nylon fabrics are shown in Figure 2.4. Ascorbic acid functioned as a reducing as well as ethylene-scavenging agent preventing oxidation of Cu NP to CuO NP, and CTAB controlled the size of the NP. The greenish tinged fabrics showed higher tensile strength and lower bending length and presented ZOI in the disk diffusion test against *S. aureus*. Teli *et al.* [199] reported acrylamide-grafted bamboo rayon fabric to immobilize Cu NP *in situ* durable up to 50 washes.

Perelshtein *et al.* [200] prepared a CuO-coated cotton bandage sonicated in Cu acetate in an ethanol-water medium and further UV-irradiated to develop *in situ* CuO NP. Crystalline CuO in nano-form is uniformly deposited on fibers and shows effectiveness against *E. coli* and *S. aureus* suitable for use in wound dressings, bed linens, and bandages. The same group advocated outstanding launderability up to 65 cycles for the sonochemical method of coating for cotton fabrics with CuO NP, suitable for minimizing hospital-acquired cross-infections [201]. Anita *et al.* [202] described a method to produce microencapsulated CuO NP for finishing of the cotton fabric. CuO NP of size ~50 nm were produced by the wet chemical method with $CuSO_4$, sodium hydroxide, and starch and further encapsulated with sodium alginate by the ionic gelation process using a sprayer. The plain weave cotton fabrics finished with microcapsule-CuO NP using the exhaustion process followed by pad-dry-cure showed activity against *E. coli* and *S. aureus* and showed better laundering durability than only CuO NP-finished fabrics. Gabbay *et al.* [203] described a multi-soaking procedure to impregnate cotton fibers with CuO NP.

Abramova *et al.* [204] developed a sonochemical reactor in which NP guided by cativation jets formed by asymmetric collapse of bubbles are hurled at high velocities toward the fiber surface. The authors reported CuO NP-incorporated cotton fabrics coated uniformly having activity against *E. coli*, *S. aureus*, *K. pneumonia*, and MRSA. To take the NP finishing of textiles further, a pilot-scale sonochemical coating plant has been successfully developed [205]. The CuO and ZnO NP have been incorporated as a nano-coating on cotton fabrics continuously for 50 m.

2.3.4 Application of ZnO NP on Textiles

In situ generation of ZnO NP (~30 nm) on cotton fabrics assisted by sonochemical irradiation coating was reported by Perelshtein *et al.* [206]. The fabrics showed activity against *S. aureus* and *E. coli* even at 0.75% of nano-coating. Uğur et *al.* [207] developed a multilayer nano-composite film consisting of ZnO NP by LBL self-assembly on cotton fabrics. Cationization of fabrics was previously carried out using 2,3-epoxypropyl trimethylammonium chloride (EP3MAC) and further dipping in alternate solutions of cationic and anionic ZnO nano-colloids. The fabrics with 10 and 16 layers of nano ZnO showed activity against *S. aureus* after 20 washing cycles.

Rajendra *et al.* [208] reported ZnO prepared *ex situ* by the wet precipitation method using $ZnSO_4$ and NaOH as precursor salts in soluble starch for pad-dry-cure finishing of cotton fabrics. The fabrics showed better activity against *S. aureus* than *E. coli* in disk diffusion and bacterial colonies reduction test. Krishnaveni *et al.* [209] reported a chitosan–ZnO complex by the hydrothermal method via *in situ* deposition of ZnO in chitosan. The chitosan–ZnO complex cross-linked on cotton fabric by the pad-dry-cure method is suitable for applications requiring antibacterial activity and biological fluorescence labeling applications. Polyamide 6/ ZnO NP nano-composite fibers have been prepared using a melt intercalation route using 0.5–5 wt% of ZnO NP, reported by Dural Erem *et al.* [210]. The nano-composite fibers showed 99% reduction in bacterial colonies of *S. aureus* and *K. pneumoniae* with increased mechanical properties.

Cotton fabrics functionalized with flower-shaped ZnO and chalcone ((E)-1-(3-hydroxyphenyl)-3-(4-methoxyphenyl) prop-2-en-1-one), an aromatic ketone-based antibacterial agent have been developed using padding mangle and gum acacia as binding agents, as developed by Sivakumar *et al.* [211]. The fabric based on combination antimicrobial therapy which relies on synergistic effect of chalcone and ZnO NP showed >99% reduction of bacterial colonies against *E.coli*, *S. aureus*, and *P. aeruginosa*.

2.3.5 Application of other NP on Textiles

In situ synthesis of magnetite and hematite NP using ferric chloride has been deposited on polyester woven fabric using a sono-chemical synthesis route as reported by Harifi *et al.* [212]. The fabrics showed bioactivity and durable magnetic properties suitable for electromagnetic shielding and antibacterial applications. Rastgoo *et al.* [213] devised a one-pot deposition method of magnetite NPS on cotton/PET blend fabric through ultrasonic irradiation. The fabric was irradiated ultrasonically in a bath containing ferrous sulfate as a precursor and CTAB as a surfactant. Asymmetric bubble collapses generating cavitation-oxidized Fe^{2+} to Fe^{3+} with hydroxyl radicals produced due to water sonolysis. The blend showed 95% activity against *S. aureus* and 99% against *C. albicans*. Gouda *et al.* [214] reported cotton gauze fabric-finished zirconium oxide NP (2–5 nm) deposited *in situ* by successive immersion in a precursor salt solution and a reducing agent solution, followed by batching and drying. The cotton gauze fabrics were effective against *S. aureus*, *S. typhimurium*, and *C. albicans* but were ineffective against *A. flavus*. Crepe satin silk fabrics immobilized with Au NP developed *in situ* using tetrachloroauric (III) acid in aqueous media have been reported by Tang *et al.* [215]. The silk fabrics were colored red and brown according to Au nano-spheres and nano-disks, respectively, with good light fastness, UV protection, thermal conductivity, and excellent antibacterial activity against *E. coli*.

2.4 Mechanism of Action of Inorganic NP

Metal and metal oxide NP show excellent biocidal activity against different strains of bacteria, fungi, and yeast at extremely low concentrations owing to their nano-scale dimensions, unique physicochemical characteristics, and high specific surface area. Although many scientific literatures are available to support beneficial attributes, the mechanism of biocidal action of metal-based NP is poorly understood and has many schools of thought [216–217]. The intrinsic chemistry of the NP is an important parameter in determining the extent and strength of antimicrobial action and is chiefly governed by three intrinsic factors—the reduction potential of the metal, its solubility in cytoplasmic fluid, and metal donor atom selectivity. These three factors play an important role in the multi-targeted action mechanism unique of NP [56, 216–218]. The mechanism of toxicity of the nanoparticles against bacteria is schematically outlined in Figure 2.5.

2.4.1 Cell Membrane Leakage and/or Impairment

Bacterial cells possess a net negative charge due to the presence of acidic phospholipids on its outer membrane with the extent of charge varying form strain to strain. Therefore, monovalent and divalent cations are electrostatically attracted toward the outer membrane. The membrane houses transporter-specific (ABC transporters and P-type transporters) or unspecific (MIT transporters) regulating the intake or excretion of proteins and nutrients [219–220]. Cu and Zn ions are permitted to enter through ion channels as they belong to a class of essential nutrients. Cu can also cause membrane impairment through lipid peroxidation, attacking the integrity of the cell. Ag is known to cause leakage of protons through the cell by unbalancing the chemi-osmotic force across the membrane. Some cations can bind with transporters inhibiting complete assimilation of ions [220].

2.4.2 Oxidative Stress Generation through ROS

Fe is an essential element present in living cells functioning as a co-factor to enzymes, catalyzing life-sustaining enzymatic reactions and engaged to coordination ligands; however, it also catalyzes Fenton reaction to produce reactive oxygen species (ROS) according to equation (1) [221–222]:

$$Fe^{2+} + H^2O^2 \rightarrow Fe^{3+} + OH- + OH\bullet \qquad (2.1)$$

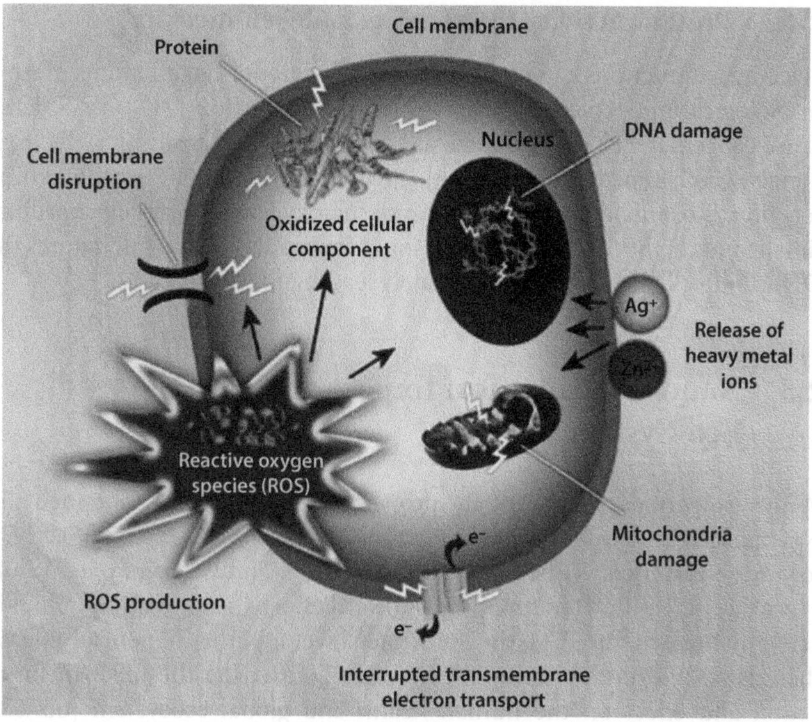

Figure 2.5 Mechanisms of toxicity of nano-particles (NP) against bacteria. NP and their ions (e.g., silver and zinc) can produce free radicals, resulting in induction of oxidative stress (i.e., reactive oxygen species; ROS). The produced ROS can irreversibly damage bacteria (e.g., their membrane, DNA, and mitochondria), resulting in bacterial death. Adapted with permission from reference [127].

Metals are characterized by their ability to contribute to redox reactions and change their oxidation states by acquiring electrons from donors. Therefore, they act as a catalyst in promoting Fenton reaction between Fe and H_2O_2 generating ROS. Metal cations such as Ag, which are non-Fenton–active, displace Fe in coordination ligands through ionic or molecular mimicry and release Fe that catalyzes Fenton reaction producing a hydroxyl ion and a hydroxyl radical. The nascent hydroxyl radical is the most potent oxidizing reactive species, readily reacting with most biomolecules. With the rise in ROS intensifying toxic oxidative stress, antioxidants inside the cell become readily depleted [223]. Thiol groups possess high affinity for metals and can get oxidized leading to the generation of more ROS. Physiological redox reactions inside the cell are thereby severely hampered [216].

2.4.3 Protein Activity Interference and Genotoxicity

Selective interactions can bind metal ions and donor ligand atoms present in biological molecules leading to protein inactivation [217]. Fe-S dehydrase clusters in cell are prone to site-specific molecular mimicry by toxic metals which replace the Fe [216]. Free Fe further catalyzes the Fenton reaction. Bio-organic complexes with sulfur, oxygen, or nitrogen coordination ligand can be formed by cupric ions. The oxidative stress denatures the DNA double helix structure, ultimately leading to cell death [224].

2.5 Nano-Toxicological Impact of NP on the Eco-System

With exceptional functional property enhancement obtained in NP-reinforced textile substrates, nano-additive-based research activity for textile materials is one of the fastest-growing industries of date. While every day, scientific literature points toward effortless fabrication of highly active nano-structured textile substrates leveraging its beneficial effects, there is only a modest amount of studies focused on the mechanism of action and effect on the human health and global ecosystem [66–68]. NP-finished products in the due course of time are bound to release NP incidentally, starting from the time of production, during washing and usage due to erosion, leading to severe malfunction of the environmental infrastructure [225]. In order to exert toxic effects to target organisms (bacteria, fungi, algae, and virus) for which it is primarily designed and fabricated, biocide metal NP above a threshold limit starts to attack non-target organisms (marine species, mammalians, and human) on incidental or accidental exposure. The interfaces between the human body and the immediate surroundings, for example, small pores in the skin and respiratory, gastrointestinal, and genital tracts, function as possible entry routes toward NP absorption into the inner system [226]. It is highly likely that toxic effects associated with the exposure of NP will attack the lowest trophic level, that is, bacteria, protozoa, fungi, yeast, and molds at the initial stage [227]. However, trophic transfer may result in the transmission of toxicity to higher strata of the energy pyramid, the human. Metal and metal oxide-based NP are already proven to exhibit adverse effects to bacteria, fungi, and algae even at very low concentrations, a characteristic having roots in its high specific surface area. NP dumped onto wastewater treatment plants based on outdated hazardous discharge control technologies are incompetent in removing nano-scale-sized impurities, disrupting the aquatic ecosystem [228–229]. Scientific studies based on the effect of Ag

NP on marine life indicate that very low concentration of NP, especially Ag NP, can be fatal to zebra fish, clams, and *Diptera* species such as *Drosophila melanogaster* [230–232]. Crustaceans and fishes are proven be most susceptible to the NP toxicity, in some cases more than the target organism.

The levels of toxicity of different NP are different due to their inherent chemistry, reactivity, and solubility. Ag NP are the most toxic of different NP used for biocidal function apart from being the most potent one too. The LC50 values (lethal concentration; concentration in water having 50% chance of causing death to aquatic life) of Ag NP reported are very low, 0.01 mg/L for crustaceans to 38 mg/L for protozoa [67]. For other NP, the LC50 values are usually below 10 mg/L, pointing toward high risks associated with nano-Ag usage. The LC50 value of CuO NP is 2–3 mg/L for crustaceans and algae and 100 mg/L for protozoa and bacteria. ZnO NP exhibit pronounced activity against algae (<0.1 mg/L) and LC50 of 10–100 mg/L against fungi, protozoa, and yeast.

Few studies are directed toward the *invivo* effect of Ag NP accumulation in lower mammalian cells of species, such as rats and mice. Sung *et al.* [233] reported the impairment of lung functions due to inflammatory lesions in lung cells of the rat in a 90-day inhalation period. Another study conducted by Kim *et al.* [234] reported the accumulation of Ag in olfactory bulbs of rat and its subsequent transfer to brain tissues. *In vitro* studies conducted on mouse stem cells and rat liver cells indicate mitochondrial dysfunction and membrane leakage. The effects of nano-colloidal Ag of diameter 2–3 nm and 30 nm, and concentration 100 ppm on the skin condition of healthy rabbits were analyzed by Lee *et al.* [63]. Quite contrarily, the authors reported total skin innoxiousness for 2–3 nm NP and rare skin erythema for 30 nm NP, making 2–3 nm NP better suitable for textile applications. Gouda *et al.* [170] reported no clinical signs of skin erythema on adult albino rabbits patched with Ag oxide and zirconium oxide NP-impregnated cotton gauze fabrics for 4 days.

Limited study has been done to draw conclusion on the effect of human cells. Studies on topical application of Ag NP on human fibroblastic sarcoma cells (tumor cells) and carcinoma cells (cancer cells) show reduced levels of cellular glutathione indicating ROS generation and oxidative stress build-up [235]. In a study by Asharani *et al.* [236], the toxicity exerted by Ag NP was researched using human lung fibroblast cells and glioblastoma cells. The authors found evidences of interruption of mitochondrial respiration chain with oxidative stress generation, disruption of ATP synthesis, and DNA damage. Ag NP exerted cytotoxic effects on both types of cells in an equal manner, giving us food for thought about serious and long-term effects of Ag NP above the threshold dose. Although some literature could be found on toxicity of Ag NP, the toxicity data on CuO and ZnO NP are relatively rare. However, Kangwansupamonkon *et al.* [197] advocated non-toxicity of apatite-coated titania NP on normal human dermal fibroblast cells through MTT assay in the cytotoxicity test having cell viability over 70%.

In the present scenario, no safety organization committee has been set up for regulating synthesis and safe usage of NP. Different metal and metal oxide NP, even though possesses unique physico-chemical properties and has been classified as new class of materials according to material scientists worldwide, are chemically similar to their bulk counterparts and therefore carry commercially the same CAS number [67]. As a result, the NP are currently judged by the nano-toxicological effect on aquatic life based on the scale set for bulk materials, which is insufficient to provide substantial evidence pertaining to their non-toxic nature. While a legislative body for the regulation of NP is the need of the hour, a fundamental elucidation and understanding toward the path of activity against living cells and its effect on human health is equally important. Till then, control of usage of NP with lower dose-dependent concentrations in applications directly in contact with human skin is to be considered for bringing down risk factors to a negligible level.

2.6 Conclusion

A current glance at the literature suggests most researches are focused on antimicrobial textile materials based on Ag NP by wet processing, physical reduction, or green synthesis. However, other metal NP such as TiO_2, Cu, CuO, and ZnO NP are also effective for fabricating antimicrobial textiles. The size and concentration of the NP plays an important role in demonstrating the strength of the antimicrobial effect with small NP of size 1–15 nm for Ag and Cu NP demonstrating the best activity at >50 ppm concentration (1–2 wt%) . The size of ZnO particles reported was higher, in the range of 30–40 nm but required less concentration (1 wt%) for successful inhibition. The antimicrobial activity of titania NP relying on its photo-catalytic activity provides best results in the presence of UV conditions alone, restricting its applications in the textile industry. Bacterial strains of *S. aureus*, *E. coli*, and *K. pneumoniae* are the most tested micro-organisms along with the fungal strain of *C. albicans*. However, most studies lack research on the effect of NP-reinforced textiles on cytotoxicity and allergic reactions to human subjects. A better understanding of NP release from textiles during washing or wear and tear and its subsequent action to the immediate environment is the need of the hour for a successful management of risk factors associated with the application and usage of NP.

Acknowledgment

The authors would like to acknowledge the Ministry of Human Resource Development (MHRD), India, for providing financial assistance.

References

1. Studer, H., Antimicrobial protection for polyolefine fibers. *Chem. Fibers Int.*, 47, 373, 1997.
2. Chen, H.L., Burns, L.D., Environmental analysis of textile products. *Cloth. Text. Res. J.*, 24, 248, 2006.
3. Ren, X., Kou, L., Kocer, H.B., Zhu, C., Worley, S.D., Broughton, R.M., Huang, T.S, Antimicrobial coating of an N-halamine biocidal monomer on cotton fibers via admicellar polymerization. *Colloids Surf. A.*, 317, 711, 2008.
4. Purwar, R., Joshi, M., Recent developments in antimicrobial finishing of textiles - A review, *AATCC Rev.*, 4, 22, 2004.
5. Mao, J. W., Durable freshness for textiles. *AATCC Rev.*, 1, 28, 2001.
6. Sundarrajan, S., Chandrasekaran, A.R., Ramakrishna, S., An update on nanomaterials-based textiles for protection and decontamination. *J. Am. Ceram. Soc.*, 93, 3955, 2010.
7. Antimicrobial coatings market by type (silver, copper, & others), application (indoor air/HVAC, medical, mold remediation, building & construction, food & beverages, textiles, & others) and geography (North America, Europe, Asia-Pacific, and ROW)—global trends and forecasts to 2018. marketsandmarkets.com, Dallas, TX, USA
8. Joshi, M., Bhattacharyya, A., Nanotechnology–a new route to high-performance functional textiles. *Text. Prog.*, 43, 155, 2011.
9. Joshi, M., Ali, S.W., Rajendran, S., Antibacterial finishing of polyester/cotton blend fabrics using neem (Azadirachta indica): A natural bioactive agent. *J. Appl. Polym. Sci.*, 106, 793, 2007.
10. Payne, J., From medical textiles to smell-free socks. *J. Soc. Dyers Colour*, 113, 48, 1997.
11. Joshi, M., Purwar, R., Udakhe, J.S., Sreedevi, R., Antimicrobial nanocomposite compositions, fibers and films, US Patent 9192625, assigned to IIT, 2015.
12. Vigo, T., Antibacterial fiber treatment and disinfection. Text. Res. J. 51, 454, 1981.
13. Joshi, M., Butola, B.S., Saha, K., Advances in topical drug delivery system: Micro to nanofibrous structures. *J. Nanosci. Nanotech.*, *14*, 853, 2014.
14. Saha, K., Butola, B., Joshi, M., Drug-loaded polyurethane/clay nanocomposite nanofibers for topical drug-delivery application. *J. Appl. Polym. Sci.*, 131, 40230, 2014.

15. Butola, B.S., Mohammad, F., Silver nanomaterials as future colorants and potential antimicrobial agents for natural and synthetic textile materials. RSC Adv., 6, 44232, 2016.
16. Szostak-Kotowa, J., Biodeterioration of textiles. *Int. Biodeterior. Biodegrad.*, 53, 165, 2004.
17. Ehrlich, H., Microbes and metals. *Appl Microbiol Biotechnol.*, 48, 687, 1997.
18. Gulrajani, M.L., Gupta, D., Periyasamy, S., Muthu, S.G., Preparation and application of silver nanoparticles on silk for imparting antimicrobial properties. *J. Appl. Polym. Sci.*, 108, 614, 2008.
19. Gregersen, T., Rapid method for distinction of Gram-negative from Gram-positive bacteria. *Eur. J. Appl. Microbial. Biotechnol.*, 5, 123, 1978.
20. Scott, E.A., Bloomfield, S.F., The survival and transfer of microbial contamination via cloths, hands and utensils. *J. Appl. Bacteriol.* 68, 271, 1990.
21. Mullard, A., Microbiology: the inside story. *Nature News*, 453, 578, 2008.
22. Gerba, C.P., Rose, J.B., Hass, C.N., Sensitive populations: who is at greatest risk? *Int. J. Food Microbiol.*, 30, 113, 1996.
23. Firstenberg-Eden, R.U.T.H., Attachment of bacteria to meat surfaces: a review. *J. Food Prot.*, 44, 602, 1981.
24. Hayat, R., Ali, S., Amara, U., Khalid, R., Ahmed, I., Soil beneficial bacteria and their role in plant growth promotion: a review. *Ann. Microbiol.*, 60, 579, 2010.
25. Zayed, G., Can the encapsulation system protect the useful bacteria against their bacteriophages. *Plant Soil*, 197, 1998.
26. Saito, T., Selection of useful probiotic lactic acid bacteria from the *Lactobacillus acidophilus* group and their applications to functional foods. *Anim. Sci. J.*, 75, 1, 2004.
27. Corsetti, A., Gobbetti, M., Balestrieri, F., Paoletti, F., Russi, L., Rossi, J., Sourdough lactic acid bacteria effects on bread firmness and staling. *J. Food Sci.*, 63, 347, 1998.
28. Eikmeier, H., Rehm, H.J., Production of citric acid with immobilized *Aspergillus niger*. *Appl. Microbial. Biotechnol.*, 20, 365, 1984.
29. Adeleye, I.A., Eruba, S., Ezeani, C.J., Isolation and characterisation of antibiotic producing microorganisms in composted Nigerian soil. *J. Environ. Biology*, *25*, 313, 2004.
30. Ara, K., Hama, M., Akiba, S., Koike, K., Okisaka, K., Hagura, T., Tomita, F., Foot odor due to microbial metabolism and its control. *Can. J. Microbiol.*, *52*, 357, 2006.
31. Grice, E.A., Segre, J.A., The skin microbiome. *Nat. Rev. Microbiol.*, 9, 244, 2011.
32. Cogen, A.L., Nizet, V., Gallo, R.L., Skin microbiota: a source of disease or defence? *Brit. J. Dermatol.*, 158, 442, 2008.
33. Rosenthal, M., Goldberg, D., Aiello, A., Larson, E., Foxman, B., Skin microbiota: microbial community structure and its potential association with health and disease. *Infect. Genet. Evol.*, 11, 839, 2011.

34. Fredricks, D.N., Microbial ecology of human skin in health and disease. *J. Invest. Dermatol. Symp. Proc.*, 6, 167, 2001.
35. Callewaert, C., De Maeseneire, E., Kerckhof, F.M., Verliefde, A., Van de Wiele, T., Boon, N., Microbial odor profile of polyester and cotton clothes after a fitness session. *Appl. Environ. Microbiol.*, 80, 6611, 2014.
36. Täubel, M., Rintala, H., Pitkäranta, M., Paulin, L., Laitinen, S., Pekkanen, J., Nevalainen, A., The occupant as a source of house dust bacteria. *J. Allergy Clin. Immunol.*, 124, 834, 2009.
37. Ventola, C.L., The antibiotic resistance crisis: Part 1. Causes and threats. *Pharm. Ther.*, 40, 277, 2015.
38. Simoncic, B., Tomsic, B., Structures of novel antimicrobial agents for textiles-a review. *Text. Res. J.*, 80, 1721, 2010.
39. Gao, Y., Cranston, R., Recent advances in antimicrobial treatments of textile. *Text. Res. J.*, 78, 60, 2008.
40. Wang, X., Wang, C., The antibacterial finish of cotton via sols containing quaternary ammonium salts, *J. Sol-Gel Sci. Technol.*, 50, 15, 2009.
41. Schindler, W.D., Hauser, P.J., *Chemical Finishing of Textiles*, Woodhead Publishing Ltd, Cambridge, 2004.
42. Heywood, D. (Ed.), *Textile Finishing*, Society of Dyers and Colourists, Bradford, 2003.
43. Gupta, D., Bhaumik, S., Antimicrobial treatment for textiles. *Indian J. Fibre Text. Res.*, 32, 254, 2007.
44. Shahid, M., Mohammad, F., Green chemistry approaches to develop antimicrobial textiles based on sustainable biopolymers–A review. *Ind. Eng. Chem. Res*, 52, 5245, 2013.
45. Joshi, M., Ali, S. W., Purwar, R., Rajendran, S., Ecofriendly antimicrobial finishing of textiles using bioactive agents based on natural products. *AATCC Rev.*, 9, 46, 2009.
46. Shahid, M., Mohammad, F., High-energy radiation induced sustainable coloration and functional finishing of textile materials. *Ind. Eng. Chem. Res.*, 54, 3727, 2015.
47. Shahid, M., Mohammad, F., Thermodynamics, kinetics and multifunctional finishing of textile materials with colorants extracted from natural renewable sources. *ACS Sustain. Chem. Eng.*, 5, 7451–7466, 2017.
48. Shahid, M., Mohammad, F., Perspectives for natural product based agents derived from industrial plants in textile applications–a review. *J. Clean. Prod.*, 57, 2, 2013.
49. Shahid, M., Mohammad, F., emerging green technologies and environment friendly products for sustainable textiles. *Roadmap to Sustainable Textiles and Clothing*, 63, 63–82, 2014.
50. Kango, S., Kalia, S., Celli, A., Njuguna, J., Habibi, Y., Kumar, R., Surface modification of inorganic nanoparticles for development of organic–inorganic nanocomposites—a review. *Prog. Polym. Sci.*, 38, 1232, 2013.

51. Rai, M., Yadav, A., Gade, A., Silver nanoparticles as a new generation of antimicrobials. *Biotechnol. Adv.*, 27, 76, 2009.
52. Sathyanarayanan, M.B., Balachandranath, R., Genji Srinivasulu, Y., Kannaiyan, S.K., Subbiahdoss, G., The effect of gold and iron-oxide nanoparticles on biofilm-forming pathogens. *ISRN Microbiology, 2013.*
53. Saraf, R., Cost effective and monodispersed zinc oxide nanoparticles synthesis and their characterization. *Int. J. Adv. Appl. Sci.*, 2, 85, 2013.
54. Shi, H., Magaye, R., Castranova, V., Zhao, J., Titanium dioxide nanoparticles: a review of current toxicological data. *Part. Fibre Toxicol.*, 10, 15, 2013.
55. Zille, A., Almeida, L., Amorim, T., Carneiro, N., Esteves, M.F., Silva, C.J., Souto, A.P., Application of nanotechnology in antimicrobial finishing of biomedical textiles. *Mater. Res. Express*, 1, 032003, 2014.
56. Palza, H., Antimicrobial polymers with metal nanoparticles. *Int. J. Mol. Sci.*, 16, 2099, 2015.
57. Dastjerdi, R., Montazer, M., A review on the application of inorganic nanostructured materials in the modification of textiles: focus on anti-microbial properties. *Colloids Surf. B*, 79, 5, 2010.
58. Lam, Y.L., Kan, C.W., Yuen, C.W.M., Developments in functional finishing of cotton fibres–wrinkle-resistant, flame-retardant and antimicrobial treatments. *Text. Prog.*, 44, 175, 2012.
59. Moritz, M., Geszke-Moritz, M., The newest achievements in synthesis, immobilization and practical applications of antibacterial nanoparticles. *Chem. Eng. J.*, 228, 596, 2013.
60. Radetić, M., Functionalization of textile materials with silver nanoparticles. *J. Mater. Sci.*, 48, 95, 2012.
61. Windler, L., Height, M., Nowack, B., Comparative evaluation of antimicrobials for textile applications. *Environ. Int.*, 53, 62, 2013.
62. Roy, A., Joshi, M., Butola, B.S., Srivastava, A.K., Silver-loaded HDPE/clay nanocomposites with antibacterial property: A potential replacement for commodity polyethylene plastic. *Polym. Compos.* 2017.
63. Lee, H.J., Jeong, S.H., Bacteriostasis and skin innoxiousness of nanosize silver colloids on textile fabrics. *Text. Res. J.*, 75, 551, 2005.
64. Filon, F. L., Mauro, M., Adami, G., Bovenzi, M., Crosera, M., Nanoparticles skin absorption: New aspects for a safety profile evaluation. *Regul. Toxicol. Pharmacol.*, 72, 310, 2015.
65. Nel, A., Xia, T., Madler, L., Li, N., Toxic potential of materials at the nanolevel. *Sci.*, 311, 622, 2006.
66. Marambio-Jones, C., Hoek, E.M., A review of the antibacterial effects of silver nanomaterials and potential implications for human health and the environment. *J. Nanopart. Res.*, 12, 1531, 2010.
67. Bondarenko, O., Juganson, K., Ivask, A., Kasemets, K., Mortimer, M., Kahru, A. Toxicity of Ag, CuO and ZnO nanoparticles to selected environmentally relevant test organisms and mammalian cells in vitro: a critical review. *Arch. Toxicol.*, 87, 1181, 2013.

68. Tolaymat, T.M., El Badawy, A.M., Genaidy, A., Scheckel, K.G., Luxton, T.P., Suidan, M., An evidence-based environmental perspective of manufactured silver nanoparticle in syntheses and applications: a systematic review and critical appraisal of peer-reviewed scientific papers. *Sci. Total Environ.*, 408, 999, 2010.
69. Tran, Q.H., Nguyen, V.Q., Le, A.T., Silver nanoparticles: synthesis, properties, toxicology, applications and perspectives. *Adv. Nat. Sci.: Nanosci. Nanotechnol.* 4, 033001, 2013.
70. Kim, J.S., Kuk, E., Yu, K.N., Kim, J.H., Park, S.J., Lee, H.J., Kim, S.H., Park, Y.K., Park, Y.H., Hwang, C.Y., Kim, Y.K., Lee, Y.S., Jeong, D.H., Cho, M.H., Antimicrobial effects of silver nanoparticles. *Nanomed. Nanotechnol. Biol. Med.* 3, 95, 2007.
71. Morones, J.R., Elechiguerra, J.L., Camacho, A., Holt, K., Kouri, J.B., Ramírez, J.T., Yacaman, M.J, The bactericidal effect of silver nanoparticles. *Nanotechnol.*, 16, 2346, 2005.
72. Viana, M.M., Mohallem, N.D.S., Miquita, D.R., Balzuweit, K., Silva-Pinto, E., Preparation of amorphous and crystalline Ag/TiO_2 nanocomposite thin films. *Appl. Surf. Sci.*, 265, 130, 2013.
73. Chopra, I., The increasing use of silver-based products as antimicrobial agents: a useful development or a cause for concern? *J. Antimicrob. Chemother.*, 59, 587, 2007.
74. Davies, R.L., Etris, S.F., The development and functions of silver in water purification and disease control. *Catal. Today*, 36, 107, 1997.
75. Matsumura, Y., Yoshikata, K., Kunisaki, S.I., Tsuchido, T., Mode of bactericidal action of silver zeolite and its comparison with that of silver nitrate. *Appl. Environ. Microbiol.*, 69, 4278, 2003.
76. Fox, C.L., Silver Sulfadiazine—A new topical therapy for pseudomonas in burns therapy of pseudomonas infection in burns. *Arch. Surg.*, 96, 184, 1968
77. Becker, R.O., Silver ions in the treatment of local infections. *Met. Based Drugs*, 6, 311, 1999.
78. Teli, M.D., Sheikh, J., Nanosilver containing grafted bamboo rayon as antibacterial material. *Fibers Polym.*, 13, 1280, 2012.
79. Zhang, G., Liu, Y., Gao, X., Chen, Y., Synthesis of silver nanoparticles and antibacterial property of silk fabrics treated by silver nanoparticles. *Nanoscale Res. Lett.*, 9, 216, 2014.
80. Roy, A., Butola, B.S., Joshi, M., Synthesis, characterization and antibacterial properties of novel nano-silver loaded acid activated montmorillonite. *Appl. Clay Sci.*, 146, 278, 2017.
81. Sun, Y., Xia, Y., Shape-controlled synthesis of gold and silver nanoparticles. *Science*, 298, 2176, 2002.
82. Lee, K.J., Lee, Y.I., Shim, I.K., Jun, B.H., Cho, H.J., Joung, J.W., Large-scale synthesis of polymer-stabilized silver nanoparticles. *Sol. St. Phen.*, 124, 1189, 2007.
83. Bönnemann, H., Brijoux, W., Brinkmann, R., Fretzen, R., Joussen, T., Köppler, R., Richter, J., Preparation, characterization, and application of

fine metal particles and metal colloids using hydrotriorganoborates. *J. Mol. Catal.*, 86, 129, 1994.

84. Balan, L., Malval, J.P., Schneider, R., Burget, D., Silver nanoparticles: new synthesis, characterization and photophysical properties. *Mater. Chem. Phys.*, 104, 417, 2007.
85. Chen, X., Schluesener, H.J., Nanosilver: a nanoproduct in medical application. *Toxicol. Lett.*, 176, 1, 2008.
86. Hebeish, A., El-Naggar, M.E., Fouda, M.M.G., Ramadan, M.A., Al-Deyab, S.S., El-Rafie, M.H., Highly effective antibacterial textiles containing green synthesized silver nanoparticles *Carbohydr. Polym.*, 86, 936, 2011.
87. Bar, H., Bhui, D.K., Sahoo, G.P., Sarkar, P., De, S.P., Misra, A., Green synthesis of silver nanoparticles using latex of Jatropha curcas. *Colloids Surf. A.*, 339, 134, 2009.
88. Ahmed, S., Ahmad, M., Swami, B.L., Ikram, S. Green synthesis of silver nanoparticles using Azadirachta indica aqueous leaf extract. *J. Radiat. Res. Appl. Sci.*, 9, 1, 2016.
89. Kumar, D.A., Palanichamy, V., Roopan, S.M., Green synthesis of silver nanoparticles using Alternanthera dentata leaf extract at room temperature and their antimicrobial activity. *Spectrochim. Acta Mol. Biomol. Spectrosc.*, 127, 168, 2014.
90. Sharma, V.K., Yngard, R.A., Lin, Y., Silver nanoparticles: green synthesis and their antimicrobial activities. *Adv. Colloid Interface Sci.*, 145, 83, 2009.
91. Chau, J.L.H., Hsu, M.K., Hsieh, C.C., Kao, C.C., Microwave plasma synthesis of silver nanopowders. *Mater. Lett.*, 59, 905, 2005.
92. Jiang, H., Manolache, S., Wong, A.C.L., Denes, F.S., Plasma-enhanced deposition of silver nanoparticles onto polymer and metal surfaces for the generation of antimicrobial characteristics. *J. Appl. Polym. Sci.*, 93, 1411, 2004.
93. Rodriguez-Sanchez, L., Blanco, M.C., Lopez-Quintela, M.A., Electrochemical synthesis of silver nanoparticles. *J. Phys. Chem. B*, 104, 9683, 2000.
94. Pant, H.R., Bajgai, M.P., Nam, K.T., Seo, Y.A., Pandeya, D.R., Hong, S.T., Kim, H.Y., Electrospun nylon-6 spider-net like nanofiber mat containing TiO_2 nanoparticles: A multifunctional nanocomposite textile material. *J. Hazard. Mater.*, 185, 124, 2011.
95. Visai, L., De Nardo, L., Punta, C., Melone, L., Cigada, A., Imbriani, M., Arciola, C.R., Titanium oxide antibacterial surfaces in biomedical devices. *Int. J. Artif. Organs.*, 34, 929, 2011.
96. Campoccia, D., Montanaro, L., Arciola, C.R., A review of the biomaterials technologies for infection-resistant surfaces. *Biomater.*, 34, 8533, 2013.
97. Hashimoto, K., Irie, H., Fujishima, A., TiO_2 photocatalysis: a historical overview and future prospects. *Jpn. J. Appl. Phys.*, 44, 8269, 2005.
98. Foster, H.A., Ditta, I.B., Varghese S., Steele, A., Photocatalytic disinfection using titanium dioxide: spectrum and mechanism of antimicrobial activity. *Appl. Microbiol. Biotechnol.*, 90, 1868, 2011.

99. Kikuchi, Y., Sunada, K., Iyoda, T., Hashimoto, K., Fujishima, A., Photocatalytic bactericidal effect of TiO_2 thin films: dynamic view of the active oxygen species responsible for the effect. *J. Photochem. Photobiol. A*, 106, 51, 1997.
100. Xue, C., Wu, J., Lan, F., Liu, W., Yang, X., Zeng, F., Xu, H., Nano titanium dioxide induces the generation of ROS and potential damage in HaCaT cells under UVA irradiation. *J Nanosci Nanotechnol.*, 10, 8500, 2010.
101. Kaida, T., Kobayashi, K., Adachi, M., Suzuki, F., Optical characteristics of titanium oxide interference film and the film laminated with oxides and their applications for cosmetics. *J. Cosmet. Sci.*, 55, 219, 2004.
102. Windler, L., Lorenz, C., Goetz, N.V., Hungerbühler, K., Amberg, M., Heuberger, M., Nowack, B., Release of titanium dioxide from textiles during washing. *Environ. Sci. Technol.*, 46, 8181, 2012.
103. Perelshtein, I., Applerot, G., Perkas, N., Grinblat, J., Gedanken, A., A one-step process for the antimicrobial finishing of textiles with crystalline TiO_2 nanoparticles. *Chem. Eur. J.*, 18, 4575, 2012.
104. Woodruff, R.S., Li, Y., Yan, J., Bishop, M., Jones, M.Y., Watanabe, F, Chen, T., Genotoxicity evaluation of titanium dioxide nanoparticles using the Ames test and Comet assay. *J. Appl. Toxicol.*, 32, 934, 2012.
105. Chen, T., Yan, J. Li, Y., Genotoxicity of titanium dioxide nanoparticles. *J. Food Drug Anal.*, *22*, 95, 2014.
106. Reddy, M.P., Venugopal, A., Subrahmanyam, M., Hydroxyapatite-supported Ag–TiO_2 as *Escherichia coli* disinfection photocatalyst. *Water Res.*, 41, 379, 2007.
107. Sung-Suh, H.M., Choi, J.R., Hah, H.J., Koo, S.M., Bae, Y.C., Comparison of Ag deposition effects on the photocatalytic activity of nanoparticulate TiO_2 under visible and UV light irradiation. *J. Photochem. Photobiol.*, 163, 37, 2004.
108. Domek, M.J., LeChevallier, M.W., Cameron, S.C., McFeters, G.A., Evidence for the role of copper in the injury process of coliform bacteria in drinking water. *Appl. Environ. Microbiol.* 48, 289, 1984.
109. Goudouri, O., Kontonasaki, E., Lohbauer, U., Boccaccini, A., Antibacterial properties of metal and metalloid ions in chronic periodontitis and peri-implantitis therapy. *Acta Biomater.* 10, 3795, 2014,
110. Uauy, R., Olivares, M., Gonzalez, M., Essentiality of copper in humans. *Am. J. Clin. Nutr.*, 67, 952, 1998.
111. Calabrese, E.J., Canada, A.T., Sacco, C. Trace elements and public health. *Annu. Rev. Public Health*, 6, 131, 1985.
112. Borkow, G., Gabbay, J., Putting copper into action: Copper-impregnated products with potent biocidal activities. *FASEB J.*, 18, 1728, 2004.
113. Gabbay, J., Borkow, G., Mishal, J., Magen, E., Zatcoff, R., Shemer-Avni, Y., Copper oxide impregnated textiles with potent biocidal activities. *J. Ind. Text.*, 35, 323, 2006.
114. Hawbaker, J.A., Speer, V.C., Hays, V.W., Catron, .D.V., Effect of copper sulfate and other chemotherapeutics in growing swine rations. *J. Anim. Sci.* 20, 163, 1961.

115. Satyaparameshwar, K., Reddy, T.R., Kumar, N.V., Study of carbohydrate metabolism in selected tissues of freshwater mussel, *Lamellidens marginalis* under copper sulphate toxicity. *J. Environ. Biol.* 27, 39, 2006.
116. Ren, G., Hu, D., Cheng, E.W.C., Vargas-Reus, M.A., Reip, P., Allaker, R.P., Characterisation of copper oxide nanoparticles for antimicrobial applications. *Int. J. Antimicrob. Agents*, 33, 587, 2009.
117. Umer, A., Naveed, S., Ramzan, N., Rafique, M.S., Selection of a suitable method for the synthesis of copper nanoparticles. *Nano*, *7*, 1230005, 2012.
118. Gawande, M.B., Goswami, A., Felpin, F.X., Asefa, T., Huang, X., Silva, R., Varma, R.S., Cu and Cu-based nanoparticles: synthesis and applications in catalysis. *Chem. Rev*, 116, 3722, 2016.
119. Yallappa, S., Manjanna, J., Sindhe, M., Satyanarayan, N., Pramod, S., Nagaraja, K., Microwave assisted rapid synthesis and biological evaluation of stable copper nanoparticles using T. arjuna bark extract. *Spectrochim. Acta Mol. Biomol. Spectrosc.*, 110, 108, 2013.
120. Cheirmadurai, K., Biswas, S., Murali, R., Thanikaivelan, P., Green synthesis of copper nanoparticles and conducting nanobiocomposites using plant and animal sources. *RSC Adv.*, 4, 19507, 2014.
121. Qi, L., Ma, J., Shen, J., Synthesis of copper nanoparticles in nonionic water-in-oil microemulsions. *J. Colloid Interface Sci.*,186, 498, 1997.
122. Ohde, H., Hunt, F., Wai, C.M., Synthesis of silver and copper nanoparticles in a water-in-supercritical-carbon dioxide microemulsion. *Chem. Mater.*,13, 4130, 2001.
123. Haas, I., Shanmugam, S., Gedanken, A., Pulsed sonoelectrochemical synthesis of size-controlled copper nanoparticles stabilized by poly (N-vinylpyrrolidone). *J. Phys. Chem. B*, 110, 16947, 2006.
124. Varshney, R., Bhadauria, S., Gaur, M.S., A review: biological synthesis of silver and copper nanoparticles. *Nano Biomed. Eng.*, 4, 99–106, 2012.
125. Varshney, R., Bhadauria, S., Gaur, M.S., Pasricha, R., Characterization of copper nanoparticles synthesized by a novel microbiological method. *J. Mineral. Met. Mat. Soc.*, 62, 102, 2010
126. Salavati-Niasari, M., Davar, F., Synthesis of copper and copper (I) oxide nanoparticles by thermal decomposition of a new precursor. *Mater. Lett.*, 63, 441, 2009.
127. Hajipour, M.J., Fromm, K.M., Ashkarran, A.A., de Aberasturi, D.J., de Larramendi, I.R., Rojo, T., Mahmoudi, M., Antibacterial properties of nanoparticles. *Trends Biotechnol.*, 30, 499, 2012.
128. Malachová, K., Praus, P., Rybková, Z., Kozák, O., Antibacterial and antifungal activities of silver, copper and zinc montmorillonites. *Appl. Clay Sci.*, 53, 642, 2011.
129. Xie, Y., He, Y., Irwin, P.L., Jin, T., Antibacterial activity and mechanism of action of zinc oxide nanoparticles against *Campylobacter jejuni*. *Appl. Environ. Microbiol.* 77, 2325, 2011.

130. Reddy, K.M., Feris, K., Bell, J., Wingett, D.G., Hanley, C., Punnoose, A., Selective toxicity of zinc oxide nanoparticles to prokaryotic and eukaryotic systems. *Appl. Phys. Lett.*, 90, 213902, 2007.
131. Jin, T., Sun, D., Su, J.Y., Zhang, H., Sue, H.J., Antimicrobial efficacy of zinc oxide quantum dots against *Listeria monocytogenes, Salmonella enteritidis,* and *Escherichia coli* O157: H7. *J. Food Sci.*, 74, 46, 2009.
132. He, L., Liu, Y., Mustapha, A., Lin, M., Antifungal activity of zinc oxide nanoparticles against *Botrytis cinerea* and *Penicillium expansum. Microbiol. Res.*, 166, 207, 2011.
133. Sugarman, B., Zinc and infection. *Rev. Infect. Dis.*, 5, 137, 1983.
134. Atmaca, S., Guel, K., Cicek, R., The effect of zinc on microbial growth. *Turk. J. Med. Sci.*, 28, 595, 1998.
135. Radke, L.L., Hahn, B.L., Wagner, D.K., Sohnle, P.G., Effect of abscess fluid supernatants on the kinetics of *Candida albicans* growth. *Clin. Immunol Immunopathol.*, 73, 344, 1994.
136. Söderberg, T., Sunzel, B., Holm, S., Elmros, T., Antibacterial effect of zinc oxide in vitro. *J. Plast. Surg. Hand Surg.*, 24, 193, 1990.
137. Vasile, B.S., Oprea, O., Voicu, G., Ficai, A., Andronescu, E., Teodorescu, A., Holban, A., Synthesis and characterization of a novel controlled release zinc oxide/gentamicin–chitosan composite with potential applications in wounds care. *Int. J. Pharm.*, 463, 161, 2014.
138. Espitia, P.J.P., Soares, N.D.F.F., dos Reis Coimbra, J.S., de Andrade, N.J., Cruz, R.S., Medeiros, E.A.A., Zinc oxide nanoparticles: synthesis, antimicrobial activity and food packaging applications. *Food Bioprocess. Technol.*, 5, 1447, 2012.
139. Ansari, S.A., Husain, Q., Qayyum, S., Azam, A., Designing and surface modification of zinc oxide nanoparticles for biomedical applications. *Food Chem. Toxicol.*, 49, 2107, 2011.
140. Wang, Z.L., Zinc oxide nanostructures: growth, properties and applications. J. *Phys. Condens. Matter.*, 16, R829, 2004.
141. Zhang, Y., Nayak, T.R., Hong, H., Cai, W., Biomedical applications of zinc oxide nanomaterials. *Curr. Mol. Med.*, 13, 1633, 2013.
142. Tayel, A.A., El-Tras, W.F., Moussa, S., El-Baz, A.F., Mahrous, H., Salem, M.F., Brimer, L., Antibacterial action of zinc oxide nanoparticles against foodborne pathogens. *J. Food Saf.*, 31, 211, 2011.
143. Roselli, M., Finamore, A., Garaguso, I., Britti, M.S., Mengheri, E., Zinc oxide protects cultured enterocytes from the damage induced by *Escherichia coli. J. Nutr.*, 133, 4077, 2003.
144. Butola, B.S., Mohammad, F., Silver nanomaterials as future colorants and potential antimicrobial agents for natural and synthetic textile materials. *RSC Adv.*, 6, 44232-44247, 2016.
145. Chernousova, S., Epple, M., Silver as antibacterial agent: ion, nanoparticle, and metal. *Angew. Chem. Int. Ed.*, 52, 1636, 2013.

146. Geranio, L., Heuberger, M., Nowack, B., The behavior of silver nanotextiles during washing *Environ. Sci. Technol.*, 43, 8113, 2009.
147. Emam, H.E., Manian, A.P., Široká, B., Duelli, H., Redl, B., Pipal, A., Bechtold, T., Treatments to impart antimicrobial activity to clothing and household cellulosic-textiles–why "Nano"-silver? *J. Clean. Prod.*, 39, 17, 2013.
148. Nateghi, M.R., Hajimirzababa, H., Effect of silver nanoparticles morphologies on antimicrobial properties of cotton fabrics. *J. Text I.*, 105, 806, 2014.
149. Perelshtein, I., Applerot, G., Perkas, N., Guibert, G., Mikhailov, S., Gedanken, A., Sonochemical coating of silver nanoparticles on textile fabrics (nylon, polyester and cotton) and their antibacterial activity. *Nanotechnol.*, 19, 245705, 2008.
150. Liu, H., Lee, Y.Y., Norsten, T.B., Chong, K. In situ formation of anti-bacterial silver nanoparticles on cotton textiles. *J. Ind. Text.*, 44, 198, 2014.
151. Lee, H.J., Yeo, S.Y., Jeong, S.H. Antibacterial effect of nanosized silver colloidal solution on textile fabrics. *J. Mater. Sci.*, 38, 2199, 2003.
152. Rajavel, K., Gomathi, R., Pandian, R., Rajendra Kumar, R.T., In situ attachment and its hydrophobicity of size-and shape-controlled silver nanoparticles on fabric surface for bioapplication. *Inorg. Nano-Met. Chem.*, 47, 1196, 2017.
153. Shateri-Khalilabad, M., Yazdanshenas, M.E., Fabrication of superhydrophobic, antibacterial, and ultraviolet-blocking cotton fabric. *J. Text I.*, 104, 861, 2013.
154. Yazdanshenas, M.E., Shateri-Khalilabad, M., One-step synthesis of superhydrophobic coating on cotton fabric by ultrasound irradiation. *Ind. Eng. Chem. Res.*, 52, 12846, 2013.
155. Montazer, M., Shamei, A., Alimohammadi, F., Synthesizing and stabilizing silver nanoparticles on polyamide fabric using silver-ammonia/PVP/UVC. *Prog. Org. Coat.*, 75, 379, 2012.
156. El-Rafie, M., Ahmed, H.B., Zahran, M.K., Characterization of nanosilver coated cotton fabrics and evaluation of its antibacterial efficacy. *Carbohydr. Polym*, 107, 174, 2014.
157. Hanh, T.T., Thu, N.T., Hien, N.Q., An, P.N., Loan, T.T.K., Hoa, P.T., Preparation of silver nanoparticles fabrics against multidrug-resistant bacteria. *Radiat. Phys. Chem.*, 121, 87, 2016.
158. Hebeish, A., Hashem, M., El-Hady, M.A., Sharaf, S., Development of CMC hydrogels loaded with silver nano-particles for medical applications. *Carbohydr. Polym.*, 92, 407, 2013.
159. Tang, B., Li, J., Hou, X., Afrin, T., Sun, L., Wang, X., Colorful and antibacterial silk fiber from anisotropic silver nanoparticles. *Ind. Eng. Chem. Res.*, 52, 4556, 2013. .
160. Montazer, M., Shamei, A., Alimohammadi, F., Stabilized nanosilver loaded nylon knitted fabric using BTCA without yellowing. *Prog. Org. Coat.*, 74, 270, 2012.

161. Yuranova, T., Rincon, A.G., Bozzi, A., Parra, S., Pulgarin, C., Albers, P., Kiwi, J., Antibacterial textiles prepared by RF-plasma and vacuum-UV mediated deposition of silver. *J. Photochem. Photobiol. Chem.*, 161, 27, 2003.
162. Zhang, Y., Peng, H., Huang, W., Zhou, Y., Zhang, X., Yan, D., Hyperbranched poly (amidoamine) as the stabilizer and reductant to prepare colloid silver nanoparticles in situ and their antibacterial activity. *J. Phys. Chem. C*, 112, 2330, 2008.
163. Yeo, S.Y., Jeong, S.H., Preparation and characterization of polypropylene/silver nanocomposite fibers. *Polym. Int.*, 52, 1053, 2003.
164. Majumdar, A., Butola, B.S., Thakur, S., Development and performance optimization of knitted antibacterial materials using polyester–silver nanocomposite fibres. *Mater. Sci. Eng. C*, 54, 26, 2015.
165. Perera, S., Bhushan, B., Bandara, R., Rajapakse, G., Rajapakse, S., Bandara, C., Morphological, antimicrobial, durability, and physical properties of untreated and treated textiles using silver-nanoparticles. *Colloids Surf. A*, 436, 975, 2013.
166. Joshi, M., Khanna, R., Shekhar, R., Jha, K., Chitosan nanocoating on cotton textile substrate using layer-by-layer self-assembly technique. *J. Appl. Polym. Sci.*, 119, 2793, 2011.
167. Dubas, S.T., Kumlangdudsana, P., Potiyaraj, P., Layer-by-layer deposition of antimicrobial silver nanoparticles on textile fibers. *Colloids Surf. A*, 289, 105, 2006.
168. Sataev, M.S., Koshkarbaeva, S.T., Tleuova, A.B., Perni, S., Aidarova, S.B., Prokopovich, P., Novel process for coating textile materials with silver to prepare antimicrobial fabrics. *Colloids Surf. A*, 442, 146, 2014.
169. Budama, L., Çakır, B.A., Topel, Ö., Hoda, N., A new strategy for producing antibacterial textile surfaces using silver nanoparticles. *Chem. Eng. J.*, 228, 489, 2013.
170. Gouda, M., Nano-zirconium oxide and nano-silver oxide/cotton gauze fabrics for antimicrobial and wound healing acceleration. *J. Ind. Text.*, 41, 222, 2012.
171. Dastjerdi, R., Montazer, M., Shahsavan, S., A new method to stabilize nanoparticles on textile surfaces. *Colloids Surf. A*, 345, 202, 2009.
172. Nischala, K., Rao, T.N., Hebalkar, N., Silica–silver core–shell particles for antibacterial textile application. *Colloids Surf. B*, 82, 203, 2011.
173. Tarimala, S., Kothari, N., Abidi, N., Hequet, E., Fralick, J., Dai, L.L., New approach to antibacterial treatment of cotton fabric with silver nanoparticle–doped silica using sol–gel process. *J. Appl. Polym. Sci.*, 101, 2938, 2006.
174. Parikh, D.V., Fink, T., Rajasekharan, K., Sachinvala, N.D., Sawhney, A.P.S., Calamari, T. A., Parikh, A.D., Antimicrobial silver/sodium carboxymethyl cotton dressings for burn wounds. *Text. Res. J.*, 75, 134, 2005.
175. Degoutin, S., Jimenez, M., Casetta, M., Bellayer, S., Chai, F., Blanchemain, N., Martel, B., Anticoagulant and antimicrobial finishing of non-woven polypropylene textiles. *Biomed. Mater.*, 7, 035001, 2012.

176. Dastjerdi, R., Mojtahedi, M.R.M., Shoshtari, A.M., Khosroshahi, A., Investigating the production and properties of Ag/TiO_2/PP antibacterial nanocomposite filament yarns. *J. Text. I.*, 101, 204, 2010.
177. Montazer, M., Behzadnia, A., Pakdel, E., Rahimi, M.K., Moghadam, M.B., Photo induced silver on nano titanium dioxide as an enhanced antimicrobial agent for wool. *J. Photochem. Photobiol. B*, 103, 207, 2011.
178. Lu, Z., Meng, M., Jiang, Y., Xie, J., UV-assisted in situ synthesis of silver nanoparticles on silk fibers for antibacterial applications. *Colloids Surf. A*, 447, 1, 2014.
179. Montazer, M., Alimohammadi, F., Shamei, A., Rahimi, M.K., In situ synthesis of nano silver on cotton using Tollens' reagent. *Carbohydr. Polym.*, 87, 1706, 2012.
180. Deng, X., Nikiforov, A., Vujosevic, D., Vuksanovic, V., Mugoša, B., Cvelbar, U., Leys, C., Antibacterial activity of nano-silver non-woven fabric prepared by atmospheric pressure plasma deposition. *Mater. Lett.*, 149, 95, 2015.
181. Radetić, M., Ilić, V., Vodnik, V., Dimitrijević, S., Jovančić, P., Šaponjić, Z., Nedeljković, J. M., Antibacterial effect of silver nanoparticles deposited on corona-treated polyester and polyamide fabrics. *Polym. Adv. Technol.*, 19, 1816, 2008.
182. Chen, Y.H., Hsu, C.C., He, J.L., Antibacterial silver coating on poly (ethylene terephthalate) fabric by using high power impulse magnetron sputtering. *Surf. Coat. Technol.*, 232, 868, 2013.
183. Gorenšek, M., Gorjanc, M., Bukošek, V., Kovač, J., Jovančić, P., Mihailović, D., Functionalization of PET fabrics by corona and nano silver. *Text. Res. J.*, 80, 253, 2010.
184. Ali, S.W., Rajendran, S., Joshi, M., Synthesis and characterization of chitosan and silver loaded chitosan nanoparticles for bioactive polyester. *Carbohydr. Polym.*, 83, 438, 2011.
185. Ali, S.W., Joshi, M., Rajendran, S. Synthesis and characterization of chitosan nanoparticles with enhanced antimicrobial activity. *Int. J. Nanosci.*, 10, 979, 2011.
186. Lu, Y.H., Chen, Y.Y., Lin, H., Wang, C., Yang, Z.D., Preparation of chitosan nanoparticles and their application to Antheraea pernyi silk. *J. Appl. Polym. Sci.*, 117, 3362, 2010.
187. Mahltig, B., Tatlises, B., Fahmi, A., Haase, H., Dendrimer stabilized silver particles for the antimicrobial finishing of textiles. *J. Text. I.*, 104, 1042, 2013.
188. Velmurugan, P., Cho, M., Lee, S.M., Park, J.H., Bae, S., Oh, B.T., Antimicrobial fabrication of cotton fabric and leather using green-synthesized nanosilver. *Carbohydr. Polym.*, 106, 319, 2014.
189. Ravindra, S., Mohan, Y.M., Reddy, N.N., Raju, K.M., Fabrication of antibacterial cotton fibres loaded with silver nanoparticles via "Green Approach". *Colloids Surf. A*, 367, 31, 2010.
190. Balakumaran, M.D., Ramachandran, R., Jagadeeswari, S., Kalaichelvan, P.T., In vitro biological properties and characterization of nanosilver coated

cotton fabrics–an application for antimicrobial textile finishing. *Int. Biodeter. Biodegr.*, 107, 48, 2016.

191. Balamurugan, M., Saravanan, S., Soga, T., Coating of green-synthesized silver nanoparticles on cotton fabric. *J. Coat. Technol. Res.*, 14, 735, 2017
192. Crick, C.R., Bear, J.C., Kafizas, A., Parkin, I.P., Superhydrophobic photocatalytic surfaces through direct incorporation of titania nanoparticles into a polymer matrix by aerosol assisted chemical vapor deposition. *Adv. Mater.*, 24, 3505, 2012.
193. Perelshtein, I., Applerot, G., Perkas, N., Grinblat, J., Gedanken, A., A one-step process for the antimicrobial finishing of textiles with crystalline TiO_2 nanoparticles. *Chem. Eur. J.*, 18, 4575, 2012.
194. Sadr, F.A., Montazer, M., *In situ* sonosynthesis of nano TiO_2 on cotton fabric. *Ultrason. Sonochem.*, 21, 681, 2014.
195. Daoud, W.A., Xin, J.H., Zhang, Y.H., Surface functionalization of cellulose fibers with titanium dioxide nanoparticles and their combined bactericidal activities. *Surf. Sci.*, 599, 69, 2005.
196. Arain, R. A., Khatri, Z., Memon, M.H., Effect of silver colloid on chemical and physical properties of cotton fabric for the application of antimicrobial textiles. *Adv. Mater. Res.*, 576, 272, 2012.
197. Kangwansupamonkon, W., Lauruengtana, V., Surassmo, S., Ruktanonchai, U., Antibacterial effect of apatite-coated titanium dioxide for textiles applications. *Nanomed. Nanotechnol. Biol. Med.*, 5, 240, 2009.
198. Komeily-Nia, Z., Montazer, M., Latifi, M., Synthesis of nano copper/nylon composite using ascorbic acid and CTAB. *Colloids Surf. A*, 439, 167, 2013.
199. Teli, M.D., Sheikh, J., Modified bamboo rayon-copper nanoparticle composites as antibacterial textiles. *Int. J. Biol. Macromol.*, 61, 302, 2013.
200. Perelshtein, I., Applerot, G., Perkas, N., Wehrschuetz-Sigl, E., Hasmann, A., Guebitz, G., Gedanken, A., CuO–cotton nanocomposite: Formation, morphology, and antibacterial activity. *Surf. Coat. Technol.*, 204, 54, 2009.
201. Perelshtein, I., Ruderman, Y., Perkas, N., Beddow, J., Singh, G., Vinatoru, M., Gedanken, A., The sonochemical coating of cotton withstands 65 washing cycles at hospital washing standards and retains its antibacterial properties. *Cellulose*, 20, 1215, 2013.
202. Anita, S., Ramachandran, T., Rajendran, R., Koushik, C.V., Mahalakshmi, M., A study of the antimicrobial property of encapsulated copper oxide nanoparticles on cotton fabric. *Text. Res. J.*, 81, 1081, 2011.
203. Gabbay, J., Borkow, G., Mishal, J., Magen, E., Zatcoff, R., Shemer-Avni, Y., Copper oxide impregnated textiles with potent biocidal activities. *J. Ind. Text.*, 35, 323, 2006.
204. Abramova, A., Gedanken, A., Popov, V., Ooi, E.H., Mason, T.J., Joyce, E.M., Bayazitov, V., A sonochemical technology for coating of textiles with antibacterial nanoparticles and equipment for its implementation. *Mater. Lett.*, 96, 121, 2013.

205. Abramov, O.V., Gedanken, A., Koltypin, Y., Perkas, N., Perelshtein, I., Joyce, E., Mason, T. J., Pilot scale sonochemical coating of nanoparticles onto textiles to produce biocidal fabrics. *Surf. Coat. Technol.*, 204, 718, 2009.
206. Perelshtein, I., Applerot, G., Perkas, N., Wehrschetz-Sigl, E., Hasmann, A., Guebitz, G.M., Gedanken, A., Antibacterial properties of an in situ generated and simultaneously deposited nanocrystalline ZnO on fabrics. *ACS Appl. Mater. Interfaces*, 1, 361, 2008.
207. Uğur, Ş.S., Sarıışık, M., Aktaş, A.H., Uçar, M.Ç., Erden, E., Modifying of cotton fabric surface with nano-ZnO multilayer films by layer-by-layer deposition method. *Nanoscale Res. Lett.*, 5, 1204, 2010.
208. Rajendra, R., Balakumar, C., Ahammed, H.A.M., Jayakumar, S., Vaideki, K., Rajesh, E., Use of zinc oxide nano particles for production of antimicrobial textiles. *Int. J. Eng. Sci. Technol.*, 2, 202, 2010.
209. Krishnaveni, R., Thambidurai, S., Industrial method of cotton fabric finishing with chitosan–ZnO composite for anti-bacterial and thermal stability. *Ind. Crops Prod.*, 47, 160, 2013.
210. Dural Erem, A., Ozcan, G., Skrifvars, M., Antibacterial activity of PA6/ZnO nanocomposite fibers. *Text. Res. J.*, 81, 1638, 2011.
211. Sivakumar, P.M., Balaji, S., Prabhawathi, V., Neelakandan, R., Manoharan, P.T., Doble, M., Effective antibacterial adhesive coating on cotton fabric using ZnO nanorods and chalcone. *Carbohydr. Polym.*, 79, 717, 2010.
212. Harifi, T., Montazer, M., In situ synthesis of iron oxide nanoparticles on polyester fabric utilizing color, magnetic, antibacterial and sono-Fenton catalytic properties. *J. Mater. Chem. B*, 2, 272, 2014.
213. Rastgoo, M., Montazer, M., Malek, R.M., Harifi, T., Rad, M.M., Ultrasound mediation for one-pot sonosynthesis and deposition of magnetite nanoparticles on cotton/polyester fabric as a novel magnetic, photocatalytic, sonocatalytic, antibacterial and antifungal textile. *Ultrason. Sonochem.*, 31, 257, 2016.
214. Gouda, M., Nano-zirconium oxide and nano-silver oxide/cotton gauze fabrics for antimicrobial and wound healing acceleration. *J. Ind. Text.*, 41, 222, 2012.
215. Tang, B., Sun, L., Kaur, J., Yu, Y., Wang, X., In-situ synthesis of gold nanoparticles for multifunctionalization of silk fabrics. *Dyes Pigm.*, 103, 183, 2014.
216. Lemire, J.A., Harrison, J.J., Turner, R.J., Antimicrobial activity of metals: mechanisms, molecular targets and applications. *Nat. Rev. Microbiol.*, 11, 371, 2013.
217. Prabhu, S., Poulose, E.K., Silver nanoparticles: mechanism of antimicrobial action, synthesis, medical applications, and toxicity effects. *Int. Nano Lett.*, 2, 32, 2012.
218. Workentine, M.L., Harrison, J.J., Stenroos, P.U., Ceri, H., Turner, R.J., Pseudomonas fluorescens' view of the periodic table. *Environ. Microbiol*, 10, 238, 2008.

219. Nies, D.H., Microbial heavy-metal resistance. *Appl. Microbiol. Biotechnol.*, 51, 730, 1999.
220. Nies, D.H., Efflux-mediated heavy metal resistance in prokaryotes. *FEMS Microbiol. Rev.*, 27, 313, 2003.
221. Pomposiello, P.J., Demple, B., Global adjustment of microbial physiology during free radical stress. *Adv. Microbiol. Physiol.*, 46, 319, 2012.
222. Keyer, K., Imlay, J.A., Superoxide accelerates DNA damage by elevating free-iron levels. *Proc. Natl. Acad. Sci.*, 93, 13635, 1996.
223. Macomber, L., Imlay, J.A., The iron-sulfur clusters of dehydratases are primary intracellular targets of copper toxicity. *Proc. Natl. Acad. Sci.*, 106, 8344, 2009.
224. Warnes, S.L., Highmore, C.J., Keevil, C.W., Horizontal transfer of antibiotic resistance genes on abiotic touch surfaces: implications for public health. *MBio. Am. Soc. Microbiol.*, 3, 00489, 2012.
225. Benn, T.M., Westerhoff, P., Nanoparticle silver released into water from commercially available sock fabrics. *Environ. Sci. Technol.*, 42, 4133, 2008.
226. Chen, X., Schluesener, H.J., Nanosilver: a nanoproduct in medical application. *Toxicol. Lett.*, 176, 1, 2008
227. Choi, O., Hu, Z., Size dependent and reactive oxygen species related nanosilver toxicity to nitrifying bacteria. *Environ. Sci. Technol.*, 42, 4583, 2008.
228. Moore, M.N., Do nanoparticles present ecotoxicological risks for the health of the aquatic environment? *Environ. Int.*, 32, 967, 2006.
229. Griffitt, R.J., Luo, J., Gao, J., Bonzongo, J.C., Barber, D.S., Effects of particle composition and species on toxicity of metallic nanomaterials in aquatic organisms. *Environ. Toxicol. Chem.*, 27, 1972, 2008.
230. Asharani, P.V., Wu, Y.L., Gong, Z., Valiyaveettil, S., Toxicity of silver nanoparticles in zebrafish models. *Nanotechnology*, 19, 255102, 2008
231. Yeo, M.K., Kang, M.S., Effects of nanometer sized silver materials on biological toxicity during zebrafish embryogenesis. *Bull. Korean Chem. Soc.*, 29, 1179, 2008.
232. Ahamed, M., Posgai, R., Gorey, T.J., Nielsen, M., Hussain, S.M., Rowe, J.J., Silver nanoparticles induced heat shock protein 70, oxidative stress and apoptosis in Drosophila melanogaster. *Toxicol. Appl. Pharmacol.*, 242, 263, 2010.
233. Sung, J.H., Ji, J.H., Yoon, J.U., Kim, D.S., Song, M.Y., Jeong, J., Kim, T.S., Lung function changes in Sprague-Dawley rats after prolonged inhalation exposure to silver nanoparticles. *Inhal. Toxicol.* 20, 567, 2008.
234. Kim, K.J., Sung, W.S., Moon, S.K., Choi, J.S., Kim, J.G., Lee, D.G., Antifungal effect of silver nanoparticles on dermatophytes. *J. Microbiol. Biotechnol.*, 18, 1482, 2008.
235. Arora, S., Jain, J., Rajwade, J.M., Paknikar, K.M., Cellular responses induced by silver nanoparticles: in vitro studies. *Toxicol. Lett.*, 179, 93, 2008.
236. AshaRani, P.V., Low Kah Mun, G., Hande, M.P., Valiyaveettil, S., Cytotoxicity and genotoxicity of silver nanoparticles in human cells. *ACS Nano*, 3, 279, 2008.

3

Nano-Zinc Oxide: Prospects in the Textile Industry

N. Vigneshwaran*, V. Prasad, A. Arputharaj, A.K. Bharimalla and P.G. Patil

ICAR-Central Institute for Research on Cotton Technology, Matunga, Mumbai, India

Abstract

Zinc oxide (ZnO), an n-type semiconductor with a band gap of 3.37 eV, is a very interesting multifunctional material having promising applications in solar cells, sensors, displays, gas sensors, varistors, piezoelectric devices, electro-acoustic transducers, photodiodes, and UV light-emitting devices. Other applications include sunscreens, gas sensors, antireflection coatings, and catalysis. ZnO shows both improved quantum yield and tunable excitonic emission energy at nanometer-scale regions. Nanoparticles of zinc oxide (nano-ZnO) score over other nanomaterials for textile applications in terms of cost-effectiveness, transparent nature, and excellent antibacterial and UV-blocking properties. The antibacterial mechanism of nano-ZnO is due to the induction of intercellular reactive oxygen species, disruption of the cell membrane, and oxidative stress, while the UV-blocking property is due to its band gap energy that coincides with that of UV rays. In spite of various advantages, the bottleneck for widespread use in textile finishing is the lack of affinity between textile fibers and nano-ZnO. Conventionally, nano-ZnO is applied onto textiles using binders to hold them on to the surface of fibers. But the wash fastness problem reduced their potency in the due course of time. Our research group extensively worked on both binder-based application and *in situ* application of nano-ZnO onto cotton textiles. *In situ* process of nano-ZnO application improved the wash fastness without the use of any binders. The literature reported that the cytotoxicity of nano-ZnO on mouse dermal fibroblast cells depended mainly on the concentration and duration of exposure. The potential toxicity may arise due to the release of zinc ions

**Corresponding author*: Vigneshwaran.N@icar.gov.in

Shahid-ul-Islam and B.S. Butola (eds.) Nanomaterials in the Wet Processing of Textiles, (113–134) © 2018 Scrivener Publishing LLC

that induces local skin effects (e.g., irritation and contact dermatitis) and absorption of toxic or sensitizing metals. This chapter covers the overall protocols for synthesis and application of nano-ZnO onto textiles and the related performance and toxicity issues.

Keywords: Antibacterial property, *in situ* synthesis, nano-ZnO, pad-dry-cure method, self-cleaning property, UV protection

3.1 Introduction

Nanoparticles are engineered materials at the molecular/atomic level (1–100 nm size range) that show size-related properties significantly different from their bulk counterparts. This distinction allows their potential use in diversified sectors such as medicines, sensors, composites, pulp & paper, and textiles. Many recent reviews are available for synthesis, characterization, and application of nano-ZnO in diversified fields [1–3] and their toxicity attributes [4–6].

Textiles can provide a suitable substrate for the growth of micro-organisms by providing suitable humidity and temperature in contact to the human body. The recent increase in public concern about hygiene has been driving many research works for the creation of an anti-microbial finish on the surface of textiles. However, using many anti-microbial agents has been avoided because of their possible harmful or toxic effects. The use of inorganic nanoparticles would be a good alternative in comparison with that of traditional harmful/toxic chemicals [7]. Nanomaterials can induce stain repellence, wrinkle-freeness, static elimination, and electrical conductivity to fibers without compromising their comfort and flexibility. Nanomaterials also offer a wider application potential to create connected garments that can sense and respond to external stimuli via electrical, color, or physiological signals [8].

3.2 Synthesis of Nano-ZnO

Nano-ZnO can occur in one-(1D), two-(2D), and three-dimensional (3D) structures [2]. Nano-ZnO is synthesized into different forms as depicted in Figure 3.1. Researchers adopted various methods for the preparation of nano-ZnO from different precursors. These methods can be broadly classified into three major categories, that is, chemical, biological, and physical methods [9] (Figure 3.2).

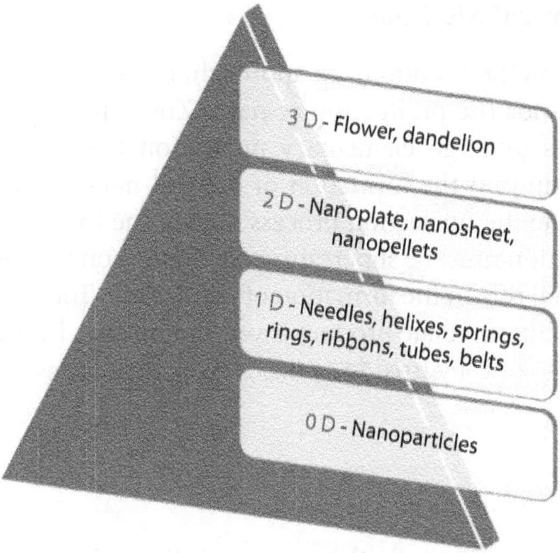

Figure 3.1 Different forms of nano-ZnO.

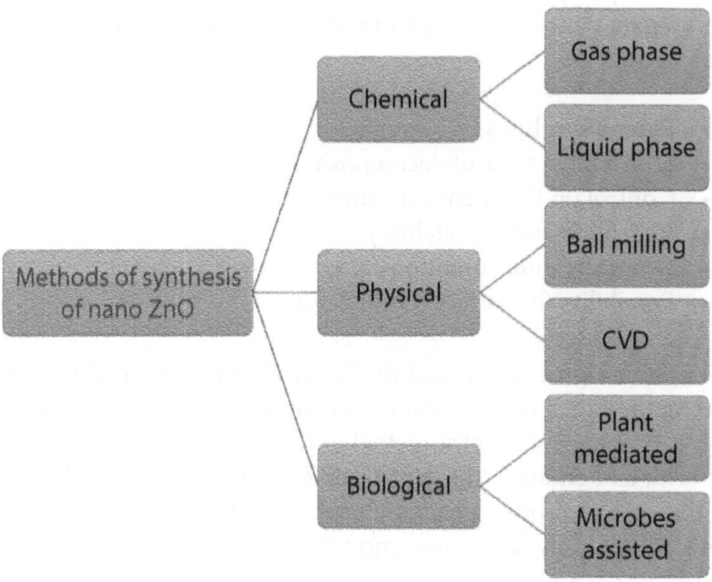

Figure 3.2 Methods for the synthesis of nano-ZnO.

3.2.1 Chemical Methods

The chemical method, bottom-up approach, is the mostly attempted synthesis method for the production of nano-ZnO. Initially, freshly formed ions/molecules undergo elementary nucleation followed by the growth processes, leading to the formation of defined nanoparticles of different sizes. To initiate the nucleation process during the formation of nanoparticles, the solution must be supersaturated. So the concentration of precursors plays a vital role in the formation of nano-ZnO. The solubility product of the formed intermediate product is very important. In the case of nano-ZnO synthesis, zinc hydroxide ($Zn(OH)^2$) is formed as an intermediate product.

3.2.1.1 Sol–Gel Method

Sol–gel is defined as a colloidal system of dispersion of one phase into another where the molecule particles dispersed in a medium have at least in one direction a dimension roughly between 1 nm and 1 mm. This method was first developed by Spanhel *et al.* [10] and later refined by Meulenkamp [11]. The sol–gel approach appears to be one of the most promising methods to prepare nano-ZnO. Some of the most important advantages of the sol–gel method are:

- Ease of synthesis
- Low temperature of decomposition
- Control on the chemical composition
- Reliability and repeatability
- Relatively mild conditions of synthesis
- Possibility for surface modification

The different precursors used for the synthesis of nano-ZnO using the sol–gel method include, zinc acetate dehydrate & ethanol; zinc 2-ethyl hexanoate, 2-propanol, and tetra methyl ammonium; zinc acetate dihydrate, oxalic acid, and ethanol. The growth of ZnO from the zinc salt precursor using the sol–gel process generally undergoes four stages, such as solvation, hydrolysis, polymerization, and transformation into ZnO.

3.2.1.2 Chemical Precipitation Method

In this method, zinc salts such as zinc acetate, zinc chloride, etc., are allowed to undergo reaction with a reducing agent such as NaOH, NH_4OH, etc.

Formed soluble or insoluble precipitate is afterward thoroughly washed with an aqueous/alcoholic solvent system and calcined at high temperatures to produce the nano-ZnO with desired morphology. To control the size and to avoid agglomeration of nanoparticles in the chemical precipitation method, capping agents or stabilizers are used. In the electrostatic stabilization method, anionic or cationic surfactants are added in the precursors to avoid aggregation of nano-ZnO particles during the formation stage. Rodriguez-Paez *et al.* [12] synthesized zinc oxide nanoparticles with different morphologies by controlling different parameters of the precipitation process such as solution concentration, pH, and washing medium.

Our research group synthesized nano-ZnO stabilized by soluble starch using water as a solvent [13]. Here, the steric hindrance principle is applied by soluble starch to stabilize the nano-ZnO. The precursors used are zinc nitrate and NaOH in the presence of soluble starch. The resultant suspension absorbed strongly at 361 nm due to the quantum confinement effect. The average size of ZnO nanoparticles was estimated to be 38 ± 3 nm using TEM and was confirmed by x-ray diffraction analysis. Thilagavathi and Geetha [14] also adopted the co-precipitation method for the synthesis of nano-ZnO using zinc acetate dihydrate and NaOH.

3.2.1.3 Hydrothermal Method

In this method, synthesis takes place in an autoclave, where the mixture of reactants is heated gradually to a temperature of 100–300 °C and left for a longer period. Due to heating followed by cooling, crystal nuclei are formed, which then grow. This method has many benefits, including the possibility of carrying out the synthesis at low temperatures. Zhang *et al.* [15] produced flower-like hexagonal nano-ZnO crystals by the organic free hydrothermal method.

3.2.1.4 Microwave Method

Microwave irradiation is an efficient and distinct heating method, and has unique features such as short reaction time, rapid volumetric heating, energy saving, and high reaction rate. In contrast to the conventional heating, in the microwave process, the heat is generated internally within the material, instead of originating from external heating sources, so an inverted temperature gradient is produced during the microwave heating process. On comparing conventional and microwave-assisted syntheses of nano-ZnO, it was reported that the presence of surfactant, microwave heating, and OH^- sources have major influences on the morphology and size of the resulting nano-ZnO products [16]. Many mechanisms have been proposed for the formation of nano-ZnO rods using microwave-assisted synthesis and in general, it includes a two-step process, namely, nucleation and growth. The alkalinity of the precursors determines the aspect ratio of the formed nano-ZnO rods [17].

3.2.1.5 Microemulsion Method

Microemulsions are thermodynamically stable, isotropic liquid mixtures of oil, water, and surfactant. In contrast to ordinary emulsions, microemulsions form upon simple mixing of the components and do not require the high shear conditions. The interfacial tension between phases, amount of energy required for formation, droplet sizes, and visual appearance are only a few of the differences seen when comparing emulsions to microemulsions. This method also allows attaining variation in the ZnO morphology, which can be achieved by changing the surfactant amount in the microemulsion formulation. Higher surfactant concentrations and calcination temperatures not exceeding 400 °C promote formation of rod-like ZnO microemulsion products [18]. Direct synthesis of pure monodispersed nano-ZnO of different shapes by basic hydrolysis of zinc nitrate with suitable control of the reactant concentrations in the water in oil microemulsion of Triton X-100/hexanol/cyclohexane was reported to produce hexagonal wurtzite nano-ZnO [19].

3.2.1.6 Sonochemical Method

It is based on acoustic cavitation resulting from the continuous formation, growth, and implosive collapse of bubbles in a liquid. Pholnak *et al.* [20] synthesized hexagonal ZnO nanotubes from the reaction between $Zn(NO_3)_2 \cdot 6H_2O$ and $C_6H_{12}N_4$ under an ultrasonic irradiation at 35 kHz for 1 h. Compared to the control experiments without sonication, the ultrasonic treatment proved to be effective in promoting active ZnO nuclei and growth of hexagonal wurtzite ZnO tubes from active nanorods. Another report revealed the antibacterial activity of nano-ZnO prepared by the sonication process against *Lactobacillus plantarum* after 24-h exposure at 1000 ppm [21].

3.2.1.7 Gas Phase Synthesis

In gas phase synthesis, one of the highly used techniques is the spray pyrolysis method, in which aerosol droplets of the precursor zinc salt are produced via flame heating. To synthesize ZnO nanoparticles, at first the precursor solution is atomized by a nebulizer under the pressure of air (7 bar). The spray-pyrolysis method has many advantages, which are high purity of synthesized powders, regular shape of particles, a better control of stoichiometry and chemical regularity in coupled oxide, and continuous working [22].

3.2.2 Physical Method

Nano-ZnO synthesis using physical methods include high energy ball milling (HEBM), melt mixing, physical vapor deposition, laser ablation, and sputter deposition. Physical methods of synthesis of nano-ZnO involve using the principle of top-down approach. In most physical/mechanical processes, the production rates of nano-ZnO are very high and are mostly used for the industrial processes. HEBM is used to produce nano-ZnO from its microcrystalline powder. For the synthesis of nanoparticles, 99.9% pure ZnO powder is milled for 2 to 50 hours [23]. In this technique, starting powder particles are trapped between highly kinetic colliding balls and the inner surface of the vial, which causes repeated deformation, rewelding, and fragmentation of premixed powders resulting in the formation of fine, dispersed particles in the grain-refined matrix.

A completely different approach, mechano-chemical procedure, is followed for synthesis of nano-ZnO by planetary ball mill using chemical precursors by the bottom-up approach [24]. Here, the morphology and crystallinity of the product could be controlled by altering the mechanical agitation and subsequent thermal treatment.

3.2.3 Green Synthesis of Nano-ZnO

Recently, green routes are mostly encouraged to the synthesis of nano-ZnO because of lesser pollution load, energy efficiency, and cost-effectiveness. The development of cost-effective, simple, and eco-friendly routes for the synthesis of nanoparticles is very important. A wide variety of biogenic sources have been put to trial by researchers as a source of green agents to facilitate the synthesis process. In addition, environmentally benign solvents are also being used these days in order to promote green synthesis [25]. A number of natural substances such as plants, fungi, algae, bacteria, and viruses are used to synthesize the nano-ZnO. Nano-ZnO could be synthesized using tomato extract [26], potato extract [27], and *Corymbia citriodora* leaf extract [28]. A recent review [29] provides updated information on green synthesis of nano-ZnO. Microbes are considered as eco-friendly factories of nanoparticle synthesis. *Aspergillus aeneus* isolate NJP12 showed highest potential for extracellular synthesis of ZnO under ambient conditions. The results revealed the synthesis of spherical nanoparticles coated with protein molecules which served as stabilizing agents [30].

3.3 Application of Nano-ZnO onto Textiles

Textile industries have developed many different kinds of functional finishes to textile materials during the last few decades. The traditional finishing technologies include exhaustion, padding, low add-on processes, foam application, printing, coating, electro-deposition, etc. The novel means of applications have also been invented to apply specialty chemicals, such as micro encapsulation and by using cyclodextrins. Functional finishes mainly involve a surface modification that can be obtained by means of chemical modification, by the application of a surface layer, or by more environment-friendly treatments such as the use of enzymes or physical modification (based mainly on plasma technology). The increasing demands for improved or even new properties of textiles, especially technical textiles as high performance products, can be observed worldwide. Some of the methods that have been used extensively for the application of nano-ZnO for multi-functionality of textiles are discussed below.

3.3.1 Sonochemical Method

The application of nano-ZnO on polymers including textiles through ultrasonic irradiation has been proven to be an effective method. The metal oxide nanoparticles have a large surface area and can be evenly and thinly spread on the surface of fabrics. This makes them a better alternative to the conventional chemicals such as triclosan, quaternary ammonium salts, and other synthetic organic compounds that now dominate the antimicrobial market [31]. The generation of hybrid antimicrobial coatings on cotton fabrics by means of a one-step simultaneous sonochemical deposition of nano-ZnO and chitosan was demonstrated to provide an excellent antibacterial activity against the pathogens, *Staphylococcus aureus* and *Escherichia coli* [32]. Simultaneous sonochemical dyeing and coating of textiles with antibacterial metal oxide nanoparticles in a one-step process was shown to have antibacterial behavior not influenced by the presence of the dyes [33].

3.3.2 Pad-Dry-Cure Method

The pad-dry-cure method is a widely used method in textile industries for chemical finishing. The active ingredient is fixed to the textile substrate with the help of the polymer resin or crosslinking agents to improve the finish durability. Since the nano-ZnO has no affinity for the textile substrate, this method is commonly used to achieve the durable effect. Our research group used this method for imparting antibacterial and UV protection activities using the nano-ZnO stabilized by soluble starch [13, 34]. The nano-ZnO-impregnated cotton fabrics showed excellent antibacterial activity against two representative bacteria, *S. aureus* (Gram positive) and *Klebsiella pneumoniae* (Gram negative). Also, nano-ZnO impregnation enhanced the protection of cotton fabrics against UV radiation in comparison with the untreated cotton fabrics. The nano-ZnO-impregnated fabrics were found to retain more than 80% antibacterial and UV-protection functions even after 25 wash cycles.

ZnO/carboxymethyl chitosan bionano-composite having the mean sizes of ZnO and carboxymethyl chitosan particles, ≈28 nm and ≈100 nm, respectively, were used as a finishing agent for the cotton fabric to impart UV protection and antibacterial properties to cotton fabrics by the pad-dry-cure method. Thus, finished cotton fabrics exhibited very good antibacterial property against Gram-positive and Gram-negative bacteria and also had a good UV protection activity [35].

The inorganic–organic hybrid polymers filled with ZnO nano particles-chitosan could be prepared through the sol–gel method and applied on the cotton fabric through the pad-dry-cure method [36]. The antimicrobial investigation results against *E. coli* and *M. luteus* showed the enhanced antibacterial activity against selected microorganisms. The antibacterial activity of textiles treated by nano-ZnO-chitosan particles increases by decreasing the molecular weight of chitosan as a low molecular weight in the composite improves the movement of chains due to low viscosity. The application of Ag-ZnO nanoparticles on cotton fabrics and cotton–polyester-blended fabrics using the pad-dry-cure method was reported by Ibănescu *et al.* [37]. The antimicrobial activity of the treated fabric was tested against the Gram-negative bacterium *E. coli* and Gram-positive *Micrococcus luteus*. The results showed substantial increase in the antimicrobial property with reasonably good washing durability.

3.3.3 *In Situ* Synthesis

The synthesis of nanomaterials inside the substrate is an entirely new means of functional finishing onto a textile substrate. Nano-ZnO could be directly synthesized inside the mesopores present on the surface of cotton fibers and the size got restricted by the size of those pores. Hence, the need of stabilizer is nullified. ZnO nanowires have been grown on cotton fabric to impart self-cleaning, superhydrophobicity, and ultraviolet (UV) blocking

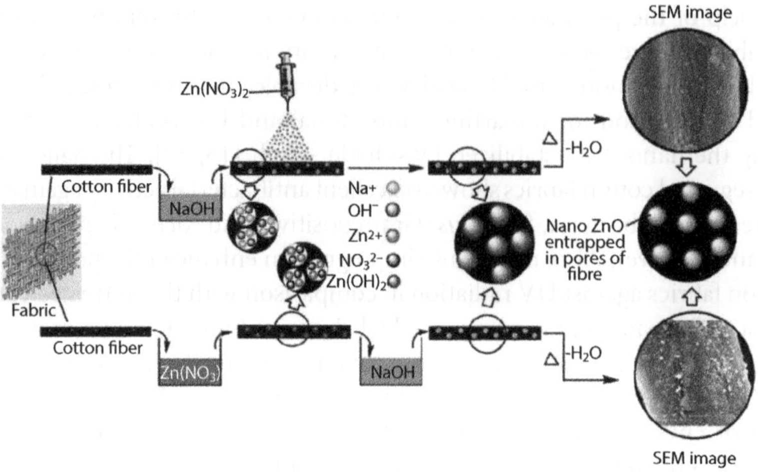

Figure 3.3 Schematic diagram showing the *in situ* synthesis of nano-ZnO on the surface of cotton fabrics. Reproduced with permission from [39].

properties. ZnO nanowires were grown by the microwave-assisted hydrothermal method and subsequently functionalized with steric acid to obtain a water contact angle of 150°, showing their superhydrophobic nature stable up to four washings. Self-cleaning activity of the ZnO nanowire-coated cotton fabric showed considerable degradation of methylene blue under UV light irradiation [38]. Our research group demonstrated the *in situ* synthesis of nano-ZnO onto 100% cotton fabrics by the spraying or dipping process resulting in durable antibacterial and UV protection finishes [39]. In this study, the nano-sized pore structure of cellulosic fibrils acted as a nucleation site for the formation of nano-ZnO from the precursors, zinc nitrate, and sodium hydroxide (Figure 3.3). Both the processes resulted in excellent antibacterial activity (>98%) against two representative pathogens, *S. aureus* and *K. pneumoniae* even after 50 wash cycles. The UV protection factor (UPF) was maintained above the minimum accepted level of 50+ till 50 wash cycles. The spraying process resulted in three times less uptake of nano-ZnO than that of the dipping process, without significant reduction in functional properties. The water absorbency and color of the terry cotton fabrics remain unaffected in the developed process.

3.3.4 Layer-by-Layer Assembly

Layer-by-layer assembly is a novel method for the fabrication of composite thin films onto textile materials. The sequential adsorption of oppositely charged polycations and polyanions on solid surfaces leads to the build-up of polyelectrolyte multilayer films. The multilayer deposition is carried out by immersing the substrate into the cationic and anionic solutions, alternately the taken for washing. A wide range of functional molecules can be incorporated within the film, including nanoparticles, dyes, proteins, and other supramolecular species. The multi-layer deposition of the nano-ZnO particle on fabrics has been reported in many literatures and the antibacterial activity, UV-protective properties, and durability of finished fabrics have been evaluated and reported to be satisfactory [40].

3.3.5 Plasma Coating of Surfaces

Adding a monomer to the gas in the plasma chamber, ultra-thin layers of polymers are deposited onto the plasma surface of virtually any material. The plasma surface coating generates invisible layers that render the surface of a hydrophobic material wettable, add a diffusion barrier to a porous material, or change plasma surface properties in many other ways. Low-pressure oxygen plasma created by an electrodeless radiofrequency discharge was applied to modify the properties of cellulosic fibers in order to improve adsorption properties toward nano-ZnO and to achieve excellent ultraviolet (UV)-protective properties of the cotton fabric [41]. The results showed that longer plasma treatment time causes higher adsorption of nanoparticles due to the increased concentration of oxygen-containing groups on the fiber surface and because of the increased surface roughness of fibers.

3.4 Properties of Nano-ZnO-Finished Textiles

The multifunctional properties of nano-ZnO finished textiles are given in Table 3.1.

3.4.1 Antibacterial Activity

The antibacterial activity and toxicity mechanism of nano-ZnO were reviewed by various authors [3, 42, 43]. The major means of antibacterial activity include the generation of reactive oxygen species (ROS), release of Zinc ions, abrasive nature of nano-ZnO, and other unknown reactions between the nanoparticles and the microbial structures. Figure 3.4 shows the possible mechanisms by which the nano-ZnO could kill or stop the growth of a bacterium. Also, it is hypothesized that the various modes of actions may act synergistically to produce a better antibacterial activity.

Table 3.1 Multifunctional properties of nano-ZnO-finished textiles.

Functional property	Corresponding physical phenomenon
Antibacterial	Generation of ROS Release of Zinc ions Abrasive nature of nano-ZnO
UV protection	Band gap absorbance in the UV region
Self-cleaning property	Photocatalytic activity
Gas sensing	Change in electrical resistivity due to adsorption of molecules on the surface of nano-ZnO
pH Sensing	The sensing mechanism is the polarization-induced bound surface charge by interaction with the polar molecules in liquids
Super hydrophobicity	Surface-modified nano-ZnO with silanes or fluorines
Rapid collection of carriers in solar cell	An array of oriented single crystalline nano-ZnO helps in rapid collection of carriers generated throughout the device as the nanorods provide direct path from the point of photogeneration to the conducting substrate. This effectively reduces the electron recombination losses.

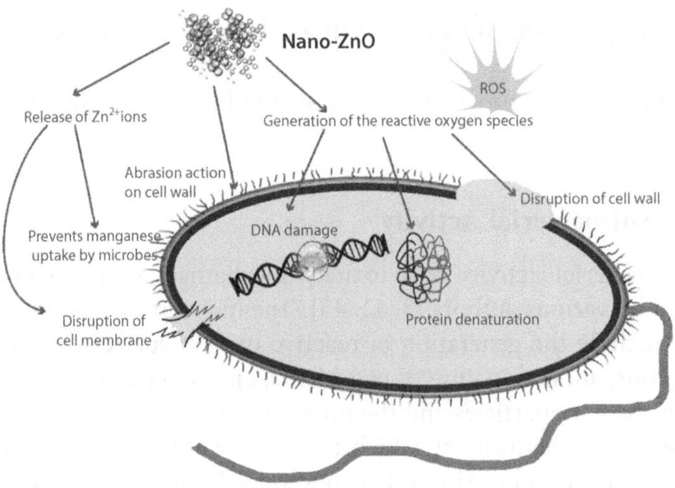

Figure 3.4 Different mechanisms for the antibacterial activity of nano-ZnO.

3.4.1.1 Generation of ROS

The ROS generation by nano-ZnO is depicted by the equations (1) to (5) [3].

$$ZnO + h\nu \rightarrow e^- + h^+ \tag{1}$$

$$h^+ + H_2O \rightarrow {}^\bullet OH + H^+ \tag{2}$$

$$e^- + O_2 \rightarrow {}^\bullet O^-_2 \tag{3}$$

$${}^\bullet O_2 + H^+ \rightarrow HO^\bullet_2 \tag{4}$$

$$HO^\bullet_2 + H^+ + e^- \rightarrow H_2O_2. \tag{5}$$

3.4.1.2 Release of Zinc Ions (Zn^{2+})

The antibacterial activity of nano-ZnO is also due to the release of zinc ions in the medium bacteria that has a significant effect on the active transport inhibition, amino acid metabolism, and enzyme system disruption. The toxicity due to soluble zinc ions was earlier proved by the use of a microbial sensor [44]. But, the release of zinc ions is affected by various parameters such as size, morphology, and porosity of nano-ZnO; concentration and capping of the nano-ZnO; and pH and chemical nature of the medium where the microbes are present. The complexation of zinc ions by the components of the broth (medium where the microbes are cultivated) increased the solubility of the zinc in the liquid medium, thereby increasing their efficacy [45].

3.4.1.3 Abrasive Nature of Nano-ZnO

The abrasiveness of nano-ZnO is also proposed to be an important parameter that decides their antibacterial activity in addition to the formation of surface oxygen species [46]. The chemical interactions between hydrogen peroxide and membrane proteins, and chemical interactions between other unknown chemical species generated due to the presence of nano-ZnO particles with the lipid bilayer also contributes to the antibacterial activity [47].

The preparation of hybrid nanoparticles (nano-ZnO along with chitosan) and application on cotton textiles displayed higher antibacterial activity by 48% and 17% against *S. aureus* and *E. coli*, respectively [32]. The presence of chitosan, a biopolymer, also improved the durability of the antimicrobial effect of the coatings in addition to improving the biocompatibility in comparison with that of nano-ZnO-coated cotton textiles.

3.4.2 UV Protection

Though sunlight is very much essential for human health, the ultraviolet (UV) rays in the sunlight pose a major problem to the skin resulting in various skin diseases. Dermatologists warn against excessive sun exposure and focus on the prevention measures such as sun creams and protective clothing. The UV rays from sun undergo various processes as depicted in Figure 3.5. The UV rays can get absorbed, reflected, scattered, transmitted, or re-emitted by fluorescence. Among these various phenomena, the transmitted UV rays could hit our skin inducing the possible damages described above. The UV rays are classified into three regions: UVA – 320 to 400 nm that causes increased aging effects on the skin; UVB – 280 to 320 nm that causes skin burn; and UVC – below 280 nm that causes skin cancer. The parameters that decide the amount of UV rays passing through include the structure of fabric, porosity/density of weave or knit, dye components, wetness, and chemical nature of fibers. The UV protection factor (UPF) states for how long someone wearing UV-protective clothing can stay in the sun without suffering skin damage.

Three standard documents that pertain to the testing and labeling of UV-protective textile products have been published by the American Society for Testing and Materials (ASTM) and the American Association of Textile Chemists and Colorists (AATCC). The titles of these documents

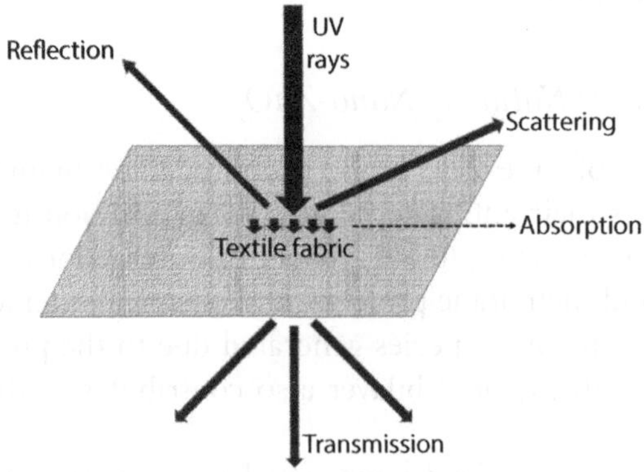

Figure 3.5 Interaction of UV rays with a textile fabric.

are ASTM D 6544 "Standard Practice for the Preparation of Textiles Prior to UV Transmission Testing", AATCC 183 "Test Method for Transmittance or Blocking of Erythemally Weighted Ultraviolet Radiation Through Fabrics", and ASTM 6603 "Standard Guide to Labeling of UV-protective Textiles" [48]. The Australian/New Zealand Standard (4399) specifies the Evaluation and Classification of sun-protective clothing. British Standards include BS EN 13758-1 (Textiles. Solar UV protective properties. Method of test for apparel fabrics) and BS EN 13758-2 (Textiles. Solar UV protective properties. Classification and marking of apparel). Determination of the UPF according to UV STANDARD 801 is based on maximum radiation intensity and the sun spectrum in Melbourne/Australia on the 1st of January of a year (i.e., at the height of Australian summer) and the most sensitive skin type for the wearer. As per the International standards, the textiles are classified according to the UPF as given in Table 3.2.

Cotton textiles treated with oxygen low pressure RF plasma for different periods of time and functionalized with 3% of ZnO nanoparticles showed higher adsorption of nanoparticles due to the increased concentration of oxygen-containing groups on the fiber surface and because of the increased surface roughness of fibers. Thus, treated cotton fabrics gained excellent UV protective properties with UPF rating 50+ [41]. ZnO nanostructures (bundle/flower-like particles) homogenously formed on the fibers' surface of a cotton fabric exhibited the UPF values more than 100 [49].

3.4.3 Self-Cleaning Property

The self-cleaning property of a semiconductor/metal oxide can be explained by its photocatalytic property [50, 51]. Nano-ZnO absorbs photons from UV rays and excites the electrons in the valence band of ZnO to the conduction band, thereby leaving a positive hole in the valence band. The photo-generated holes and electrons can initiate oxidation–reduction reaction, which is able to form the hydroxyl radicals and reactive oxygen

Table 3.2 Classification of textiles based on UPF.

UPF	UV protection	Approximate percentage of UV blocked
15–24	Good	93.3–95.8
25–39	Very Good	96–97.4
40–50+	Excellent	97.5–98

species that are directly responsible for the oxidative degradation of organic compounds [52]. Figure 3.6 shows the degradation of MB stains by nano-ZnO-coated cotton fabrics. The intensities of MB stains on the nano-ZnO fabrics significantly decreased due to the photocatalytic degradation. Also, the photocatalytic degradation properties clearly depended on the concentration of nano-ZnO available on the surface of coated fabric; the rate of photocatalytic degradation increased with the increasing nano-ZnO (from 3 to 9 wt%). Another research team demonstrated the use of nano-ZnO for degradation of both MB and congo red stains by direct *in situ* growth of vertically aligned one-dimensional nano-ZnO onto textile surfaces [53].

Synthetic fabrics such as polyester on treatment with nano-ZnO demonstrated excellent antibacterial and self-cleaning effect [54]. The authors

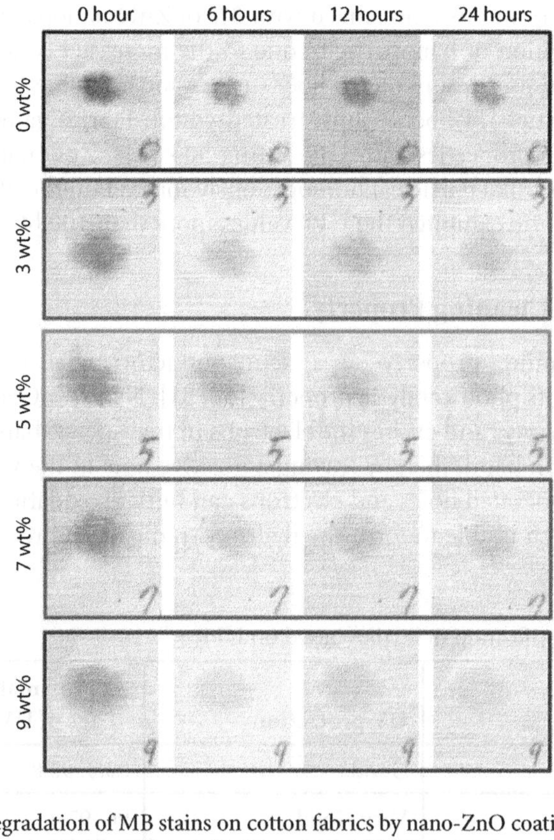

Figure 3.6 Degradation of MB stains on cotton fabrics by nano-ZnO coating. The X-axis represents the concentration of nano-ZnO coating and Y-axis represents the time of irradiation using a solar simulator. Reproduced with permission from [52].

demonstrated that alkali pre-treatment resulted in enhanced pick-up of nano-ZnO and hence, increased performance of functional properties.

3.4.4 Biosensing

The unique properties of nano-ZnO and the ease of its synthesis make this material extremely interesting for diversified applications. The direct *in situ* growth of vertically aligned one-dimensional (1D) nano-ZnO on the surfaces of conductive textiles (composed of 55% silver and 45% nylon) proved to be an excellent substrate for the fabrication of biosensor to detect ALD [2-methyl-2-(methylthio) propionaldehyde-O-(methylcarbamoyl) oxime], acarbamates family of pesticides. These pesticides are widely used in the protection of fruits and vegetables such as onion, potato, nut, cotton, tobacco, and sugar beet. The addition of ALD in the system resulted in an inhibition of immobilized enzyme activity that was quantified by monitoring the variation of electrode potential [53].

3.4.5 Super Hydrophobicity

The wetting states and the closely related wetting behavior vary with the surface structures and surface energies. For super hydrophobicity, the water contact angle needs to be more than 150° (see Table 3.3). In general, the nano-ZnO structures are modified with low surface energy materials such as silanes and fluorines to eliminate the influence of polarity of ZnO materials on wetting property [55]. In a different approach, Ce-doped nano-ZnO prepared via the chemical precipitation method and applied onto the surface of cotton fabrics was demonstrated to impart super hydrophobicity with a contact angle of 148° while that of pure nano-ZnO-treated was 103° [56]. Super-hydrophobic cotton fabrics prepared by fabricating rough surfaces using SiO_2 nanoparticles and ZnO nanorod arrays with a subsequent *n*-dodecyltrimethoxysilane (DTMS) modification showed a water contact

Table 3.3 Classification of textile surfaces based on the water contact angle.

Water contact angle	Classification	Effect of water droplet on fabric surface
<90°	Hydrophilic	Spreads out
>90°	Hydrophobic	Beads-up
>150°	Super hydrophobic	Rolls out

angle more than 150°. The lower roll-off angle of a nano-ZnO-treated cotton fabric, which is attributed to the discontinuous three-phase contact line, showed better water-repellent properties [57].

3.5 Conclusion

Nano-ZnO imparts durable multi-functional properties to textile materials and hence, a promising candidate for futuristic material in the field of textile finishing. The development of green syntheses protocols, zero-discharge textile finishing methodology, and safer handling and disposal techniques will sustain the use of this material (nano-ZnO) in the textile industry.

References

1. Shi, L.-E., Li, Z.-H., Zheng, W., Zhao, Y.-F., Jin, Y.-F., Tang, Z.-X., Synthesis, antibacterial activity, antibacterial mechanism and food applications of ZnO nanoparticles: a review. *Food Addit. Contam. Part A*, 31, 173–186, 2014.
2. Kołodziejczak-Radzimska, A., Jesionowski, T., Zinc oxide—From synthesis to application: A review. *Materials*, 7, 2833, 2014.
3. Sirelkhatim, A., Mahmud, S., Seeni, A., Kaus, N.H.M., Ann, L.C., Bakhori, S.K.M., Hasan, H., Mohamad, D., Review on zinc oxide nanoparticles: antibacterial activity and toxicity mechanism. *Nano-micro Lett.*, 7, 219–242, 2015.
4. Ma, H., Williams, P. L., Diamond, S. A., Ecotoxicity of manufactured ZnO nanoparticles – A review. *Environ. Pollut.*, 172, 76–85, 2013.
5. Bondarenko, O., Juganson, K., Ivask, A., Kasemets, K., Mortimer, M., Kahru, A., Toxicity of Ag, CuO and ZnO nanoparticles to selected environmentally relevant test organisms and mammalian cells in vitro: a critical review. *Arch. Toxicol.*, 87, 1181–1200, 2013.
6. Baun, A., Hartmann, N.B., Grieger, K., Kusk, K.O., Ecotoxicity of engineered nanoparticles to aquatic invertebrates: a brief review and recommendations for future toxicity testing. *Ecotoxicology*, 17, 387–395, 2008.
7. Dastjerdi, R., Montazer, M., A review on the application of inorganic nano-structured materials in the modification of textiles: Focus on anti-microbial properties. *Colloids Surf. B Biointerfaces*, 79, 5–18, 2010.
8. Yetisen, A. K., Qu, H., Manbachi, A., Butt, H., Dokmeci, M. R., Hinestroza, J. P., Skorobogatiy, M., Khademhosseini, A., Yun, S. H., Nanotechnology in textiles. *ACS Nano*, 10, 3042–3068, 2016.
9. Naveed Ul Haq, A., Nadhman, A., Ullah, I., Mustafa, G., Yasinzai, M., Khan, I., Synthesis approaches of zinc oxide nanoparticles: The dilemma of ecotoxicity. *J. Nanomater.*, 2017, 14, 2017.

10. Spanhel, L., Anderson, M.A., Semiconductor clusters in the sol-gel process: quantized aggregation, gelation, and crystal growth in concentrated zinc oxide colloids. *J. Am. Chem. Soc.*, 113, 2826–2833, 1991.
11. Meulenkamp, E. A., Synthesis and growth of ZnO nanoparticles. *J. Phys. Chem. B*, 102, 5566–5572, 1998.
12. Rodríguez-Paéz, J. E., Caballero, A. C., Villegas, M., Moure, C., Durán, P., Fernández, J. F., Controlled precipitation methods: formation mechanism of ZnO nanoparticles. *J. Eur. Ceram. Soc.*, 21, 925–930, 2001.
13. Vigneshwaran, N., Kumar, S., Kathe, A., Varadarajan, P., Prasad, V., Functional finishing of cotton fabrics using zinc oxide–soluble starch nanocomposites. *Nanotechnology*, 17, 5087, 2006.
14. Thilagavathi, T., Geetha, D., Low-temperature hydrothermal synthesis and characterization of ZnO nanoparticles. *Indian J. Phys.*, 87, 747–750, 2013.
15. Hui, Z., Deren, Y., Xiangyang, M., Yujie, J., Jin, X., Duanlin, Q., Synthesis of flower-like ZnO nanostructures by an organic-free hydrothermal process. *Nanotechnology*, 15, 622, 2004.
16. Erten-Ela, S., Cogal, S., Icli, S., Conventional and microwave-assisted synthesis of ZnO nanorods and effects of PEG400 as a surfactant on the morphology. *Inorg. Chim. Acta*, 362, 1855–1858, 2009.
17. Uma Sangari, N., Chitra Devi, S., Synthesis and characterization of nano ZnO rods via microwave assisted chemical precipitation method. *J. Solid State Chem.*, 197, 483–488, 2013.
18. Yıldırım, Ö.A., Durucan, C., Synthesis of zinc oxide nanoparticles elaborated by microemulsion method. *J. Alloys Compd.*, 506, 944–949, 2010.
19. Sarkar, D., Tikku, S., Thapar, V., Srinivasa, R. S., Khilar, K.C., Formation of zinc oxide nanoparticles of different shapes in water-in-oil microemulsion. *Colloids Surf. Physicochem. Eng. Aspects*, 381, 123–129, 2011.
20. Pholnak, C., Sirisathitkul, C., Harding, D.J., Characterizations of octahedral zinc oxide synthesized by sonochemical method. *J. Phys. Chem. Solids*, 72, 817–823, 2011.
21. Shi, L.-E., Fang, X.-J., Zhang, Z.-L., Zhou, T., Jiang, D., Wu, H.-H., Tang, Z.-X., Preparation of nano-ZnO using sonication method and its antibacterial characteristics. *Int. J. Food Sci. Tech.*, 47, 1866–1871, 2012.
22. Lee, S.D., Sang-Hun, N., Myoung-Hwa, K., Kim, Y.D., Jin-Hyo, B., Synthesis of ZnO nanoparticles by spray-pyrolysis method and their photocatalytic effect, in: 2010 3rd International Nanoelectronics Conference (INEC), pp. 572–573. 2010.
23. Salah, N., Habib, S.S., Khan, Z.H., Memic, A., Azam, A., Alarfaj, E., Zahed, N., Al-Hamedi, S., High-energy ball milling technique for ZnO nanoparticles as antibacterial material. *Int. J. Nanomed.*, 6, 863–869, 2011.
24. Manzoor, U., Siddique, S., Ahmed, R., Noreen, Z., Bokhari, H., Ahmad, I., Antibacterial, structural and optical characterization of mechano-chemically prepared ZnO nanoparticles. *PLoS ONE* 11(5): e0154704, 1–12, 2016.

25. Shanker, U., Jassal, V., Rani, M., Kaith, B.S., Towards green synthesis of nanoparticles: From bio-assisted sources to benign solvents. A review. *Int. J. Environ. Anal. Chem.*, 96, 801–835, 2016.
26. Sutradhar, P., Saha, M., Green synthesis of zinc oxide nanoparticles using tomato (Lycopersicon esculentum) extract and its photovoltaic application. *J. Exp. Nanosci.*, 11, 314–327, 2016.
27. Buazar, F., Bavi, M., Kroushawi, F., Halvani, M., Khaledi-Nasab, A., Hossieni, S.A., Potato extract as reducing agent and stabiliser in a facile green one-step synthesis of ZnO nanoparticles. *J. Exp. Nanosci.*, 11, 175–184, 2016.
28. Zheng, Y., Fu, L., Han, F., Wang, A., Cai, W., Yu, J., Yang, J., Peng, F., Green biosynthesis and characterization of zinc oxide nanoparticles using *Corymbia citriodora* leaf extract and their photocatalytic activity. *Green Chem. Lett. Rev.*, 8, 59–63, 2015.
29. Agarwal, H., Venkat Kumar, S., Rajeshkumar, S., A review on green synthesis of zinc oxide nanoparticles – An eco-friendly approach. *Resource-Efficient Technologies*, https://doi.org/10.1016/j.reffit.2017.03.002 In press, Corrected proof. Available online 18 April 2017
30. Jain, N., Bhargava, A., Tarafdar, J.C., Singh, S.K., Panwar, J., A biomimetic approach towards synthesis of zinc oxide nanoparticles. *Appl. Microbiol. Biotechnol.*, 97, 859–869, 2013.
31. Perelshtein, I., Perkas, N., Gedanken, A., Ultrasonic coating of textiles by antibacterial and antibiofilm nanoparticles. In: *Handbook of Ultrasonics and Sonochemistry*, Springer Singapore, Singapore, 2016, pp. 967–993.
32. Petkova, P., Francesko, A., Fernandes, M.M., Mendoza, E., Perelshtein, I., Gedanken, A., Tzanov, T., Sonochemical coating of textiles with hybrid ZnO/chitosan antimicrobial nanoparticles. *ACS Appl. Mater. Interfaces*, 6, 1164–1172, 2014.
33. Perelshtein, I., Lipovsky, A., Perkas, N., Tzanov, T., Gedanken, A., Sonochemical co-deposition of antibacterial nanoparticles and dyes on textiles. *Beilstein J. Nanotechnol.*, 7, 1–8, 2016.
34. Yadav, A., Prasad, V., Kathe, A., Raj, S., Yadav, D., Sundaramoorthy, C., Vigneshwaran, N., Functional finishing in cotton fabrics using zinc oxide nanoparticles. *Bull. Mater. Sci.*, 29, 641–645, 2006.
35. El. Shafei, A., Abou-Okeil, A., ZnO/carboxymethyl chitosan bionano-composite to impart antibacterial and UV protection for cotton fabric. *Carbohydr. Polym.*, 83, 920–925, 2011.
36. Farouk, A., Moussa, S., Ulbricht, M., Textor, T., ZnO nanoparticles-chitosan composite as antibacterial finish for textiles. *Int. J. Carbohydr. Chem.*, 2012, 1–8, 2012. doi:10.1155/2012/693629, Article ID 693629
37. Ibănescu, M., Muşat, V., Textor, T., Badilita, V., Mahltig, B., Photocatalytic and antimicrobial Ag/ZnO nanocomposites for functionalization of textile fabrics. *J. Alloys Compd.*, 610, 244–249, 2014.
38. Ates, E.S., Unalan, H.E., Zinc oxide nanowire enhanced multifunctional coatings for cotton fabrics. *Thin Solid Films*, 520, 4658–4661, 2012.

39. Prasad, V., Arputharaj, A., Bharimalla, A., Patil, P., Vigneshwaran, N., Durable multifunctional finishing of cotton fabrics by in situ synthesis of nano-ZnO. *Appl. Surf. Sci.*, 390, 936–940, 2016.
40. Uğur, Ş. S., Sarıışık, M., Aktaş, A.H., Uçar, M.Ç., Erden, E., Modifying of cotton fabric surface with nano-ZnO multilayer films by layer-by-layer deposition method. *Nanoscale Res. Lett.*, 5, 1204, 2010.
41. Jazbec, K., Sala, M., Mozetic, M., Vesel, A., Gorjanc, M., Functionalization of cellulose fibres with oxygen plasma and ZnO nanoparticles for achieving UV protective properties. *J. Nanomaterials*, 16, 25–25, 2015.
42. Ivask, A., Juganson, K., Bondarenko, O., Mortimer, M., Aruoja, V., Kasemets, K., Blinova, I., Heinlaan, M., Slaveykova, V., Kahru, A., Mechanisms of toxic action of Ag, ZnO and CuO nanoparticles to selected ecotoxicological test organisms and mammalian cells in vitro: A comparative review. *Nanotoxicology*, 8, 57–71, 2014.
43. Wang, L., Hu, C., Shao, L., The antimicrobial activity of nanoparticles: present situation and prospects for the future. *Int. J. Nanomed.*, 12, 1227–1249, 2017.
44. Kasemets, K., Ivask, A., Dubourguier, H.-C., Kahru, A., Toxicity of nanoparticles of ZnO, CuO and TiO2 to yeast Saccharomyces cerevisiae. *Toxicol. In Vitro*, 23, 1116–1122, 2009.
45. Pasquet, J., Chevalier, Y., Pelletier, J., Couval, E., Bouvier, D., Bolzinger, M.-A., The contribution of zinc ions to the antimicrobial activity of zinc oxide. *Colloids Surf. Physicochem. Eng. Aspects.*, 457, 263–274, 2014.
46. Nagarajan, P., Rajagopalan, V., Enhanced bioactivity of ZnO nanoparticles—an antimicrobial study. *Sci. Technol. Adv. Mater.*, 9, 035004, 2008.
47. Zhang, L., Jiang, Y., Ding, Y., Daskalakis, N., Jeuken, L., Povey, M., O'Neill, A.J., York, D.W., Mechanistic investigation into antibacterial behaviour of suspensions of ZnO nanoparticles against *E. coli*. *J. Nanopart. Res.*, 12, 1625–1636, 2010.
48. Hatch, K. L., American standards for UV-protective textiles. In: R. Dummer, F.O. Nestle, G. Burg (Eds.) *Cancers of the Skin: Proceedings of the 8th World Congress*, Springer, Berlin, pp. 42–47. 2002.
49. Shateri-Khalilabad, M., Yazdanshenas, M. E., Bifunctionalization of cotton textiles by ZnO nanostructures: antimicrobial activity and ultraviolet protection. *Text. Res. J.*, 83, 993–1004, 2013.
50. Fujishima, A., Honda, K., Electrochemical photolysis of water at a semiconductor electrode. *Nature*, 238, 37–38, 1972.
51. Lee, K.M., Abd Hamid, S.B., Lai, C.W., Mechanism and kinetics study for photocatalytic oxidation degradation: A case study for phenoxyacetic acid organic pollutant. *J. Nanomater.*, 2015, 10, 2015.
52. Zhu, C., Shi, J., Xu, S., Ishimori, M., Sui, J., Morikawa, H., Design and characterization of self-cleaning cotton fabrics exploiting zinc oxide nanoparticle-triggered photocatalytic degradation. *Cellulose*, 24, 2657–2667, 2017.

53. Hatamie, A., Khan, A., Golabi, M., Turner, A.P.F., Beni, V., Mak, W.C., Sadollahkhani, A., Alnoor, H., Zargar, B., Bano, S., Nur, O., Willander, M., Zinc oxide nanostructure-modified textile and its application to biosensing, photocatalysis, and as antibacterial material. *Langmuir*, 31, 10913–10921, 2015.
54. Nourbakhsh, S., Montazer, M., khandaghabadi, Z., Zinc oxide nano particles coating on polyester fabric functionalized through alkali treatment. *J. Ind. Text.*, 1, 2016. https://doi.org/10.1177/1528083716657819 Article first published online: June 30, 2016.
55. Wu, J., Chen, J., Xia, J., Lei, W., Wang, B.-p., A brief review on bioinspired ZnO superhydrophobic surfaces: Theory, synthesis, and applications. *Adv. Mat. Sci. Eng.*, 2013, 1–10, 2013. http://dx.doi.org/10.1155/2013/232681
56. Gao, D., Lyu, L., Lyu, B., Ma, J., Yang, L., Zhang, J., Multifunctional cotton fabric loaded with Ce doped ZnO nanorods. *Mater. Res. Bull.*, 89, 102–107, 2017.
57. Xu, B., Cai, Z., Wang, W., Ge, F., Preparation of superhydrophobic cotton fabrics based on SiO_2 nanoparticles and ZnO nanorod arrays with subsequent hydrophobic modification. *Surf. Coat. Technol.*, 204, 1556–1561, 2010.

4

Application of Nanomaterials in the Remediation of Textile Effluents from Aqueous Solutions

Mohammad Kashif Uddin[1,*] **and Ziaur Rehman**[2]

[1,*]*Basic Engineering Science Department, College of Engineering, Majmaah University, Al-Majmaah, Saudi Arabia*
[2]*Department of Civil and Environmental Engineering, College of Engineering, Majmaah University, Al-Majmaah, Saudi Arabia*

Abstract

Textile dyes, if present in wastewater, have hazardous effects on the life of aquatic animals and human beings. Advances in nano-science and technology have led to the evolution of pollution control. In order to treat dye-contaminated wastewater, several methods and nanomaterials have been successfully used in recent years. The small-size nanomaterials bring new opportunities for the operation of the dye-contaminated wastewater treatment technologies. In this review, the last five years' literatures (2012–2017) have been collected to highlight the successful usage of different nanomaterials to remove various types of dyes from the aqueous solution. A thorough literature survey revealed the application potential of nanomaterials in the remediation of textile effluents.

Keywords: Dyes, nanomaterials, water treatment, adsorption

4.1 Introduction

Water is a natural gift whose contamination has become a serious problem. There are many sources of water pollution, but the discharge of untreated

**Corresponding author*: mohdkashifchem@gmail.com: m.kashifuddin@mu.edu.sa

Shahid-ul-Islam and B.S. Butola (eds.) Nanomaterials in the Wet Processing of Textiles, (135–162) © 2018 Scrivener Publishing LLC

effluents from textile dyeing mills into the water bodies is a serious one. Highly concentrated colored wastewater that consists of a mixture of various dyes is highly toxic for living beings. Among the different pollutants of the aquatic ecosystem, dye molecules are highly toxic and mutagenic to both human beings and aquatic species. The molecular structure of dyes contains a complex aromatic ring, which is difficult to decompose and makes them carcinogenic [1, 2]. The discharge of dyes in an aqueous stream poses severe ecological problems to the aquatic life, food web, and to the aesthetic nature of the environment [3]. Dye absorbs and reflects the sunlight by reducing its penetration in water which causes a negative effect on the photosynthesis process, and can interfere in bacterial growth under water [1]. Dye containing wastewater is generally high in both color and organic content. Color is the undesirable physical contaminant, which can be easily recognized and is highly visible in aqueous solution, even in small amount. Organic content is the organic material present in water, which includes both humic and non-humic fractions. Organic pollutants consist of proteins, lipids, carbohydrates, cellulose, fats, lignin, and nucleic acids along with various combinations.

Textile, finishing, and dye manufacturing industries release a large quantity of wastewater containing toxic dyestuffs into the aquatic systems [4]. Dyes are widely used for coloring products in several industries such as textiles, leather, paper, rubbers, paint, tannery, pharmaceuticals, plastics, foodstuffs, cosmetics, etc. It is estimated that over 700,000 tons of dyes and pigments are produced annually worldwide, 20% of which are utilized for textile dyeing and finishing processes [5]. Textile industries are among the most polluted industries, which discharge wastewater of untreated dyestuffs in large volume and effluent composition [6, 7, 8]. Almost, 1,000 tones/year or more of dyes are discharged into wastewater by the textile industries worldwide [9]. The total dye consumption of the textile industries worldwide is in excess of 107 kg/year [10]. Effluents released from the textile industries are of synthetic origin, which contain a large variety of dyes, additives, pigments, and derivatives. Textile dyes are also potentially more toxic because they are difficult to be treated by the conventional wastewater treatment [11]. The chemical reagents used in textile industries have diversity in their chemical composition, as it ranges from inorganic compounds to polymers and organic products [12]. Dye removal from textile wastewater has been a big challenge over the last few decades. The development of effective methods to remove dyes from industrial wastewater is very important because of the potential toxicity of dyes, and thus is of high demand among the scientific community from the last few years [12].

In order to remove these dyes, many physical, chemical, and biological processes such as precipitation, separation, biodegradation, photo-degradation, chemical oxidation, chemical coagulation, foam flotation, reverse osmosis, electrochemical treatment, emulsion, ultrafiltration, photo-catalysis, ion exchange, pre-concentration, evaporation, sedimentation, aerobic or anaerobic treatment, adsorption, etc. have been developed, but most of them suffer from the economic point of view [13]. Out of these methods, the adsorption technique is superior and versatile to eliminate heavy metals from contaminated water because of its low cost, simplicity, ease of operation, efficiency in treatment, good applicability, high capacity, reliability, less energy consumption, and simplicity [14–20]. Adsorption is a mass transfer process by which a substance is transferred from the liquid phase to the surface of a solid, and becomes bound by physical and/or chemical interactions [15, 16]. In recent years, various nanomaterials have been widely synthesized, and applied in the removal of dyes from the aqueous solution [21]. Continuous development in nanomaterial research is providing effective wastewater treatment technologies to complete fresh water demand of the living beings. However, the need of new treatment methods and updated, reliable, efficient materials is still required to produce high-quality drinking water and to overcome the challenges of providing clean water in adequate amount.

Nanomaterials are of atomic size structure and smaller than 100 nm scale level possessing novel properties, higher surface area-to-volume ratio to offer high reactivity toward environmental contaminants, pollution detection and prevention, and water treatment. Nanoparticles can penetrate deeper and thus can treat wastewater effectively, which is generally not possible by conventional technologies [22]. Nanoparticles can behave as a colloid by mixing with water and they can display quantum-size effects [23–24]. Nanomaterials have been synthesized in various forms such as wires, tubes, films, particles, dots, etc. [25]. Magnetic nanoparticles are also promising because of their chemical structure, binding properties, high efficiency, rapid recovery due to high specific surface area, and low cost. In the treatment of wastewater, many efficient, eco-friendly, and cost-effective nanomaterials have been developed having unique functionalities for potential decontamination of industrial effluents, surface water, ground water, and drinking water [26, 27]. Excellent adsorption results are reported in many studies due to some key features such as high surface area, microporous/mesoporous structure, high dispersion and adsorption ability, and economical and environment-friendly nature. In this chapter, high adsorption capacity of nanoparticles toward various dyes has been reviewed by analyzing and tabulating the various optimal experimental conditions (solution pH, equilibrium contact time, amount of adsorbent, and temperature) as well as adsorption isotherms, kinetics, and thermodynamics data of different nanomaterials toward various types of dyes. The results finally concluded the excellent adsorption capacity of nanomaterials for the removal of several dyes from the aqueous solution.

4.2 Types of Dyes

Dyes can be classified based on their application and chemical structure. Dye molecules consist of group of atoms known as chromophores and the auxochromes. Chromophores have diverse functional groups responsible for producing color in dye, while the auxochromes intensify the solubility and color of dye. The most common auxochromes are amine, carboxyl, sulfonate, and hydroxyl [28, 29]. There are more than 100,000 commercially available dyes with over 7×10^5 tons of dyestuffs produced annually [1]. There are many structural varieties of dyes, such as acidic, basic, disperse, azo, diazo, and metal complex dyes. These varieties can be divided into cationic, nonionic, or anionic types. Anionic dyes are the direct, acid, and reactive dyes [30]. These are bright in color, water-soluble, and most toxic dyes, as they tend to pass untraced by conventional wastewater treatment [31]. Nonionic dyes are dispersed dyes because they do not ionize in an aqueous medium. Cationic dyes are a kind of alkaline dyes, which can be dissociated into positively charged ions in the aqueous solution and can interact with a negative group to form salt. Cationic dyes are also mostly applied to the dyeing of the polyester fiber and acrylic fiber [32].

Synthetic dyes are of the following types:

a) *Acid dyes* – Acid dyes are inexpensive, water-soluble anionic dyes that are applied to fibers such as silk, wool, nylon, and modified acrylic fibers under neutral to acidic conditions. The maximum quantity of acid dye absorbed depends on the amount of sulfuric acid (H_2SO_4) present in the solution.

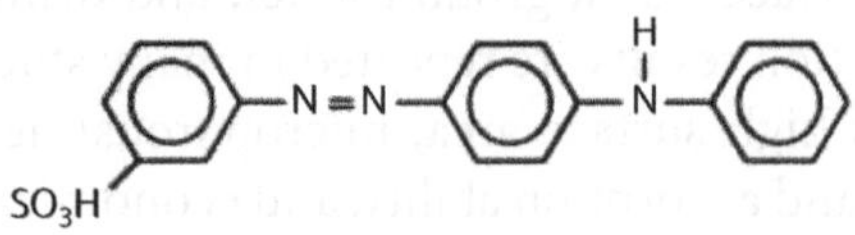

(Acid Yellow 36*)
*Source: http://textilelearner.blogspot.in/2015/01/different-types-of-dyes-with-chemical.html

b) *Basic dyes* – Basic dyes are water-soluble cationic dyes that are mainly applied to acrylic fibers, coloration of papers, and sometimes to wool and silk. Acetic acid is usually added with basic dyes to increase the uptake of the dye onto the fiber. They give good fastness and bright shades to acrylics.

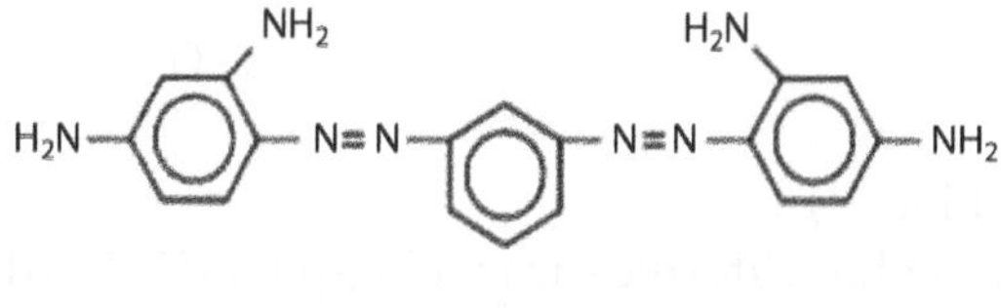

(Basic Brown 1*)
*Source: http://textilelearner.blogspot.in/2015/01/different-types-of-dyes-with-chemical.html

c) *Direct dyes* – Direct dyes are inexpensive and easy to use, but have poor fastness quality. Direct dyes are applied on cotton, paper, leather, wool, silk, and nylon in a neutral or slightly alkaline dye bath, with the addition of an electrolyte, that is sodium chloride (NaCl), sodium sulfate (Na_2SO_4), or sodium carbonate (Na_2CO_3) to accelerate the rate of uptake of dye by the fiber.

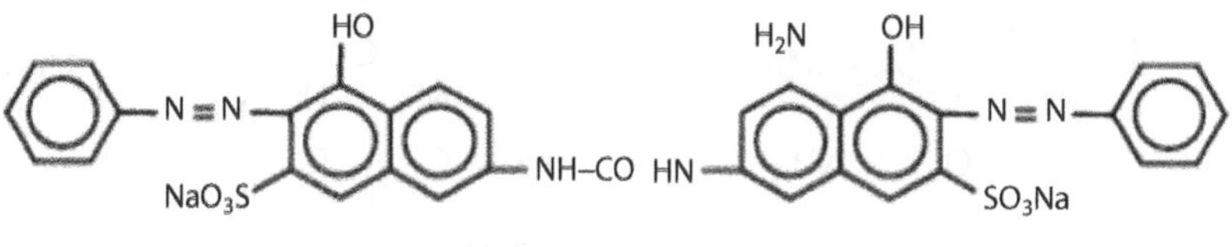

(Direct Dye 26*)
*Source: http://textilelearner.blogspot.in/2015/01/different-types-of-dyes-with-chemical.html

d) *Vat dyes* – Vat dyes are insoluble in water and incapable of dyeing fibers directly. They can become soluble by the use of a strong reducing agent, such as sodium hydrosulfide (NaHS) dissolved in sodium hydroxide (NaOH). The first synthetic vat dye was an indigo created in 1879. The color of denim is due to indigo, the original vat dye. Vat dyes are expensive because of the initial cost as well as the method of application.

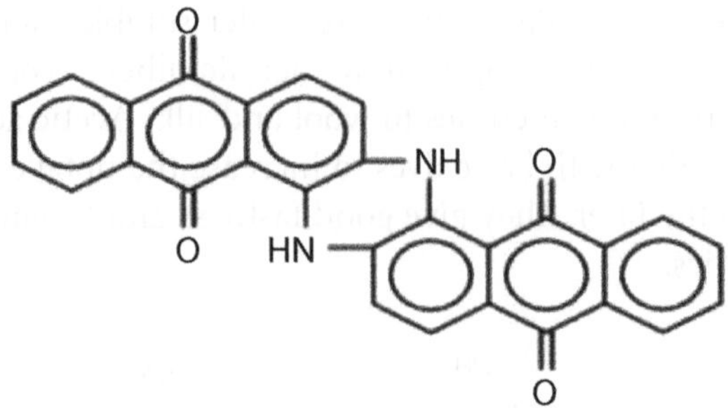

(Vat Blue 4*)
*Source: http://textilelearner.blogspot.in/2015/01/different-types-of-dyes-with-chemical.html

e) *Reactive dyes* – Reactive dyes are the best one for dyeing cotton and other cellulose fibers. They were first developed in 1956 by I.C.I., U.K. Reactive dyes have excellent fastness properties and areapplied to natural and synthetic cellulosic fibers, natural protein fibers, and polyamide fibers. Reactive dyes are retained onto the fiber by means of a chemical reaction. They have covalent bonds that attach to natural fibers to make them most permanent of dyes.

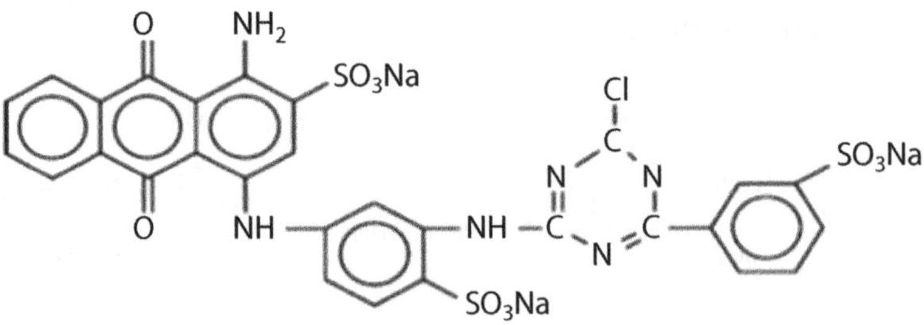

(Reactive blue 5*)
*Source: http://textilelearner.blogspot.in/2015/01/different-types-of-dyes-with-chemical.html

f) *Disperse dyes* – Disperse dyes are water-insoluble and were originally developed for the dyeing of cellulose acetate. These dyes are finely ground in the presence of a dispersing agent to be sold as a paste, spray-dried, or as a powder. These dyes have very fine particle size that provides a large surface area and high color uptake by the fiber. These are applied to the fibers that are most commonly dyed with disperse dyes such as cellulose diacetate, acrylic and nylon fibers, cellulose triacetate, and polyester fibers. Among them, polyester fibers provide satisfactory dyeing results as they are hydrophobic and have a significant crystalline content.

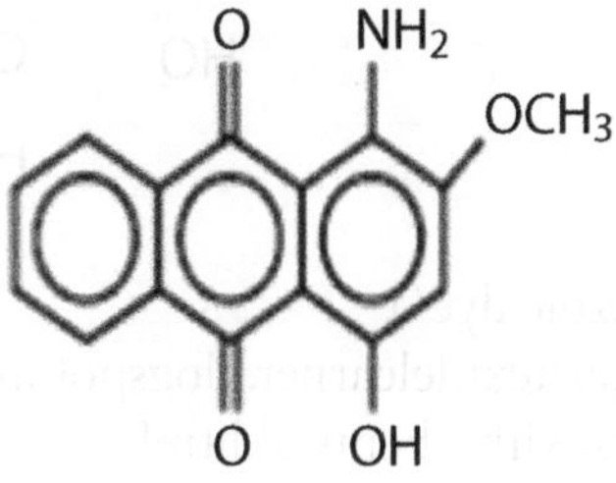

(Disperse red 4*)
*Source: http://textilelearner.blogspot.in/2015/01/different-types-of-dyes-with-chemical.html

g) *Sulfur dyes* – Sulfur dyes are inexpensive dyes that are used to dye dark-colored cotton. Natural and man-made cellulosic fibers are used to dye with sulfur dyes. Sulfur Black 1 is the largest selling dye by volume. This type of dye is affected by heating the fabric in an organic compound solution, typically a nitrophenol derivative, and sulfide or polysulfide, which then reacts with the sulfide source to form dark colors that adhere to the fabric.

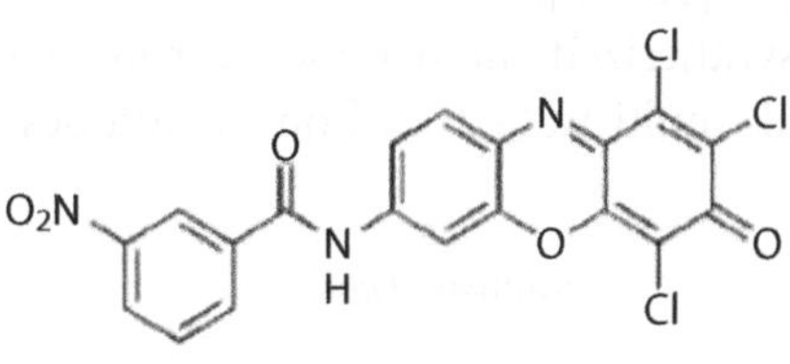

(Sulfur red 7*)
*Source: http://textilelearner.blogspot.in/2015/01/different-types-of-dyes-with-chemical.html

h) *Azoic colors* – Azoic dyeing is a technique in which an insoluble azo dye is produced directly onto or within the fiber. This type of dye can be formed by toxic chemicals and are fast to washing. Azoic colors are used mostly on cotton and nylon. Azoic colors give bright and high intensity colors, in comparison to common dyes.

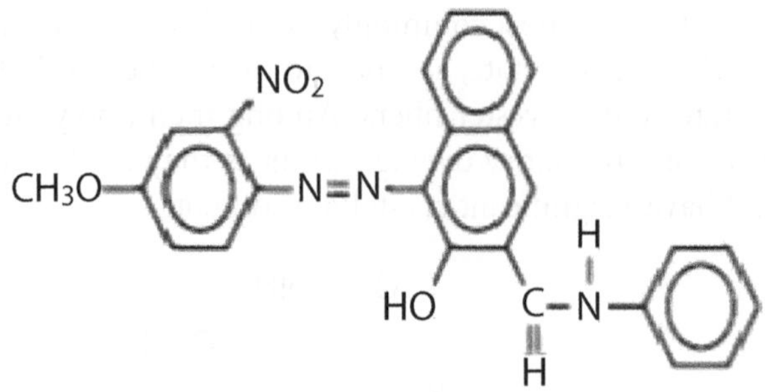

(Bluish red azoic dye*)
*Source: http://textilelearner.blogspot.in/2015/01/different-types-of-dyes-with-chemical.html

Figure 4.1 shows the classification of dyes, which is as follows:

4.3 Adsorption of Various Dyes on Nanomaterials

The discharge of industrial effluents containing textile dyes is a matter of concern because these pollutants can cause harmful effects in the environment and human health. Nano-adsorbents offer significant decontamination potential due to their unique characteristics such as high efficiency and rapid recovery due to high specific surface area. Table 4.1 listed the maximum adsorption capacities of various nano-adsorbents used to remove several dyes [33–63].

Bahgat *et al.* synthesized multi-walled carbon nanotubes (CNTs) and $NiFe_2O_4$-decorated multi-walled carbon nanotubes ($NiFe_2O_4$–CNTs) to

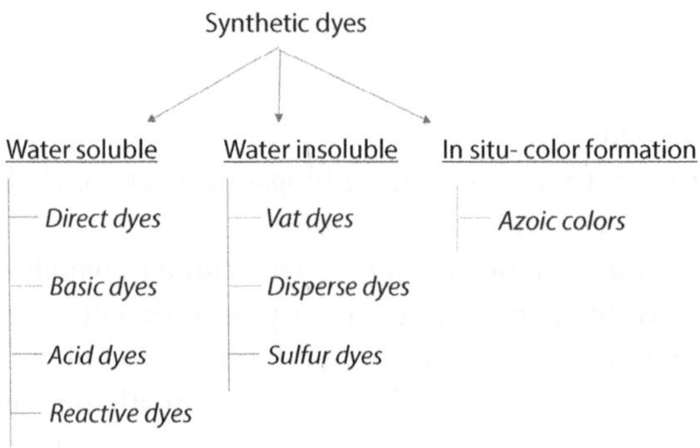

Figure 4.1 Classification of dyes.

Table 4.1 Maximum adsorption capacities of various nanomaterials toward different dyes.

Nanomaterials	Targeted dye	Adsorption capacity, qm [mg/g]	References
Nano-carbon xerogels	Methylene blue	33.00	[33]
Nano-carbon xerogels	Rhodamine B	42.00	[33]
Carbon nanotube [CNT]	Methyl green	146.00	[34]
$NiFe_2O_4$–CNTs	Methyl green	88.49	[34]
Silver-nanocomposite	Crystal violet	1.85	[35]
Graphene oxide [GO] – zeolite nanosheet	Rhodamine B	55.56	[36]
Benzene carboxylic acid derivatized graphene oxide–zeolite powder	Rhodamine B	67.56	[36]
Fe_3O_4 and reduced graphene oxide [RGO] composite	Rhodamine B	142.86	[37]
Silica-coated magnetic nanoparticles	Methyl orange	53.19	[38]
α-MnO_2 micronests composed of nanowires	Congo red	625.00	[39]
Nano-alumina	Reactive red 120	65.23	[40]
Modified nano-graphite/Fe_3O_4 composite	Methyl violet	144.71	[41]
Iron oxide magnetic nanoparticles decorated silica colloid	Methyl blue	1.44	[42]
Iron oxide magnetic nanoparticles decorated silica colloid	Methyl orange	8.39	[42]

(Continued)

Table 4.1 Cont.

Nanomaterials	Targeted dye	Adsorption capacity, qm [mg/g]	References
Alumina nanoparticles	Color black G	263.16	[43]
Multiwall carbon nanotubes	Maxilon blue	260.70	[44]
Fe_2O_3–biochar nano-composite	Methyl orange	20.53	[45]
Magnetic composite sorbent	Malachite green	159.10	[46]
Magnetic composite sorbent	Congo red	73.00	[46]
Magnetic composite sorbent	Indigo carmine	110.30	[46]
Nano zerovalent iron algal biocomposite	Malachite green	0.56	[47]
Carbon-coated Fe_3O_4 nanoparticles	Methylene blue	141.30	[48]
Bimodal porous silica microspheres decorated with polydopamine nanoparticles	Methylene blue	75.44	[49]
Carrageenan/Silica hybrid nano-adsorbents	Methylene blue	530.00	[50]
Copper oxide nanoflake	Malachite green oxalate	178.89	[51]
Copper oxide nanoflake	Methyl orange	158.83	[51]
Nickel oxide nanoflake	Malachite green oxalate	189.03	[51]
Nickel oxide nanoflake	Methyl orange	165.83	[51]
MgO nanomaterial	Remazol red RB-133	27.30	[52]

Alumina nanoparticle	Orange G	93.30	[53]
Cu_2Se nanoparticles	Methylene blue	33.30	[54]
Magnetite nanoparticles loaded Fig leaves	Crystal violet	53.47	[55]
Magnetite nanoparticles loaded Azolla	Crystal violet	30.21	[55]
Magnetite nanoparticles loaded Fig leaves	Methylene blue	61.72	[55]
Magnetite nanoparticles loaded Azolla	Methylene blue	25.00	[55]
Zero valent iron nanoparticles	Drimarene dye S-RB [Reactive Red 198]	5339.30	[56]
Magnetically separable porous iron-oxide nanocomposite	Methylene blue	298.00	[57]
Graphene oxide/magnesium oxide nanocomposite	Methylene blue	833.00	[58]
Acid/base bifunctional carbonaceous nanomaterial	Methylene blue	536.60	[59]
Acid/base bifunctional carbonaceous nanomaterial	Acid red 18	46.12	[59]
Organosilica nanoparticle	Phenol red	175.44	[60]
Poly[styrene-co-methacrylic acid]-coated magnetite nanoparticle	Crystal violet	416.66	[61]
Poly[styrene-co-methacrylic acid]-coated magnetite nanoparticle	Rhodamine B	69.54	[61]
Magnetic $Ba_3[PO_4]_2/Fe_3O_4$-nanoparticle	Malachite green	1639.00	[62]
Nanostructured magnesium oxide particles	Reactive blue 19	250.00	[63]

evaluate the adsorption capability toward organic dye-methyl green [34]. CNTs were prepared via the chemical vapor deposition (CVD) method and functionalized using a mixture of concentrated acids, while $NiFe_2O_4$–CNTs were prepared by *in situ* chemical precipitation of metal hydroxides followed by hydrothermal processing. It was found in the kinetics study of this research that the pseudo-second-order model controlled the adsorption of methyl green onto CNTs, while it was the pseudo-first-order model for the adsorption onto $NiFe_2O_4$–CNTs. The linear correlation coefficients (R^2), equilibrium adsorption data, standard deviations of Langmuir and Freundlich isotherms, and other results of the study showed that CNTs and $NiFe_2O_4$–MWCNTs were efficient adsorbents for the removal of methyl green from wastewater.

Zhang *et al.* [64] claimed to achieve the highest adsorption capacity among various reported adsorbents in removing methyl blue (MB) by zinc oxide (ZnO) nanoparticles. The adsorption process of MB on ZnO nanoparticles was found to be selective and independent of temperature and pH in its experimental range. The adsorption kinetics followed a pseudo-second-order model and exhibited a two-stage intra-particle diffusion model. The Temkin isotherm model indicated that the adsorption process was spontaneous and uniformly distributed. However, the adsorption isotherm was only calculated and fitted well with the Temkin isotherm, but the parameters of other main and important isotherms such as Langmuir and Freundlich were not calculated in this study.

In search of a cost-effective and environment-friendly material for adsorption of toxic crystal violet dye, green synthesis of silver nanoparticles (AgNPs) using soil was carried as a novel nanocomposite [35]. In comparison with soil under same experimental conditions, this nanocomposite was found to have a higher adsorption capacity toward the crystal violet dye. The percentage dye removal was homogenous, spontaneous, and endothermic, suggesting that the used nano-adsorbent was efficient in removing crystal violet from the effluent solution.

Zinc–aluminum-layered double hydroxide (Zn–Al LDH), an innovative nano-structured inorganic adsorbent, was synthesized for the separation of reactive yellow 84 (RY84) dye from several textile wastewater effluents. The study also revealed that the coupling of Zn–Al(NO_3^-) LDH nano-adsorbent for a solid phase extraction (SPE) procedure with spectrophotometric detection exhibited a simple and low-cost technique that can be used for the determination of RY84 dye [65].

Zeolite particles functionalized with graphene oxide (GO) nanosheets and a carboxylated diazonium salt (4-carboxybenzenediazoniumtetrafluoroborate) and were applied as effective and environmentally favorable adsorbents to remove cationic dye, rhodamine B, from aqueous solution [36]. Results of this study indicated that benzene carboxylic acid derivatized GO-zeolite powders showed higher adsorption capacities compared to pristine natural zeolite and GO-coated zeolite powders.

Simply made and cost-effective nano-alumina were found to be quite efficient for the removal of reactive red (RR) 120 dye ($C_{44}H_{24}C_{12}N_{14}O_{20}S_6Na_6$) from the aqueous solution [40]. It was found in this study that with an increase of the adsorbent dose, the dye removal efficiency increased, while the amount of dye adsorbed per unit mass (mg/g) decreased.

The deposition–precipitation method was used to synthesize and develop nano-graphite/Fe_3O_4 composite (NG/FC) for the removal of methyl violet (MV) from aqueous solution [41]. The Langmuir isotherm model and the pseudo-second-order kinetic model fitted the experimental data quite well. High stability and remarkable regeneration ability indicated that the NG/FC composite was a promising adsorbent for the removal of MV from wastewater.

Chi *et al.* [42] explored the potential environmental engineering application of silica colloid–polyelectrolyte–iron oxide nanocomposite. The magnetic and catalytic properties of the nanocomposite were synthesized via layer-by-layer assembly to adsorb organic dyes, methylene blue (MB) and methyl orange (MO), from aqueous environment. The electrostatic interactions between the dyes and as-synthesized nanomaterials verified that silica colloid–polyelectrolyte–iron oxide composite was a superior adsorbent due to its catalytic properties. 100% dye removal efficiency was recorded on silica–PDDA–IOMNPs nanocomposite neglecting the charge of dye involved. This nanocomposite can be repeatedly used for at least five times for the water treatment purpose that made it a versatile nanoagent for environmental engineering application. The adsorption isotherm data for silica–PDDA–IOMNPs nanocomposite in MO and MB removal were well fitted with the Langmuir model suggesting that the adsorption occurred with monolayer coverage. Dye interaction on the adsorption process and the catalytic degradation mechanism by Fenton and Fenton-like reaction were described appropriately by the pseudo-second-order kinetic model.

An alumina nanoparticle was developed using the combustion synthesis method and was utilized as an adsorbent for the removal color black G (CBG) from wastewater [43]. It was noted that the maximum adsorption of CBG was obtained at a pH value of 2. The kinetic data obtained during the experiments are better fitted with the pseudo-first-order model for CBG (R^2 = 0.971). The Langmuir isotherm model described the phenomenon for the removal of CBG using the alumina nanoparticles well. The Microsoft solver technique was used to obtain the optimum values of initial concentration, pH, contact time, and adsorbent dosage for maximum adsorption capacity.

A low-temperature hydrothermal route has been successfully developed to synthesize multiwall carbon nanotubes (MWCNTs) at 190 °C with catalyst Pd by using polyethylene glycol (PEG) as the carbon source [44]. The adsorption property of maxilon blue (GRL) from aqueous solution onto MWCNTs was studied by batch experiments as a function of mass dosage, pH of solution, initial dye concentration, and temperature. Equilibrium isotherm data were tested using Langmuir, Freundlich, and Temkin models. Among them, the Freundlich equilibrium isotherm fitted well the experimental data, indicating the homogeneity of the adsorbent surface sites. The adsorption of GRL dye onto MWCNTs was a spontaneous and endothermic process, which indicated that the adsorption was favored at high temperature.

A Fe_2O_3–biochar nanocomposite (Fe_2O_3–BC) was prepared from $FeCl_3$-impregnated pulp and paper sludge (PPS) by pyrolysis at 750 °C to be used as an adsorbent for the treatment of methyl orange (MO)-containing wastewater [45]. The study also reported the comparison between the adsorption capacities of Fe_2O_3–BC with unactivated biochar (BC). MO adsorption followed pseudo-second-order kinetics for both BC and Fe_2O_3–BC with R^2 values of 0.996 and 0.999, respectively. Gibbs free energy calculations confirmed the adsorption was energetically favorable and spontaneous with a high preference for adsorption on both adsorbents. The hybrid nature of the nanocomposites was responsible for the efficient removal of MO (97%) from contaminated wastewater. However, regeneration capacity of the adsorbents and the effect of competition on the adsorption of different dyes still need to be investigated.

Another novel nano-zerovalent iron S. swartzii (nZVI-SS) biocomposite was synthesized for the removal of malachite green (MG) from polluted wastewater [66]. The column experimental data obtained at different conditions were analyzed using three different models, namely, Adam–Bohart model, Thomas model, and Yoon–Nelson model, which provided a good breakthrough curve prediction. However, the results obtained from the Thomas model and the Yoon–Nelson model were more satisfactory. The packed bed investigation on the biosorption of MG onto nZVI-SS biocomposite revealed the importance of bed height, flow rate, and inlet solute concentration on MG biosorption. The increase in bed height resulted in the extension of the breakthrough time, whereas increase in flow rate down-regulated the breakthrough time.

Carbon-coated Fe_3O_4 nanoparticles (Fe_3O_4@CNPs) were synthesized by using ultra small citrus pectin as the carbon source with an average size of 7 nm and a specific surface area up to 58.72 $m^2\,g^{-1}$ to remove methylene blue from contaminated water samples. It was concluded that the synthesized Fe_3O_4@CNPs had great potential for methylene blue removal from wastewater. Citrus pectin was efficiently utilized in this study [48].

In another study, a new type of porous silica microspheres was successfully synthesized by a modified two-step sol-emulsion-gel method, which offered fast and efficient synthesis of the porous bimodal silica particles. These bimodal meso/macro-porous silica microspheres (MSM) were then coated with polydopamine (PDA) nanoparticles of 39 nm in size. The high specific surface area (612.3 m^2 g^{-1}), fast mass transfer (0.9–2.67 10^{-3} mL min^{-1} mg) of synthesized adsorbent, and abundant functional groups of PDA were the main reasons for the high removal of methylene blue (MB) from the aqueous solution. The MB adsorption was highly dependent on solution pH, influent MB concentration, and flow rate of MB solution. Maximum adsorption capacity could reach up to 83.8 mg g^{-1} at pH 10, which was larger than that of most of other reported adsorbents. Thomas and Adams–Bohart models showed a good agreement with the experimental data. The regeneration process was conducted using 0.1 M HCl solution and the regenerated adsorbent could be reused for the adsorption process successfully [49].

Magnetic iron oxide nanoparticles coated with κ-carrageenan/silica organic/inorganic hybrid shells were synthesized and used as novel adsorbents for the magnetically assisted removal of methylene blue (MB) from water [50]. These hybrid materials were enriched with ester sulfate groups due to extensive grafting of κ-carrageenan over the siliceous domains by using a new surface modification method. The MB removal efficiency over six consecutive adsorption/desorption cycles was above 97%, which demonstrated the reusability potential of these hybrid sorbents.

Highly efficient and environmentally benign clay mineral, Fe(III)–montmorillonite [Fe(III)–Mt], was explored for the adsorption of methylene blue (MB) dye from the aqueous solution [67]. The Fe(III)–Mt was interacted with the MB dye solution at different pH, temperature, and solid-to-liquid ratio. The MB dye removal was rapid at basic pH and increased with temperature up to 40 °C. A complete reduction (100%) was occurred in about 7 min at pH 7 and 10 while at pH 3 in about 10 min. This study revealed that Fe(III)–Mt had the potential to be used as an adsorbent to remove cationic pollutants effectively and rapidly from industrial wastewater.

The CuO and NiO nano-flakes were synthesized by a hydrothermal reaction and used as a potential adsorbent for malachite green oxalate (MGO) and methyl orange (MO) (51). XRD, SEM, and TEM characterized the structure, morphology, and surface properties of the nanoparticles. The adsorption of MGO increased with an increase in the pH, while

MO adsorption showed an opposite trend. However, the full potential of CuO and NiO for removing MGO and MO from real industry wastewater samples demands further studies.

Three MgO nanomaterials (MgOS, MgON, and MgOU) were synthesized by sol–gel and two different precipitation methods, respectively, and used for the sorption of remazol red (RB-133) dye from the aqueous solution. The complete removal efficiency of the dye on MgOS, MgON, and MgOU adsorbents was attained in less time of 11, 40, and 60 min, respectively. Hence, the MgOS adsorbent was found to possess the highest removal efficiency of the dye from aqueous solutions. The Langmuir isotherm model was found suitable to describe adsorption, while the dye adsorption followed pseudo-first-order kinetics. Evaluation of thermodynamic parameters revealed that the adsorption process was exothermic and spontaneous. The results showed that potentially lowering capital and operational costs of MgOs adsorbent was a promising one for the removal of synthetic dye in wastewater treatment [52].

The efficiency of an organo-palygorskite-Fe_3O_4 nanomaterial was recently investigated in the removal of two anionic dyes, methyl orange and indigo carmine, from aqueous solution [68]. The production of palygorskite-Fe_3O_4 nanoparticles organophilized with cetyl trimethylammonium bromide (CTAB) was proven by different characterizations. The adsorbed anionic dye on organo-palygorskite-Fe_3O_4 might be completely recovered by the action of a magnet, showing an excellent adsorptive property. The negative values calculated for the free energies of adsorption indicated that the adsorption of methyl orange and indigo carmine on the active sites of organo-palygorskite-Fe_3O_4 surface was spontaneous.

Orange G (OG), an anionic dye, was removed from aqueous solutions by the application of alumina nanoparticles [53]. Alumina nanoparticles were successfully synthesized via the sol–gel technique and the particle size was found to be in the range of 30–35 nm. The formation of nanoscale alumina particles was analyzed by the TEM and SEM techniques. The effect of various important parameters on dye removal was examined in this study and it was found that adsorption was highly pH dependent and a maximum removal of 98.4% was observed at pH 2.5. The mechanism of the adsorption process was also interpreted with the help of diffusion models. The Weber–Morris and diffusion models depicted that the mechanism of the removal process was controlled by both the film and pore diffusion but external diffusion controlled the overall rate of the sorption process.

Cobalt ferrite nanoparticles (CFO NPs) were synthesized by a facile and polyethylene glycol (PEG)-assisted hydrothermal method [69]. The as-synthesized samples showed highly selective adsorption characteristics for organic dyes: methyl orange (MO) and methyl blue (MB), Congo red (CR). PEG/$CoFe_2O_4$ nanocomposites had higher adsorption properties for congo red dye, in comparison to methyl orange and methyl blue. This study confirmed that CFO NPs were an excellent selective adsorbent to remove organic dyes from wastewater.

Different types of cationic dyes: methylene blue (MB) and rhodamine B (RB), and anionic dyes: methyl orange (MO) and eosin Y (EY), were removed from aqueous solutions using Cu_2Se nanoparticles (Cu_2SeNPs) by the adsorption method in a study [54]. The driving force for the adsorption of dyes had electrostatic and π–π interactions between Cu_2SeNPs and dyes. Cu_2SeNPs showed great potential as an adsorbent for dyes removal due to its good stability, functionalization, and reusability. Furthermore, Cu_2SeNPs can be recycled from selenium nanoparticles adsorbing copper.

Superparamagnetic rGO-Co_3O_4 nanocomposite removed rhodamine B, methyl orange, and rose Bengal (5–15 lM) in less than 1 min at neutral pH from polluted water [70]. The magnetic properties of the nanocomposite allowed simple separation from the liquid phase by application of an external magnet (1T). The results of the study evidenced that rGO-Co_3O_4 had potential capability for the removal of heavy metal ions and organic dyes from wastewater and other catalytic transformations in the field of environmental remediation.

The surface of magnetite nanoparticle-loaded fig leaves (MNLFL) and magnetite nanoparticle-loaded Azolla (MNLA) was utilized as natural cheap sources of adsorbents to adsorb cationic dyes: crystal violet and methylene blue. The kinetic studies of adsorption were tested for pseudo-first-order, pseudo-second-order, intraparticle diffusion, and Elovich models. At optimum conditions, a pseudo-second-order kinetic model best described the adsorption of the crystal violet and methylene blue on the surface of MNLFL and MNLA adsorbents. The Langmuir isotherm fitted more than the Freundlich and Temkin isotherm in equilibrium data [55].

A two-step process utilizing surface modification of gas-phase-synthesized iron-oxide nanoparticles and a subsequent polymerization process synthesized a new type of porous iron-oxide/polymer nanocomposite [57]. Iron-oxide/polymer composite adsorbent with a large surface area was tested for the removal of methylene blue (MB). The super-paramagnetic properties of the nanocomposite adsorbent allowed its easy separation from water by a simple magnet, which enabled the development of a clean and safe process for water pollution remediation. This work contributed toward the development of easy handling and environment-friendly adsorbent and adsorption process, as the novel mesoporous magnetic

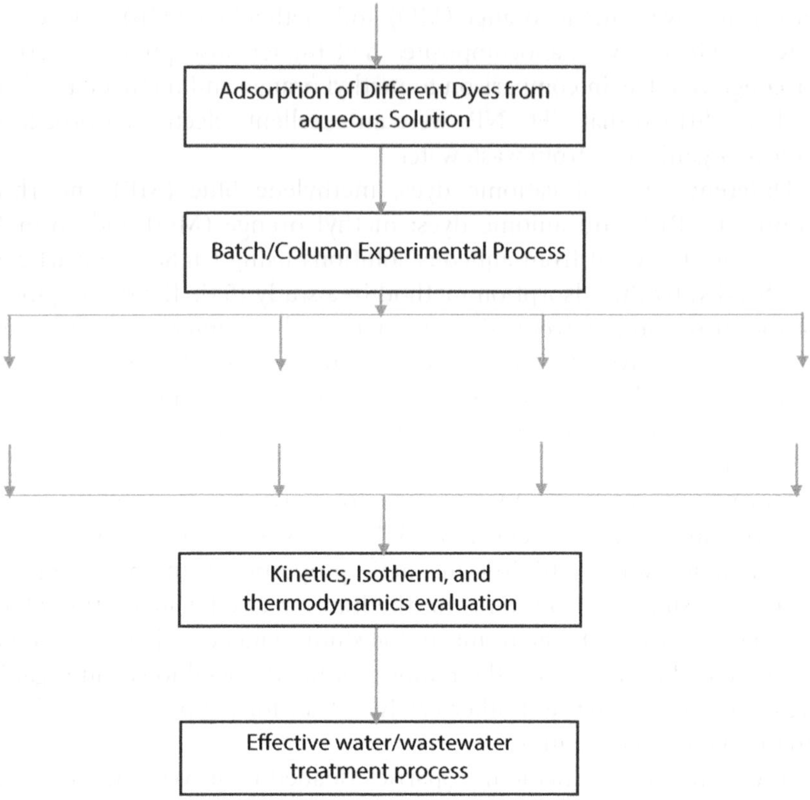

Figure 4.2 Pictorial diagram of the chapter.

iron-oxide/polymer adsorbent was a promising low-cost adsorbent for the large-scale removal of methylene blue (MB) from aqueous solutions.

Graphene oxide/magnesium oxide nanocomposites (GO/MgO NCs) were synthesized by the formation of chemical bonding between MgO and GO, and applied as an adsorbent for the removal of MB from the aqueous solution [71]. The prepared hybrid composite materials were characterized using scanning electron microscopy, transmission electron microscopy, X-ray diffraction, Fourier transform infrared spectrum, X-ray photoelectron spectroscopy, and thermogravimetric analysis. The results revealed that electrostatic attraction could be the dominant mechanism of adsorption between GO/MgO NCs and MB for pH above pHpzc, whereas for pH below pHpzc, other adsorption mechanisms such as hydrogen bonding and π–π interaction might attribute to adsorption. The high adsorption capacity of GO/MgO composites made it a promising adsorbent for water and wastewater treatment.

New types of antimicrobial capped ZnO nanoparticles were prepared recently at low temperature in water without calcination using PIL based on a PAMPSA/VP copolymer [72]. Crosslinked 2-acrylamido-2-methyl propane sulfonic acid-co-acrylonitrile nanocomposites were then prepared by ZnO nanoparticles to apply as an adsorbent for the removal of a harmful organic pollutant: ethylene blue dye. The ZnO composites achieved high adsorption removal rate for 3000 mg/L of MB from water during 20 min.

The adsorption and kinetic studies were carried out on three textile dyes, namely Reactive Blue 222 (RB 222), Reactive Red 195 (RR 195), and Reactive Yellow145 (RY 145) using PRO-BEN [a bentonite modified with a new cationic proline polymer (L-prolineepichlorohydrin polymer)] in a study [73]. A higher concentration of dyes had increased the adsorption efficiency, while the increase in pH, salt content, and the temperature had a negative effect on adsorption. The adsorption process was exothermic and spontaneous. The second-order kinetic model had proved to describe the rate of adsorption for the range of initial concentrations. The mechanism of adsorption followed intraparticle diffusion as well as external mass transfer, indicating that PRO-BEN was effective in adsorbing anionic dyes, and could be employed as an industrial solution to effluent remediation.

Fe_3O_4/Poly(styrene-co-methacrylic acid) (St-co-MAA) particles with different particle sizes (20 and 255 nm) were synthesized by mini emulsion polymerization via two routes and used as adsorbents for the removal of crystal violet (CV) and Rhodamine B (RB) from the water solution. The effects of various factors on the adsorption capacity, such as contact time, pH of dyes solution, and initial dyes concentration were investigated. Adsorption kinetics of Fe_3O_4/Poly(St-co-MAA) were well explained by the pseudo-second-order model, suggesting a chemical adsorption process. The dye-adsorbed magnetic Poly(St-co-MAA) could be easily desorbed and reused for at least four cycles with a little decrease in adsorption capacity [61].

Table 4.2 comprises the kinetics and thermodynamics adsorption parameters, and Figure 4.2 shows the pictorial diagram of the chapter.

4.4 Conclusion

It can be noted in most of the reported works that equilibrium experiments and adsorption studies were conducted by the batch technique to report maximum adsorption capacities of nanomaterials toward targeted dye. Adsorption is one of the most important processes of metal uptake with lots of properties that take place at the mineral–solute interface and thus

Table 4.2 Adsorption parameters of different dyes onto various nanomaterials.

Nanomaterials	**Targeted dye**	**Pseudo-first-order model parameters k_1 (min^{-1}); R^2**	**Pseudo-second-order model parameters k_2 (g/mg min); R^2**	**Thermodynamic parameters Δh^0 (kJ/mol); ΔH^0 (kJ/mol); ΔS^0 (J/mol K)**	**References**
Carbon nanotube [CNT]	Methyl green	0.014; 0.988	0.0002; 0.988	−5.09; 38.90; 45.33	[34]
$NiFe_2O_4$-CNTs	Methyl green	0.001; 0.981	9.31×10^{-6}; 0.90	−7.43; 20.83; 94.57	[34]
α-MnO_2 micronests	Congo red	13.35×10^{-3}; 0.951	11.11×10^{-4}; 0.999	−2.58; 6.89; 0.031	[39]
Nano-graphite/Fe_3O_4 composite	Methyl violet	0.089; 0.890	0.003; 0.004	−26.71; 19.20; 154.01	[41]
Copper-oxide nanoflakes	Malachite green oxalate	0.036; 0.940	0.100; 0.990	−7.43; −4.11; 15.60	[51]
Copper-oxide nanoflakes	Methyl orange	0.039; 0.900	0.070; 0.990	−8.01; −16.62; 79.64	[51]
Nickel-oxide nanoflakes	Malachite green oxalate	0.044; 0.923	0.068; 0.998	−6.52; −16.21; 73.74	[51]
Nickel-oxide nanoflakes	Methyl orange	0.040; 0.839	0.078; 0.996	−6.35; −9.97; 52.90	[51]
MgOs nanomaterial	Remazol Red RB-133	0.430; 0.987	0.022; 0.993	−113.10; −655.20; −1.76	[52]

Cuprous selenide nanoparticles (Cu_2SeNPs)	Methylene blue	0.015; 0.974	0.050; 0.997	−17.62; 3.48; 70.75	[54]
Amino acid proline-based polymer nanocomposite	Reactive blue 222 (RB 222)	0.026; 0.981	0.124; 0.999	−71.28; −37.806; −112.33	[73]
Amino acid proline-based polymer nanocomposite	Reactive red 195 (RR 195)	0.027; 0.979	0.135; 1.000	−31.40; −18.297; −43.98	[73]
Amino acid proline-based polymer nanocomposite	Reactive yellow (RY 145)	0.025; 0.887	0.167; 1.000	−30.16; −17.717; −41.74	[73]

of great importance in the removal capacity of toxic dyes by nanomaterials. The performance adsorption process in the reported studies has been checked by examining the effect of various factors such as pH, contact time, initial influent concentration, temperature, adsorbent dosage, etc. To determine the adsorption rate and adsorption efficiency of nanomaterials in removing toxic dye effluents from water and wastewater, various parameters such as adsorption isotherms, kinetics, and thermodynamics have been demonstrated. Desorption and regeneration studies are important factors and thus implied by various researchers in the reported studies to determine the application, recovery, and reuse of nano-adsorbents. To characterize the nanomaterials, several instrumentation methods were used such as FTIR, SEM, TEM, XRD, TGA, etc. in almost all the reported studies. It can be concluded that the use of nanotechnology to address water problems is the promising application to solve technical problems in removing water pollutants. Nanomaterials are of increased interest for research and development around the globe based on their capacity for selective adsorption of dye molecules from aqueous solution to solve water problems.

References

1. Zollinger, H. *Colour Chemistry-Synthesis: Properties and Application of Organic Dyes and Pigments*. New York: VCH, 1987.
2. Zeng, Q.F., Fu, J., Zhou, Y., Shi, Y.T., Zhu, H.L., Photooxidation degradation of reactive brilliant red K-2BP in aqueous solution by ultraviolet radiation/sodium hypochlorite. *CLEAN – Soil, Air, Water* 37, 574–580, 2009.
3. Mohan, D., Kunwar, P.S., Gurdeep S., Kundan K., Removal of dyes from wastewater using Flyash, a low-cost adsorbent. *Ind. Eng. Chem. Res.* 41, 3688–3695, 2002.
4. Vautier, M., Guillard, C., Herrmann, J.M., Photocatalytic degradation of dyes in water. Case study of indigo and of indigo carmine. *J. Catal.* 201, 46–59, 2001.
5. Gessner, T., Mayer, U., *Ullmann's Encyclopedia of Industrial Chemistry. Part A27. Triarylmethane and Diarylmethane Dyes*. 6th edition. New York: Wiley-VCH, 2001.
6. Sen, S., Demirer, G.N., Anaerobic treatment of real textile wastewater with a fluidized bed reactor. *Water Res.* 37, 1868–1878, 2003.
7. Dos Santos, A.B., *Reductive Decolourisation of Dyes by Thermophilic Anaerobic Granular Sludge*. PhD-Thesis. Wageningen University, The Netherlands, 2005.
8. Shanmugam, S., Muthiah, P., Harinarayan, J.P., Chiya, A.B., Swaminathan, G., Box Behnken design based optimization of solar induced photo catalytic decolourization of textile dye effluent. *Cent. Eur. J. Eng.* 3, 135–144, 2013.

9. Marc, R., Asian textile dye makers are a growing power in changing market. *Chem. Eng. News Bureau* 73, 10–12, 1996.
10. Wong, Y.C., Szeto, Y.S., Cheung, W.H., McKay, G., Equilibrium studies for acid dye adsorption onto chitosan, *Langmuir*, 19, 7888–7894, 2003.
11. Chatterjee, D., Patnam, V., Sikdar, A., Joshi, P., Misra, R., Rao, N.N., Kinetics of the decoloration of reactive dyes over visible light-irradiated TiO_2 semiconductor photocatalyst. *J. Hazard. Mater.* 156, 435–441, 2008.
12. Bharathi, K.S., Ramesh, S.T., Removal of dyes using agricultural waste as low-cost adsorbents: a review. *Appl. Water Sci.* 3, 773–790, 2013.
13. Hossein, A.Z., Ebrahim, G., Zeynab, T., Zinc–aluminum layered double hydroxide as a nano-sorbent for removal of Reactive Yellow 84 dye from textile wastewater effluents. *J. Iran Chem. Soc.* 10, 1103–1112, 2013.
14. Rao, R.A.K., Ikram, S., Uddin, M.K., Removal of Cr (VI) from aqueous solution on seeds of *Artimisia absinthium* (novel plant material). *Desal. Water Treat.* 54, 3358–3371, 2015.
15. Rao, R.A.K., Kashifuddin, M., Pottery glaze--An excellent adsorbent for the removal of Cu (II) from aqueous solution. *Acta Geochimica* 31, 136, 2012.
16. Rao, R.A.K., Kashifuddin, M., Adsorption properties of coriander seed powder (Coriandrum sativum): extraction and pre-concentration of Pb (II), Cu (II) and Zn (II) ions from aqueous solution. *Ads. Sci. Tech.* 30, 127–146, 2012.
17. Rao, R.A.K., Kashifuddin, M., Kinetics and isotherm studies of Cd (II) adsorption from aqueous solution utilizing seeds of bottlebrush plant (*Callistemon chisholmii*). *Appl. Water Sci.* 4, 371–383, 2014.
18. Rao, R.A.K., Kashifuddin, M., Adsorption studies of Cd (II) on ball clay: Comparison with other natural clays. *Arab. J. Chem.* 9, S1233–S1241, 2016.
19. Rao, R.A.K., Ikram, S., Uddin, M.K., Removal of CD (II) from aqueous solution by exploring the biosorption characteristics of gaozaban (Onosma bracteatum). *J. Env. Chem. Eng.* 2, 1155–1164, 2014.
20. Rao, R.A.K., Rehman, F., Kashifuddin, M., Removal of Cr (VI) from electroplating wastewater using fruit peel of Leechi (Litchi chinensis). *Desal. Water Treat.* 49, 136–146, 2012.
21. Uddin, M.K., Bushra R., Synthesis and characterization of composite cation-exchange material and its application in removing toxic pollutants, *Enhancing Cleanup of Environmental Pollutants*, Springer, 297–311, 2017.
22. Gautam, P., Madathil, D., Nair, A.N.B., Nanotechnology in waste water treatment: a review. *Inter. J. Chem. Tech. Res.* 5, 2303–2308, 2013.
23. Alivisatos, A.P., Semiconductor clusters, nanocrystals, and quantum dots. *Science*, 271, 933–937, 1996.
24. Amin, M.T., Alazba, A.A., Manzoor, U., A review of removal of pollutants from water/wastewater using different types of nanomaterials. *Adv. Mater. Sci. Eng.* 1-24, 2014.
25. Edelstein, A.S., Cammaratra, R.C., *Nanomaterials: Synthesis, Properties and Applications*, CRC Press, Second Edition, 616, 1996.

26. Khan, M.A., Uddin, M.K., Bushra, R., Ahmad, A., Nabi, S.A., Synthesis and characterization of polyaniline Zr (IV) molybdophosphate for the adsorption of phenol from aqueous solution. *React. Kinet. Mech. Catal.* 113 (2), 499–517, 2014.
27. Anjum, M., Miandad, R., Waqas, M., Gehany, F., Barakat, M.A., Remediation of wastewater using various nano-materials. *Arab. J. Chem.* https://doi.org/10.1016/j.arabjc.2016.10.004, 2016.
28. Christie, R., *Colour Chemistry*. Cambridge, UK: The Royal Chemical Society, 2001.
29. Dos Santos, A.B., Cervantes, F.J., van Lier, J.B., Review paper on current technologies for decolourisation of textile wastewaters: Perspectives for anaerobic biotechnology. *Biores. Tech.* 98, 2369–2385, 2007.
30. Mishra, G., Tripathy, M., A critical review of the treatments for decolourization of textile effluent. *Colourage* 40, 35–38, 1993.
31. Robinson, T., McMullan, G., Marchant, R., Nigam, P., Remediation of dyes in textile effluent: a critical review on current treatment technologies with a proposed alternative. *Biores. Tech.* 77, 247–255, 2001.
32. Cationic dyes, https://www.chemicalbook.com/ProductCatalog_EN/161120.htm, 2016.
33. Girgis, B.S., Attia, A.A., Fathy, N.A., Potential of nano-carbon xerogels in the remediation of dye-contaminated water discharges, *Desalination* 265, 169–176, 2011.
34. Bahgat, M., Farghali, M.M., El Rouby, W., Khedr, M.., Mohassab-Ahmed, M.Y., Adsorption of methyl green dye onto multi-walled carbon nanotubes decorated with Ni nanoferrite. *Appl. Nanosci.* 3, 251–261, 2013.
35. Satapathy, M.K., Banerjee, P., Das, P., Plant-mediated synthesis of silver-nanocomposite as novel effective azo dye adsorbent. *Appl. Nanosci.* 5, 1–9, 2015.
36. Yu, Y., Murthy, B.N., Shapter, J.S., Constantopoulos, K.A., Voelcker, N.H., Ellis, A.V., Benzene carboxylic acid derivatized graphene oxide nanosheets on natural zeolites as effective adsorbents for cationic dye removal. *J. Hazard. Mater.* 260, 330–338, 2013.
37. Qin, Y., Long, M., Tan, B., Zhou, B., RhB adsorption performance of magnetic adsorbent Fe_3O_4/RGO composite and its regeneration through a Fenton-like reaction. *Nano-Micro Lett.* 6, 125–135, 2014.
38. Shariati-Rad, M., Irandoust, M., Amri, S., Feyzi, M., Ja'fari, F., Magnetic solid phase adsorption, preconcentration and determination of methyl orange in water samples using silica coated magnetic nanoparticles and central composite design. *Int. Nano Lett.* 4, 91–101, 2014.
39. Zhang, W., Zhao, W., Zhou, Z., Yang, Z., Facile synthesis of α-MnO_2 micronests composed of nanowires and their enhanced adsorption to Congo red. *Front. Chem. Sci. Eng.* 8, 64–72, 2014.
40. Nadafi, K., Vosoughi, M., Asadi, A., Borna, M.O., Shirmardi, M., Reactive red 120 dye removal from aqueous solution by adsorption on nano-alumina. *J. Water Chem. Tech.* 36, 125–133, 2014.

41. Li, C., Dong, Y., Yang, J., Li, Y., Huang, C., Modified nano-graphite/Fe_3O_4 composite as efficient adsorbent for the removal of methyl violet from aqueous solution. *J. Mol. Liq.* 196, 348–356, 2014.
42. Chi, B.H.X., Yeap, S.P., Ahmad, A.L., Lim, J., Layer-by-layer assembly of iron oxide magnetic nanoparticles decorated silica colloid for water remediation. *Chem. Eng. J.* 243, 68–78, 2014.
43. Bhargavi, R.J., Maheshwari, U., Gupta, S., Synthesis and use of alumina nanoparticles as an adsorbent for the removal of Zn(II) and CBG dye from wastewater. *Int. J. Ind. Chem.* 6, 31–41, 2015.
44. Alkaim, A.F., Sadik, Z., Mahdi, D.K., Alshrefi, S.M., Al-Sammarraie, A.M., Alamgir, F.M., Singh, P.M., Aljeboree, A.M., Preparation, structure and adsorption properties of synthesized multiwall carbon nanotubes for highly effective removal of maxilon blue dye. *Korean J. Chem. Eng.*, 32, 1–6, 2015.
45. Chaukura, N., Murimba, E.C., Gwenzi, W., Synthesis, characterisation and methyl orange adsorption capacity of ferric oxide–biochar nano-composites derived from pulp and paper sludge. *Appl Water Sci* 7, 2175–2186, 2017. doi: 10.1007/s13201-016-0392-5
46. Makarchuk, O.V., Dontsova, T.A., Astrelin, I.M., Magnetic nanocomposites as efficient sorption materials for removing dyes from aqueous solutions. *Nanoscale Res. Lett.* 11, 161, 2016.
47. Liu, X., Lang, M., Liu, M., Su, R., Wang, M., Qi, W., He, Z., Highly efficient catalysis of Azo dyes using recyclable silver nanoparticles immobilized on tannic acid-grafted eggshell membrane. *Nanoscale Res. Lett.* 11, 440, 2016.
48. Zhang, W., Zhang, L.Y., Zhao, X.J., Zhou, Z., Citrus pectin derived ultrasmall Fe_3O_4@C nanoparticles as a high-performance adsorbent toward removal of methylene blue. *J. Mol. Liq.* 222, 995–1002, 2016.
49. Ataei-Germi, T., Nematollahzadeh, A., Bimodal porous silica microspheres decorated with polydopamine nano-particles for the adsorption of methylene blue in fixed-bed columns. *J. Coll. Inter. Sci.* 470, 172–182, 2016.
50. Soares, S.F., Simoes, T.R., Trindade, T., Daniel-da-Silva, A.L., Highly efficient removal of dye from water using magnetic carrageenan/silica hybrid nano-adsorbents. *Water Air Soil Pollut.* 228, 87, 2017.
51. Kumar, K.Y., Archana, S., Raj, T.N.V., Prasana, B.P., Raghu, M.S., Muralidhara, H.B., Superb adsorption capacity of hydrothermally synthesized copper oxide and nickel oxide nanoflakes towards anionic and cationic dyes. *J. Sci.: Adv. Mater. Dev.*, http://dx.doi.org/10.101.6/j.jsamd.2017.05.006
52. Mahmoud, H.R., Ibrahim, S.M., El-Molla, S.A., Textile dye removal from aqueous solutions using cheap MgO nanomaterials: Adsorption kinetics, isotherm studies and thermodynamics. *Adv. Powder Tech.* 27, 223–231, 2016.
53. Banerjee, S., Dubey, S., Gautam, R.K., Chattopadhyaya, M.C., Sharma, Y.C., Adsorption characteristics of alumina nanoparticles for the removal of hazardous dye, Orange G from aqueous solutions. *Arab. J. Chem.* 2017, http://dx.doi.org/10.1016/j.arabjc.2016.12.016

54. Li, Q., Liu, Z., Huang, L., Teng, J., Bai, Y., Characterization and mechanism elucidation of dye adsorption using cuprous selenide nanoparticles from aqueous solutions. *Chem. Res. Chin. Univ.*, 32, 1010–1015, 2016.
55. Alizadeh, N., Shariati, S., Besharati, N., Adsorption of crystal violet and methylene blue on azolla and fig leaves modified with magnetite iron oxide nanoparticles. *Int. J. Environ. Res.* 11, 197–206, 2017.
56. Ahuja, N., Chopra, A.K., Ansari, A.A., Removal of colour from aqueous solutions by using zero valent iron nanoparticles. *IOSR J. Environ. Sci. Toxic. Food Techn.* 10, 4–14, 2016.
57. Sehlleier, Y.H., Hardt, S., Schulz, C., Wiggers, H., A novel magnetically-separable porous iron-oxide nanocomposite as an adsorbent for methylene blue (MB) dye. *J. Environ. Chem. Eng.* 4, 3779–3787, 2016.
58. Heidarizad, M., Şengör, S.S., Synthesis of graphene oxide/magnesium oxide nanocomposites with high-rate adsorption of methylene blue. *J. Mol. Liq.* 224, 607–617, 2016.
59. Li, K., Ma, C.F., Ling, Y., Li, M., Gao, Q., Luo, W.J., Acid/base bifunctional carbonaceous nanomaterial with large surface area: Preparation, characterization, and adsorption properties for cationic and anionic compounds. *Mat. Chem. Phys.* 162, 149–161, 2015.
60. Chen, F., Zhao, E., Kim, T., Wang, J., Hableel, G., Reardon, P.J.T., Ananthakrishna, S.J., Wang, T., Arconada-Alvarez, S., Knowles, J.C., Jokerst, J.V., Organosilica nanoparticles with an intrinsic secondary amine: An efficient and reusable adsorbent for dyes. *Appl. Mater. Interfaces* 9, 15566–15576, 2017.
61. Hayasi, M., Karimi, M., Synthesis of poly(styrene-co-methacrylic acid)-coated magnetite nanoparticles as effective adsorbents for the removal of crystal violet and Rhodamine B: a comparative study. *Polym. Bull.* 74, 1995–2016, 2017.
62. Zhang, F., Wei, Z., Zhang, W., Cui, H., Effective adsorption of malachite green using magnetic barium phosphate composite from aqueous solution. *Spectrochim. Acta A: Mol. Biomol. Spectr.* 182, 116–122, 2017.
63. Nga, N.K., Hong, P.T.T., Lam, T.D., Huy, T.Q., A facile synthesis of nano-structured magnesium oxide particles for enhanced adsorption performance in reactive blue 19 removal. *J.Coll. Inter. Sci.* 398, 210–216, 2013.
64. Zhang, F., Yang, Y., Tan, R., Song, W., Adsorption behavior and mechanism of methyl blue on zinc oxide nanoparticles. *J. Nanopart. Res.* 15, 2034, 2013.
65. Abdolmohammad-Zadeh, H., Ghorbani, E., Talleb, Z., Zinc–aluminum layered double hydroxide as a nano-sorbent for removal of Reactive Yellow 84 dye from textile wastewater effluents. *J. Iran Chem. Soc.* 10, 1103–1112, 2013.
66. Jerold, M., Joseph, D., Patra, N., Sivasubramanian, V., Fixed-bed column studies for the removal of hazardous malachite green dye from aqueous solution using novel nano zerovalent iron algal biocomposite. *Nanotechnol. Environ. Eng.* 1, 8, 2016.

67. Vinuth, M., Naik, H.S.B., Mahadevaswamy, M.M., Prabhakara, M.C., Environmentally benign Fe(III)–montmorillonite for rapid adsorption of methylene blue dye in aqueous medium under ambient conditions. *Fash. Text.* 4, 8, 2017.
68. Middea, A., Spinelli, L.S., Souza Jr, F.G., Neumann, R., Fernandes, T.L.A.P., Gomes, O.F.M., Preparation and characterization of an organo-palygorskite-Fe_3O_4 nanomaterial for removal of anionic dyes from wastewater. *Appl. Clay Sci.* 139, 45–53, 2017.
69. Wu, X., Wang, W., Li, F., Khaimanov, S., Tsidaeva, N., Lahoubi, M., PEG-assisted hydrothermal synthesis of $CoFe_2O_4$ nanoparticles with enhanced selective adsorption properties for different dyes. *Appl. Surf. Sci.* 389, 1003–1011, 2016.
70. Al Nafiey, A., Addad, A., Sieber, B., Chastanet, G., Barras, A., Szunerits, S., Boukherroub, R., Reduced graphene oxide decorated with Co3O4 nanoparticles (rGO-Co_3O_4) nanocomposite: A reusable catalyst for highly efficient reduction of 4-nitrophenol, and Cr(VI) and dye removal from aqueous solutions. *Chem. Engg J.* 322, 375–384, 2017.
71. Heidarizad, M., Sengor, S.S., Synthesis of graphene oxide/magnesium oxide nanocomposites with high-rate adsorption of methylene blue. *J. Mol. Liq.* 224, 607–617, 2016.
72. Atta, A.A., Al-Lohedan, H.A., Ezzat, A.O., Tawfik, A.M., Hashem, A.I., Synthesis of zinc oxide nanocomposites using poly (ionic liquids) based on quaternary ammonium acrylamidomethyl propane sulfonate for water treatment. *J. Mol. Liq.* 236, 38–47, 2017.
73. Raghunath, S., Anand, K., Gengan, R.M., Nayunigari, M.K., Maity, A., Sorption isotherms, kinetic and optimization process of amino acid proline based polymer nanocomposite for the removal of selected textile dyes from industrial wastewater. *J. Photochem. Photobiol. B: Bio.* 166, 189–201, 2016.

5

Chitosan–Graphene-Grafted Nanocomposite Materials for Wastewater Treatment

Mohammad Shahadat[1,2,*], Ankita Jha[1], Parveen Fatemeh Rupani[3], Asha Embrandiri[4], Shaikh Ziauddin Ahammad[1] and S. Wazed Ali[2]

[1]*Department of Biochemical Engineering and Biotechnology, Indian Institute of Technology, New Delhi, India*
[2]*Department of Textile Technology, Indian Institute of Technology, New Delhi, India*
[3]*School of Environment and Safety Engineering, Jiangsu University, Zhenjiang, China*
[4]*Department Of Environmental Studies, Kannur University, Kerala. India.*

Abstract

This chapter deals with the fabrication of cost-effective, eco-friendly chitosan (Ch)–graphene oxide (GO)-grafted nanocomposite (Ch–GO) for the removal of environmental pollutants (metal ions and dyes). Advanced characterization techniques such as SEM (scanning electron microscope), TEM (transmission electron microscopy), FTIR (Fourier Transform Infrared Spectroscopy), TGA/DTA (thermogravimetric analysis/differential thermal analysis), RAMAN, XRD (X-ray diffraction), EIS (electrochemical impedance spectroscopy), and AFM (atomic-force microscopy) analyses are employed to investigate the improvement in physicochemical and mechanical properties of nanocomposite materials. Various characterization results clearly depict the existence of hydroxy, epoxy, and carboxyl groups (–OH, –OOH, and –COOH) in the matrix of Ch–GO nanocompiste as well as adsorption of metal and dye contaminates with higher metal removal efficiency. The adsorption behavior of the Ch–GO nanocomposite is elucidated in terms of the effect of pH, concentration, and temperature to achieve optimal removal capacity of pollutants. Kinetic studies establish quick adsorption followed by pseudo-first and pseudo-second and modified order kinetics. Thermodynamics

**Corresponding author*: mdshahadat93@gmail.com

Shahid-ul-Islam and B.S. Butola (eds.) Nanomaterials in the Wet Processing of Textiles, (163–186) © 2018 Scrivener Publishing LLC

analysis reveals the measurement of free energy ($\Delta G°$), enthalpy ($\Delta H°$), and entropy ($\Delta S°$). The excellent adsorption efficiency of Ch–GO nanocomposite materials confirms their application for the treatment of organic and inorganic contaminants in water bodies to control water pollution. In upcoming years, Ch–GO nanocomposite adsorbents are anticipated to release their possible significant applications in diverse fields.

Keywords: Chitosan, graphene oxide, fabrication, characterization, metal ions, dyes, water pollution

5.1 Introduction

Discharging of dyes and metal ion containing effluents of various industries into water bodies has become the main cause of water pollution [1]. These effluents change the taste, odor, and property of water, which create a significantly bad impact to the life of aquatic flora and fauna [2]. Dye-based industries including printing, paper making, textile, and the paint industry are widely facing the problem of dye-contaminated wastewater [3]. Among the contaminated pollutants in industrial effluents, dyes are the most harmful substances which hinder penetration of sunlight into water and prohibit the photosynthesis of aquatic animals. In addition, these dye molecules affect the quality of water and result in carcinogenic influence [4]. Industrialization and urbanization have increased the discharging of metal ions, which include aluminum (Al), arsenic (As), cadmium (Cd), chromium (Cr), manganese (Mn), nickel (Ni), iron (Fe), zinc (Zn), lead (Pb), copper (Cu), and mercury (Hg) in soil, air, and water, which severely harm the environment [5–7]. These metal ions are consumed by the human through food or water, adsorbed in the body, and play a key role during cellular activity up to a certain limit [8–12]. The metal ions can be beneficial only up to a certain concentration. However, a higher concentration causes a number of physiological effects. Various metal ions together with their permissible limits according to the World Health Organization (WHO) are shown in Table 5.1.

Besides dyes and metal ions, other organic entities including insecticides, herbicides, fungicides, wood preservatives, and the use of antibiotics are also highly toxic, and affect body's soft tissues and cause a number of diseases (e.g., gastrointestinal, carcinogenicity, etc.) [22–26]. Therefore, it is important to treat industrial effluents and solid wastes before discharging into water resources. In this regard, a number of chemical and biological treatment methods (e.g., flotation, chemical precipitation, electro-coagulation-flocculation, membrane filtration, ion exchange, modified bentonite clay, natural adsorbent, etc.) are used [6, 27–38]. These methods

Table 5.1 Permissible limits of metal pollutants in water and their effect in human beings.

Metal ions	Permissible limits (mgL^{-1})	Affected body parts	References
Al	2.0×10^{-1}	Nervous system	[13]
As	5.0×10^{-2}	Nervous system (central and peripheral) and gastrointestinal	[14]
Cd	1.0×10^{-2}	Lungs	[15]
Cu	3.0	Heart disease, tooth decay, hair and skin	[16]
Cr	1.0×10^{-1}	Nose (nasal itching)	[17]
Hg	2.0×10^{-3}	CNS and eyes	[18]
Mn	1.0×10^{-1}	CNS and eye	[19]
Ni	2.0×10^{-1}	Fingers, hands, and forearms; lung and nose	[20]
Zn	3.0	Skin	[21]

are associated with high-cost, time-consuming, pre-treatment procedures as well as are inconvenient for onsite analyses. As compared to conventional methods, adsorption by hybrid ion-exchange materials has gained popularity and is being effectively used for the removal of dyes and metal ions [11, 39–41]. However, there are certain drawbacks [42]. The limitations of composite materials have been overcome by the synthesis of chitosan–graphene-grafted nanocomposite adsorbents for the treatment of wastewaters.

5.2 Chitosan–Graphene-Grafted Nanocomposite

Recently, various methods and techniques have been developed to fabricate cost-effective polysaccharide adsorbents namely starch, cellulose, alginate, and chitosan [43]. Chitosan (Ch) is a low-cost, non-toxic, highly reactive multifunctional biomaterial which is obtained by the deacetylation of chitin [44] and shows higher adsorption efficiency owing to the existence of a large number of amino groups [45]. It has been successfully applied in a number of fields such as the food industry, cosmetic, medicine, water treatment, etc. The presence of amine and hydroxyl groups makes the Ch as a potential biosorbent for the removal of metal ions in wastewaters due to the formation of stable metal chelates [46]. Chitosan is also associated with some limitations as it is soluble in the acidic medium except water and alkaline solutions. To improve mechanical strength and adsorption efficiency of Ch for the treatment of metal ions in aqueous media, efforts have been made to modify Ch with a suitable adsorbent [47]. In this regard, a new member of the carbon family, graphene (Gr), was selected for the modification of Ch owing to its unique structure (two-dimensional, one-atom-thick structure) together with high mechanical properties and high surface area [48]. One of the particular derivatives of graphene is known as graphene oxide (GO), which has attracted much attention for the adsorption of metal ions, dyes and other uses [49]. The GO consisting of various functional groups (–OH, –OOH, –COOH), which provide interfacial interaction with other adsorbent to synthesize thermally and chemically stable products as well as to improve other properties (hydrophilicity). In addition, these functional groups show strong binding affinity with metal ions and dyes molecules [49, 50]. The GO has been successfully applied in the form of effective adsorbent for the removal of metal ions (Cd(II), Co(II), and Cu(II)) from water [50, 51]. The GO-based membrane was developed using green-induced directional method [52]. The membrane was employed for the adsorption of Cu(II), Ni(II), and Cd(II) ions. The adsorption of metal ions on GO membranes was interrupted by weak ionic strength. Maximum adsorption of Cu(II), Ni(II), and Cd(II) on membrane was found to be 72.6, 83.8, and 62.3 mg g^{-1}, respectively. Adsorption isotherm followed the Langmuir isotherm model and adsorption kinetics followed the pseudo-second-order model for the adsorption of metal ions. It has been noticed that the adsorption process reached equilibrium in a shorter period of time due to larger interlayer spacing of GO membranes. Based on recycling and regeneration study, the prepared GO membranes can be used more than six times without decline in adsorption capacity.

Yang *et al.* in 2010 demonstrated the interaction between positively charged ions on Cu(II) and negatively charged GO. It was able to remove Cu(II) ions [53–55]. The results have proved that GO can also be used for the adsorption of Pb(II), Cd(II), Co(II), and U(VI). However, it is noteworthy to mention that pH is an important factor in the adsorption of heavy metals using GO because the pH-dependent functional groups on the GO surface are the potential sites for metal adsorption. Zhang *et al.* reported the treatment of arsenate with 95% removal efficiency within 4–7 pH range using a cross-linked ferric hydroxide–GO composite [56]. The fabrication of composite was done by *in situ* oxidation of ferrous sulfate using hydrogen peroxide.

A recent study carried out by Zhu *et al.* indicated efficient adsorption of As(III) from contaminated water via Fe–Fe_2O_3 core shell coated with graphene [57]. Fe_2O_3 helps to increase adsorption sites whereas Fe core maintains the magnetic separation. This finding was explored by Luo *et al.* who observed that pH plays an important role for the adsorption of metal ions. The increase in the pH range improves the adsorption rate of MnO_2 nanoparticles into Fe_3O_4–rGO materials. Therefore, MnO_2 functions as a good oxidant for the effective adsorption of As(III) and As(V). There is a great application potential foreseen. The removal of Cr(VI) from polluted waters by catalytic reduction of Cr(VI) to Cr(III) using iron-decorated grapheme sheets was reported by Humera *et al.* [58].

In another study, Zhu *et al.* [59] synthesized graphene-coated core double-shell nanoparticles for the removal of Cr(VI) from wastewater. This core was made up of crystalline iron while the outer shell was made up of iron oxide. The external shell covered an amorphous Si–S–O. Results indicated total removal of Cr(VI) in 5 min which could be separated from the solution using an external magnetic field. Metal ion induced self-assembly of a FeOOH–rGO hydrogel by the NH_4OH reduction of $FeSO_4$ and GO was fabricated by Cong *et al.* [60]. This composite proved to be effective in the removal of gasoline, Cr(VI), and Pb(II) from polluted water. Similarly, Sreeprasad *et al.* synthesized rGO–metal/metal oxide composites using a redox-like reaction between GO and the metal precursors, which resulted in the effective removal of Hg(II) from polluted water [61]. In the same vain, GO coated with TiO_2 can be used in the removal of the heavy metal ions; Zn(II), Cd(II), and Pb(II) [61]. In order to increase the adsorption capacity of graphene, Polypyrrole–rGO hybrid was synthesized via a noncovalent functionalization of graphene sheets with polypyrrole [62]. The outcome indicated the growth of polypyrrole on the graphene sheets resulting in high surface area of mesoporous nature. The adsorption capacity is also affected by the ionic strength as it determines the concentration of heavy metal ions available for adsorption. 3D materials have also been applied in the adsorption of metal contaminants from polluted water [73]. This work was further extended by Li *et al.* [63] who revealed that polypyrrole–rGO nanosheets could be used in the removal of Cr(VI) from aqueous solutions. Sacrificial MnO_2 nanoslices served as the composite material, with the polypyrrole retaining the morphology of the MnO_2 nanostructure. As with 3D graphene materials, these polypyrrole materials show an increased adsorption capacity due to a change in the structure of composite material. Vasudevan and Lakshmi [64] proved that graphene can be applied for the effective removal of phosphate ions from aqueous solutions. In this study, graphene was prepared by liquid phase exfoliation of graphite. This process establishes functional groups onto the graphene surfaces, which are vital for the phosphate adsorption. The authors concluded that this adsorption is spontaneous and chemically irreversible, with a maximum adsorption capacity of 89.4 mg g^{-1}. Correspondingly, Li *et al.* [65] have shown that exfoliated graphene can be used to adsorb fluoride ions (monolayer adsorption capacity 35.6 mg g^{-1} at pH 7 and 298 K) from the solution. The results reveal that it cannot be effectively employed in industrial settings even though it has proven to be an effective adsorbent due to its irreversibility.

Thus, GO membranes have been effectively employed for the removal of heavy and toxic metal ions. However, self-association and van der wall interaction between GO layers inhibit their use in potential application [66]. To overcome all the drawbacks associated with native Ch and GO, chitosan–graphene-supported nanocomposite materials were synthesized using different chemical routes. The present chapter deals with the synthesis of chitosan–graphene-grafted nanocomposite materials and their application for the removal of metal ions and dyes from wastewaters.

5.3 Removal and Recovery of Environmental Pollutants

5.3.1 Removal of Heavy Metals

Graphene oxide/chitosan (GO-Ch)-supported nanocomposite was fabricated for the removal of Cr(VI) ions [66]. The composite material was synthesized by the reaction of GO and Ch modified with disodium ethylene-diaminetetraacetate (EDTA-2Na) (GCE). In terms of adsorption capacity, GO-Ch-modified showed much more adsorption efficiency: 86.17 mg/g for Cr(VI) at pH 2 compared to unmodified Ch and established equilibrium in a short time (90 min). Interestingly, adsorption kinetics study covered a pseudo-second-order model together with intraparticle diffusion and film diffusion models. In between the Langmuir and Freundlich isotherms, the adsorption process was found to be best fitted for the Freundlich isotherm and demonstrated high value for the Freundlich constant (K_F) which increased with increasing temperature leading to higher adsorption. Thermodynamics studies were spontaneous and highly endothermic nature at high temperature for the adsorption of Cr(VI) on GEC. The re-generation study confirmed the reusability of the GEC nanocomposite, which showed only 5% reduction in the adsorption efficiency after seven cycles. Thus, a higher adsorption capacity along with good re-generability of GEC makes the GEC a potential material for the removal of Cr(VI) ions in contaminated water. A comparative study regarding the adsorption behavior of GEC for the removal of Cr(VI) with other reported adsorbent is shown in Table 5.2.

A novel nanocomposite was synthesized by the implementation of amino-functionalized magnetic composite, $CoFe_2O_4$/chitosan/graphene [75]. Various characterization techniques including FTIR, XRD, and BET were employed to investigate morphology and chemical makeup of the composite. In addition, the effect of adsorption dose, pH, and contact time was determined to examine adsorption behavior of the composite. The optimal adsorption capacity of Ch-GO was found to be 361.0 mg/g at 323 K at neutral pH (7.0). Adsorption kinetics was established by the pseudo-first-order equation with a higher value of R^2 (0.9976) as compared to pseudo-second-order kinetics. Adsorption isotherms were also studied at different temperatures, which confirmed the Langmuir isotherm model. The thermodynamic study revealed a spontaneous and endothermic nature of the adsorption process [76]. The synthesized nanocomposite of $CoFe_2O_4$/chitosan/graphene successfully reduced Hg(II) ions from the wastewater sample.

Table 5.2 A comparative study for adsorption capacity of Cr (VI) on different chitosan-grafted adsorbents.

Composite material	Adsorption capacity q_{max}(mg g^{-1})	Reference
GCE	86.17	[66]
Ch-bent composite	89.13	[67]
CD-E-MGO	68.41	[68]
Ch-morillonite	58.82	[69]
Ch-PVA	52.88	[70]
EDTA-Ch	51.81	[71]
Ash-Ch	36.22	[72]
$MnFe_2O_4$-Ch	35.2	[73]
Ch-cellulose	13.05	[74]

To improve the quality of wastewater, a hybrid of reduced graphene oxide (RGO)/chitosan/lysine (Ch) with PLL was developed using electropolymerization and cyclic voltammetry [77]. The pH of the solution plays an important role for the adsorption of metal ions in optimum yield; therefore, the optimal pH-4.5 was chosen due to protonation of hydrophilic groups that result in reduction in the metal ion adsorption. The synthesized electrode was effectively applied for the removal and recovery of Cd(II): 96.0–103.33%, Pb(II): 93.33–97.6%, and Cu(II): 96.66–101.33% in a tap water sample.

Triehtylenetetramine-based graphene oxide and chitosan (TGO-Ch) composites were developed by microwave irradiation and conventional heating method [78]. The adsorption of Cr(VI) was compared using both the above-mentioned methods. However, microwave irradiation resulted in higher yield. Optimal capacity of Cr(VI) was achieved (219.5 mg g^{-1}) at pH 2 with 20 mg RGO-Ch (303 K). The adsorption kinetics followed pseudo-second-order kinetics and quick equilibrium adsorption (85%) was achieved in 20 min. A thermodynamic study established an endothermic and spontaneous process. The recycling study was performed to reuse the exhausted material after the adsorption process. The solvents, HCl and NaOH, were employed to desorb adsorbed Cr(VI) on TGO-Ch. The material showed 92.25% and 82.82% removal efficiency by using NaOH (1 mol L^{-1}) after three cycles; however, 80% removal capacity was archived after five cycles which confirmed the effective removal of Cr(VI) in waste water.

Another chitosan coating attapulgite (Ch-ATP) has been fabricated by self-assembly methodology and applied for the treatment of U(VI) in waste solutions [79]. In comparison to ATP and pure chitosan, modified Ch-ATP showed improved adsorption capacity of U(VI). Batch adsorption study and spectroscopic analyses confirmed the formation of inner-sphere complexes of U(VI) on the surface of Ch-ATP owing to the bonding of newly generated functional groups (i.e., -OH and -COOH). Under acidic conditions, the composite material revealed improvement in the adsorption of U(VI) in the presence of humic substances (HS); however, the alkaline medium restricted its adsorption capacity (desorption). The coordination of U(VI) with O^{2-} and –OH groups of Ch-ATP was confirmed by XPS analysis. It was observed that after sorption of uranium, the existence of specific band at ~382 eV verified high sorption capacity of Ch-ATP. Based on good adsorption capacity, eco-friendly and cost-efficient Ch-ATP can be effectively applied for the pre-concentration of U(VI) in contaminated water and nuclear reactor discharges.

Chandra *et al.* have synthesized magnetite-reduced graphene (MRGO) nanocomposites for the removal of arsenic in wastewater [80].

Morphological investigation of the composite was done by FTIR and XRD techniques. The magnetic MRGO demonstrated superior paramagnetic behavior at room temperature and might be separated by an external magnetic flux. These composites show high binding capability toward As(III) and As(V) due to the formation of active or sorption sites in the matrix of reduced graphene. The composite material showed 99.9% arsenic removal efficiency in water.

Mercapto-grafted graphene oxide–magnetic chitosan (MC-GOCh) bio absorbent has been formulated for the pre-concentration and extraction of Hg(II) using a facile eco-friendly method [81]. Cold vapor atomic absorption spectroscopy was used for the detection of Hg(II). Maximum adsorption of mercury was achieved under optimal conditions of sorbent dose of 60 mg at pH 6.5 and surface assimilation time of 10 min by using 3 mL HCl (0.1M) and thiourea (2% w/v) for a breakthrough curve. An exceptional linear curve was obtained in 0.12–80 ng mL^{-1} range ($R^2 = 0.999$) together with a pre-concentration factor (80). The limit of detection (LOD) and limit of quantification (LOQ) were found to be 6.0×10^{-2} ng mL^{-1} and 12.0×10^{-2} ng mL^{-1}, respectively. The value of relative standard deviation (4.7%) demonstrated good repeatability together with good recoveries (95 to 100%) of metal ions Hg(II) in real water.

A porous hybrid chitosan–gelatin/graphene oxide (ChG-GO) nanocomposite has been used for the removal of metal ions with significant recoveries (97%) [82]. The material was prepared by unidirectional freeze-drying method and characterized using advanced techniques. Highly porous nature of ChG-GO was found owing to the development of new cavity in the matrix of a composite and the material showed selectivity for the detection of Cu(II) and Pb(II) ions. Physico-chemical parameters such as wet ability, absorption capacity, compressive strength, the effect of pH, concentration of GO, and metal ion along with the effect of EDTA (ethylenediaminetetraacetic acid) were determined for the optimal absorption of metal ions on the composite material. The improvement in absorption efficiency of ChG-GO was achieved (1700% to 2700%) by the enhancement in the content of GO (1% by wt.) in the matrix of monolith which increased porosity and surface area and resulted in higher adsorption capacity. The compressive strength composite monoliths increases with enhancing the content of GO in both states (dry and wet), which proved a strong interaction between the polymer matrix of chitosan and GO sheets. The composite material containing 5% GO showed higher (2.5 times) compressive strength as compared to 1% GO by weight. The adsorption efficiency of water increases due to the improvement in functional groups (–OH, –COOH) on the surface of ChG-GO composite and results in higher adsorption of metal ions. The adsorption efficiency (Q_s) of Pb(II) increases (68.5 to 99 mg/g) with the increasing content of GO (5 wt. %) in the ChG matrix. Moreover, 20% to 88% increment in the adsorption capacity of Pb(II) was found by the addition of EDTA. The ChG-GO nanocomposite showed 80% adsorption capacity after five desorption cycles which confirmed its ability for the treatment of metal ions.

Li *et al.* fabricated graphene-supported cyclodextrin-chitosan/graphene oxide nanocomposite (cycledex-Ch-GO) using a facile chemical route. The synthesized grafted nanocomposite has been used for the adsorption of Cr(VI) in an aqueous solution [83]. The magnetic property of cyclodex-Ch-GO was found due to the reaction of Fe_3O_4 with cyclodex-Ch-GO. Based on a high surface area, existence of –OH and $-NH_2$ groups, the material has high potential to extract Cr(VI) from water under the influence of a magnetic field. The adsorption of Cr (VI) ions was followed by the Langmuir isotherm together with a maximum removal capacity of Cr(VI) at low pH (6.8) while chemical kinetics was followed by pseudo-second-order kinetics. High performance of cyclodex-Ch-GO was due to its high surface area which caused optimal removal of Cr(VI) in wastewater.

An interesting work related to the removal of Cd(II), methylene blue (MB), and orange G (OG) using magnetic graphene oxide (MGO) nanocomposite has been reported by Deng *et al.* [84]. The pseudo-second-order model and the Langmuir isotherm model were found to be best fitted for maximum adsorption of Cd(II) (91.2 mg g^{-1}), MB (64.23 mg g^{-1}), and OG (20.85 mg g^{-1}), respectively. The synthesized nanocomposites were also utilized for the removal of dyes from real water samples. It has been observed that in tap water samples, the natural process capability of Cd(II) was 65.39% from ultrapure water. However, the water samples had very little interference with the performance of MGO toward MB and OG, indicating that MGO was appropriated for the removal of ionic dyes as well as Cd (II) from real water. On increasing the concentration of MB, the adsorption capacity of Cd(II) was found to decrease. The GO showed a reversible trend (improvement in adsorption efficiency) in the binary systems of Cd(II)–MB and Cd(II)–OG, respectively. Therefore, there might be chances of interference of MB and OG with an MGO composite which demonstrated improved removal potential of dye molecules in water samples.

A hybrid of $Fe_3O_{4-}SiO_2$-chitosan/GO (MS-Ch-GO) nanocomposite has been synthesized for the treatment of tetracycline (TC) in aquatic solutions [85]. Various parameters such as the effect of zeta potential, concentration of containment, and pH effect were examined. A pH range, 6.0–7.0, demonstrated the highest adsorption efficiency due to the lower electrostatics repulsion. The effect of concentration of TC and different metal ions

(Na^+, K^+, Ca^{2+}, and Mg^2) revealed a maximum adsorption of 47.5 at pH 3.0 50.2 mmol kg^{-1} and 69.2 mmol kg^{-1} at pH 6.0 for TC, while in the absence of cations, the adsorption of TC onto MS-Ch-GO was found to be 50.2 mmol kg^{-1} at pH 3.0 to 67.6 mmol kg^{-1} at pH 6.0, respectively. The effect of Cu(II) concentration (0.1, 0.2 mM) was also examined to observe the change in the adsorption efficiency of TC on MS-Ch-GO. The improvement in the adsorption capacity of TC was found in the presence of Cu(II) on increasing the pH of the solution from 3.0 to 7.0. Moreover, at pH 7.0, the MS-ChGO composite showed the highest adsorption capacity for 0.1 mM and 0.2 mM Cu(II) as 173.60 mmol kg^{-1} and 183.47 mmol kg^{-1}, respectively. Improvement in the adsorption of TC in the presence of Cu(II) might be due to the formation of a complex among Cu (II) between TC and MSCG (where Cu(II) acts as bridge). Pseudo-second-order kinetics and Freundlich isotherm models were best fitted for the optimal adsorption of TC and metal ions.

Graphene-based GO-Ch-FeOOH nanocomposite has been fabricated by the reaction of graphene oxide, chitosan, and FeS using the sol–gel chemical route [86]. The prepared material was investigated under various conditions of physico-chemical parameters, namely, sorbent mass, pH, temperature, shaking speed, and contact time. The synthesized material achieved better adsorption capacity toward the removal of Pb(II) which may be owing to the existence of new functional groups in the matrix of GO-Ch-FeOOH. The pH of solution also plays an important role for optimal adsorption of metal ions. The adsorption capacity of Pb(II) increased with increasing solution's pH (from 3 to 5.5). After a pH of 5.5, a decline in the adsorption efficiency was found which may be due to the protonation of amine group ($-NH_2$ to $-NH_3$) as a result of electrostatic repulsion between NH_3 and Pb(II). On the other hand, at higher pH, sufficient $-OH^-$ adsorbed on the surface of composite led to the decline in the adsorption capacity of Pb(II). A thermodynamic study established the applicability of the Langmuir adsorption isotherm (on the basis of higher R^2 value) compared to the Freundlich model. Thus, chitosan–graphene-based nanocomposites have a potential to treat wastewater for the removal of heavy metal ions. Chitosan-gelatin/GO (Ch-gel-GO) was used by Zhang *et al.* for the adsorption of heavy metals from aqueous solutions. In addition to this, Fan *et al.* [87] disclosed that a magnetic chitosan–rGO composite material can be exploited for the efficient removal of Cr(VI) from polluted waters.

5.3.2 Treatment of Organic Pollutant

The removal of fuchsine dye from contaminated waters using magnetic graphene nanocomposites was investigated by Wang *et al.* [88]. Results indicated high removal efficiency with fast adsorption time, rendering the composite a suitable material for purification of water. In recent studies, Fan *et al.* demonstrated that the adsorption capacity of methylene blue (MB) in polluted water can be enhanced by using a magnetic chitosan–rGO composite material [4]. Nguyen-Phan *et al.* reported that a 3D wormhole like mesoporous titanosilicate–rGO was used for the removal of methylene blue from waste water [89]. The effect of Ti dopant in an Si matrix was investigated which established that a high concentration of Ti (i.e., 20 mol%) enhanced the adsorption capacity of the composite. On the other hand, improvement in the concentration of Ti resulted in a negative effect on the photocatalytic dye degradation activity. Similar work was reported by Li *et al.* to fabricate a $Mg(OH)_2$–rGO composite by the simultaneous reduction of $MgCl_2$ and GO using NaOH [90]. $Mg(OH)_2$ helped rGO to maintain a mesoporous structure and high surface area which proved to be beneficial for the removal of MB from water. Magnetic Ni-rGO nanocomposites were fabricated by the simultaneous hydrazine reduction of GO and $NiCl_2$ [91]. A complete adsorption of organic dyes (RhB and MB) was achieved in 4 h. This work was repeated by Li *et al.*, showing that a Ni–rGO nanocomposite material can be successfully used in the removal of aromatic compounds (benzene, toluene, and dimethylbenzene) from aqueous and non-aqueous solutions.

Graphene-coated biochar has been employed for the removal of Poly aromatic hydrocarbon (PAH) [92]. As compared to native biochar, the adsorption capacity of biochar with graphene coating increased by 20 times. The P25–rGO composite material showed the potential to adsorb methylene blue from an aqueous solution by using a self-assembly process of P25 on the graphene sheets. An improvement in the adsorption capacity was achieved by using nanoribbons or nanotubes because of high surface area in nanosheets as compared to other nanostructures. Tifeng Jiao *et al.* evaluated the degradation capacity of methylene blue (MB) and Rhodamine B (RhB) from textile wastewater via GO-based composite hydrogels by self-assembly of Ch and GO in an *in situ* reduction approach [93]. The outcome of experiments indicated that prepared 3D GO-based hydrogels efficiently removed dyes and exhibited good photocatalytic performance for RhB and MB in accordance with the pseudo-second-order model. Authors concluded that chitosan molecule was incorporated to facilitate the gelation process of the GO sheets. For the removal of a pharmaceutical compound (dorzolamide) from biomedical synthetic wastewaters, a novel graphite oxide/poly(acrylic acid)-grafted chitosan nanocomposite (GO/Ch) was fabricated and used as a biosorbent [94]. GO-Ch showed a higher adsorption capacity in comparison with the parent materials (graphite oxide and poly(acrylic acid)-grafted chitosan). It implied that the reactive groups of GO and Ch interact with amino groups of dorzolamide and the abundance of carboxyl groups of GO-CSA composite was the main reason for enhancing its adsorption capacity.

A novel magnetic bioadsorbent composite made up of magnetic chitosan and graphene oxide (MChGO) was prepared by Fan *et al.* [4]. Optimization of dye was achieved at pH 3, whereas pH 12 was preferable for desorption. Kinetic experiments presented relatively "fast" adsorption together with the best fitting of modified pseudo-second-order equation. The equilibrium data were fitted to the Langmuir, Freundlich, and Langmuir–Freundlich models, with maximum adsorption capacities (391, 401, and 425 mg g^{-1}, respectively) at 25 °C, 45 °C, and 65 °C. In accordance with the thermodynamic theory, the positive values of $\Delta H°$ suggested endothermic nature of the adsorption while the negative values of $\Delta G°$ indicated high adsorption affinity for dye molecules. Synthesis of graphite oxide (GO) with magnetic chitosan (Ch) to form (GO-Ch) was also carried out to serve as an adsorbent for an industrial dye (Reactive Black 5) [95]. An effective synthesis method for the preparation of the MGCh nanocomposite material was conducted by Sheshmani *et al.* to treat Acid Orange 7 (AO7) [96]. Dye removal efficiency was highly dependent on medium pH, contact time, and adsorbent concentration. Data indicated that MGCh adsorbent has a good potential for the removal of dye in wastewater. Some important chitosan–graphene-grafted nanocomposites for water treatment are listed in Table 5.3.

5.4 Conclusion

This chapter reports facile synthesis of chitosan–graphene (Ch-GO)-supported nanocomposite materials using sol–gel chemical route and hydrothermal methods. These nanocomposites have been successfully employed for the removal of organic and inorganic pollutants. The Ch-GO-grafted material was selected on the basis of existence of different functional groups ($-NH_2$, –H, -OOH) in the matrix of a composite which provides a porous nanostructure via hydrogen bonding and other interactions. Morphological characterization confirmed the formation of a porous nanocomposite of Ch-GO. The data of metal ion adsorption and dyes removal efficiency established that Ch-GO-based nano-materials have showed good pollution-removal potential. Based on the significant finding of chitosan–graphene, it is expected that the Ch-GO-based nanocomposite has momentous prospective to open new ways for their synthesis and applications in wastewater treatment. The achievement of environmentally stable Ch-GO

Table 5.3 Chitosan–graphene-grafted nanocomposite for wastewater treatment.

S. no.	Nanocomposite material	Characterization techniques	Applications	References
1.	GO/Ch/EDTA	SEM, TEM, FTIR, XRD	Adsorption of Cr(VI)	[66]
2.	$CoFe_2O_4$/Ch/ Graphene	XRD, FTIR, BET	Removal of Hg(II)	[75]
3.	RGO-Chi/PLL	EIS, CV	Analysis of Cd(II), Pb(II), Cu(II)	[77]
4.	TGO/Ch	FTIR, BET, SEM, XRD	Removal of Cr(VI)	[78]
5.	Ch-ATP	SEM, XRD	Detection of U(VI)	[79]
6.	M-RGO	FTIR, XRD	Elimination of As(II), As(V)	[80]
7.	GO-MC	SEM, TEM, FTIR, XRD	Extraction of Hg(II)	[81]
8.	Ch-gelatin sheets/graphene oxide	XRD, TEM and SEM	Absorption of Cu(II), Pb(II)	[82]
9.	Cyclodextrin-Ch-GO	SEM, TEM, XRD	Removal of Cr(VI), Cr(III)	[83]
10.	MGO	SEM,TEM, XRD, XPS	Detection of Cd(II), (MB), and orange G (OG)	[84]
11.	Fe_3O_2-SiO_2-Ch-GO	SEM, TEM, XRD, FT-IR, XPS	Adsorption of TC and Na(I), K(I), Ca(II), Mg(II)	[85]

12.	GO-Ch-FeOOH	FTIR, XRD, SEM	Extraction of Pb(II)	[86]
13.	RGO-Ch-Ag	SEM, TEM, EDX	Removal of MB and Rhodamine B	[93]
14.	Ch-GO	SEM,FTIR, XRD	Adsorption of MB	[4]
15.	GO-Ch	SEM-EDX, FTIR XRD, XPS, DTA- DTG, VSM	Adsorption of RB5	[95]
16.	GO-Ch	SEM , FTIR, XRD, VSM,	Adsorbent of Acid Orange 7	[96]
17.	GO-Ch	BET, SEM, TEM, XRD and FTIR	Removal of Cr(VI)	[97]

nanocomposite for a large-scale industrial application still remains challenging; nonetheless, the ongoing research to optimize the adsorption capacity of nanocomposite materials will be able to achieve its goal in foreseeable future.

Acknowledgment

The authors express their appreciation to Science and Engineering Research Board (SERB, DST, Govt. of India) Fast Tract Young Scientist Scheme (SB/FT/CS-122/2014) for providing Postdoctoral Fellowship to Mohammad Shahadat.

References

1. Huber, P., Carré, B., Decolorization of process waters in deinking mills and similar applications: A review. *BioResources*, 7, 1366, 2012.
2. Elizalde-González, M.P., Hernández-Montoya, V., Removal of acid orange 7 by guava seed carbon: A four parameter optimization study. *J. Hazard. Mater.*, 168, 515, 2009.
3. Xiong, Y., Strunk, P.J, Xia, H., Zhu, H., Karlsson, H.T., Treatment of dye wastewater containing acid orange II using a cell with three-phase three-dimensional electrode. *Water Res.*, 35, 4226, 2001.
4. Fan, L., Luo, C., Li, X., Lu, F., Qiu, H., Sun, M., Fabrication of novel magnetic chitosan grafted with graphene oxide to enhance adsorption properties for methyl blue. *J. Hazard. Mater.*, 215, 272, 2012.
5. Gumpu, M.B., Sethuraman, S., Krishnan, U.M., Rayappan, J.B.B., A review on detection of heavy metal ions in water–An electrochemical approach. *Sens. Actuators B.*, 213, 515, 2015.
6. Tianlik, T., Norulaini, N.A.R.N., Shahadat, M., Yoonsing, W., Omar, A.K.M., Risk assessment of metal contamination in soil and groundwater in Asia: a review of recent trends as well as existing environmental laws and regulations. *Pedosphere*, 26, 431, 2016.
7. Teh, T., Ab Rahman, N.N.N., Shahadat, M., Wong, Y., Syakir, M.I., Omar, A.M., A comparative study of metal contamination in soil using the borehole method. *Environ. Monit. Assess.*, 188, 1, 2016.
8. Cheah, I.K., Tang, R.M., Yew, T.S., Lim, K.H., Halliwell, B., Administration of pure ergothioneine to healthy human subjects: uptake metabolism and effects on biomarkers of oxidative damage and inflammation. *Antiox. Redox Signal.*, 26, 193, 2017.
9. Al Hamouz, O.C.S., Estatie, M.K., Morsy, M.A., Saleh, T.A., Lead ion removal by novel highly cross-linked Mannich based polymers. *J. Taiwan. Inst. Chem. Eng.*, 70, 345, 2017.

10. Hamlet, S.M., Ivanovski, S., Inflammatory cytokine response to titanium surface chemistry and topography. In: *The Immune Response to Implanted Materials and Devices*. Springer, 2017, pp. 151–167.
11. Siddiqui, S.I., Removal of arsenic from water through adsorption onto metal oxide-coated material. *Mater. Res. Found.*, 15.
12. Chaudhry, S.A., Ahmed, M., Siddiqui, S.I., Ahmed, S., Fe (III)–Sn (IV) mixed binary oxide-coated sand preparation and its use for the removal of As (III) and As (V) from water: application of isotherm, kinetic and thermodynamics. *J. Mol. Liq.*, 224, 431,2016.
13. Panhwar, A.H., Kazi, T.G., Afridi, H.I., Arain, S.A., Arain, M.S., Brahaman, K.D., Arain, S.S., Correlation of cadmium and aluminum in blood samples of kidney disorder patients with drinking water and tobacco smoking: related health risk, Environ. *Geochem. Health*, 38, 265, 2016.
14. Chakraborti, D., Rahman, M.M., Chatterjee, A., Das, D., Das, B., Nayak, B., Pal, A., Chowdhury, U.K., Ahmed, S., Biswas, B.K., Fate of over 480 million inhabitants living in arsenic and fluoride endemic Indian districts: magnitude health socio-economic effects and mitigation approaches. *J. Trace Elem. Med. Biol.*, 38, 33, 2016.
15. Satarug, S., Vesey, D.A., Gobe, G.C., Health risk assessment of dietary cadmium intake: Do current guidelines indicate how much is safe. *Environ. Health Perspect.*, 125, 2017.
16. Giri, S., Singh, A.K., Human health risk assessment due to dietary intake of heavy metals through rice in the mining areas of Singhbhum Copper Belt India. *Environ. Sci. Pollut. Res.*, 1, 2017.
17. Tulchinsky, T.H., Micronutrient deficiency conditions: global health issues. *Public Health Rev.*, 32 , 2017.
18. Gibb, H., O'Leary, K.G., Mercury exposure and health impacts among individuals in the artisanal and small-scale gold mining community: a comprehensive review. *Environ. Health Perspect.* (Online), 122, 2014.
19. González-Merizalde, M.V., Menezes-Filho, J.A., Cruz-Erazo, C.T., Bermeo-Flores, S.A., Sánchez-Castillo, M.O., Hernández-Bonilla, D., Mora, A., Manganese and mercury levels in wate sediments and children living near gold-mining areas of the Nangaritza River Basin Ecuadorian Amazon. *Arch. Environ. Contam. Toxicol.*, 71, 171, 2016.
20. Hseu, Z.Y., Lai, Y.J., Nickel accumulation in paddy rice on serpentine soils containing high geogenic nickel contents in Taiwan. *Environ. Geochem. Health.*, 1, 2017.
21. Bakar, N.F.A., Zin, N.M., Kechik, T.S.M.L.T., Santhanam, J., Commercial herbal slimming products: evaluation of heavy metals and microorganism contamination at different batch production. *Jurnal Sains Kesihatan Malaysia (Malaysian Journal of Health Sciences)*, 15, 2017.
22. Donham, K.J., Thelin, A., Health effects of agricultural pesticides. In: *Agricultural Medicine: Rural Occupational and Environmental Health Safety and Prevention*. John Wiley & Sons, pp. 205, 2016.

23. Lavery, M.J., Stull, C., Wolf, R., Cutaneous signs of poisoning. *Emerg. Dermatol.* 314, 2017.
24. Andersson, D.I., Hughes, D., Microbiological effects of sublethal levels of antibiotics. *Nat. Rev. Microbiol.*, 12, 465, 2014.
25. Noman, E.A., Rahman, N.N., Shahadat, M., Nagao, H., Al-Karkhi, A.F., Al-Gheethi, A., Omar, A.K., Supercritical fluid CO_2 technique for destruction of pathogenic fungal spores in solid clinical wastes. *Clean-Soil Air Water*, 44, 1700, 2016.
26. Lah, T.N.T., Norulaini, N.A.R.N., Shahadat, M., Nagao, H., Hossain, M.S., Omar, M.S., Utilization of industrial waste for the production of cellulase by the cultivation of trichoderma via solid state fermentation. *Environ. Processes.*, 3, 803, 2016.
27. Aly, O., A new development of wastewater treatment unit for paint shop in vehicle industry. *Int. J. Sci. Basic Appl. Res. (IJSBAR)*, 21,15, 2015.
28. Oncel, M., Muhcu, A., Demirbas, E., Kobya, M., A comparative study of chemical precipitation and electrocoagulation for treatment of coal acid drainage wastewater. *J. Environ. Chem. Eng.*, 1, 989, 2015.
29. Luo, Y., Guo, W., Ngo, H.H., Nghiem, L.D., Hai, F.I., Zhang, J., Liang, S., Wang, X.C., A review on the occurrence of micropollutants in the aquatic environment and their fate and removal during wastewater treatment. *Sci. Total Environ.*, 473, 619, 2014.
30. Nabi, S., Shahadat, M., Bushra, R., Shalla, A., Ahmed, F., Development of composite ion-exchange adsorbent for pollutants removal from environmental wastes. *Chem. Eng. J.*, 165, 405, 2010.
31. Nabi, S.A., Shahadat, M., Shalla, A.H., Khan, A.M., Removal of heavy metals from synthetic mixture as well as pharmaceutical sample via cation exchange resin modified with rhodamine B: its thermodynamic and kinetic studies. *Clean-Soil Air Water*, 39 , 1120, 2011.
32. Rahman, N.N.N.A., Shahadat, M., Omar, F.M., Chew, A.W., Kadir, M.O.A., Dry Trichoderma biomass: biosorption behavior for the treatment of toxic heavy metal ions. *Desal. Water Treat.*, 57, 13106, 2016.
33. Ab Kadir, N.N., Shahadat, M., Ismail, S., Formulation study for softening of hard water using surfactant modified bentonite adsorbent coating. *Appl. Clay Sci.*, 137, 168, 2017.
34. Chaudhry, S.A., Zaidi, Z., Siddiqui, S.I., Isotherm kinetic and thermodynamics of arsenic adsorption onto iron-zirconium binary oxide-coated sand (IZBOCS): modelling and process optimization. *J. Mol. Liq.*, 229, 230, 2017.
35. Nabi, S.A., Bushra, R., Shahadat, M., Application of Zr (IV) tungstate for removal of metal ions from aqueous solutions. *Toxicol. Environ. Chem.*, 94, 468, 2012.
36. Azha, S.F., Shahadat, M., Ismail, S., Acrylic polymer emulsion supported bentonite clay coating for the analysis of industrial dye. *Dyes Pigm.*, 2017.
37. Shahadat, M., Teng, T.T., Rafatullah, M., Shaikh, Z., Sreekrishnan, T., Ali, S.W., Bacterial Bioflocculants: A review of Recent Advances and Perspectives. *Che. Eng. J.*, 2017.

38. Rupani, P.F., Embrandiri, A., Ibrahim, M.H., Shahadat, M., Hansen, S.B., Mansor, N.N.A., Bioremediation of palm industry wastes using vermicomposting technology: its environmental application as green fertilizer. *3 Biotech*, 7, 155 , 2017.
39. Bushra, R., Shahadat, M., Ahmad, A., Nabi, S., Umar, K., Oves, M., Raeissi, A., Muneer, M., Synthesis characterization antimicrobial activity and applications of polyanilineTi (IV) arsenophosphate adsorbent for the analysis of organic and inorganic pollutants. *J. Hazard. Mater.*, 264 ,481, 2014.
40. Shahadat, M., Shalla, A., Raeissi, A., Synthesis characterization and sorption behavior of a novel composite cation exchange adsorbent . *Ind. Eng. Chem. Res.*, 51, 15525, 2012.
41. Shahadat, M., Teng, T.T., Rafatullah, M., Arshad, M., Titanium-based nanocomposite materials: A review of recent advances and perspectives. *Colloids Surf.*, B 126, 121, 2015.
42. Nabi, S., Shahadat, M., Bushra, R., Shalla, S., Azam, A., Synthesis and characterization of nano-composite ion-exchanger; its adsorption behavior. *Colloids Surf.*, B 87, 122, 2011.
43. Papageorgiou, S.K., Katsaros, F., Kouvelos, E., Kanellopoulos, N., Prediction of binary adsorption isotherms of Cu2+, Cd2+ and Pb2+ on calcium alginate beads from single adsorption data. *J. Hazard. Mater.*, 162, 1347, 2009.
44. Kuang, S.P., Wang, Z.Z., Liu, J., Wu, Z.C., Preparation of triethylene-tetramine grafted magnetic chitosan for adsorption of Pb (II) ion from aqueous solutions. *J. Hazard. Mater.*, 260, 210, 2013.
45. Chatterjee, S., Woo, S.H., The removal of nitrate from aqueous solutions by chitosan hydrogel beads. *J. Hazard. Mater.*, 164 , 1012, 2009.
46. Kołodyńska, D., Chitosan as an effective low-cost sorbent of heavy metal complexes with the polyaspartic acid. *Chem. Eng. J.*, 173, 520, 2011.
47. Heidari, A., Younesi, A., Mehraban, Z., Heikkinen, H., Selective adsorption of Pb (II), Cd (II), and Ni (II) ions from aqueous solution using chitosan–MAA nanoparticles. *Int. J. Biol. Macromol.*, 61 , 251, 2013.
48. Zhang, Z., Luo, H., Jiang, X., Jiang, Z., Yang, C., Synthesis of reduced graphene oxide-montmorillonite nanocomposite and its application in hexavalent chromium removal from aqueous solutions. *RSC Adv.*, 5, 47408, 2015.
49. Fan, L., Luo, C., Sun, M., Li, X., Lu, F., Qiu, H., Preparation of novel magnetic chitosan/graphene oxide composite as effective adsorbents toward methylene blue. *Bioresour. Technol.*, 114, 703, 2012.
50. Zhao, G., Li, J., Ren, X., Chen, C., Wang, X., Few-layered graphene oxide nanosheets as superior sorbents for heavy metal ion pollution management. *Environ. Sci. Technol.*, 45, 10454, 2011.
51. Wu, W., Yang, Y., Zhou, H., Ye, T., Huang, Z., Liu, R., Kuang, Y., Highly efficient removal of Cu (II) from aqueous solution by using graphene oxide. *Water Air Soil Pollut.*, 224, 1372, 2013.
52. Tan, P., Sun, J., Hu, Y., Fang, Z., Bi, Q., Chen, Y., Cheng, J., Adsorption of Cu^{2+} Cd^{2+} and Ni 2+ from aqueous single metal solutions on graphene oxide membranes. *J. Hazard. Mater.*, 297, 251, 2015.

53. Yang, S.T., Chang, Y., Wang, H., Liu, G., Chen, S., Wang, Y., Liu, Y., Cao, A., Folding/aggregation of graphene oxide and its application in Cu 2+ removal. *J. Colloid Interface Sci.*, 351, 122, 2010.
54. Li, Z., Chen, F., Yuan, L., Liu, Y., Zhao, Y., Chai, Z., Shi, W., Uranium (VI) adsorption on graphene oxide nanosheets from aqueous solutions. *Chem. Eng. J.*, 210, 539, 2012.
55. Zhao, G., Wen, T., Yang, X., Yang, S., Liao, J., Hu, J., Shao, D., Wang, X., Preconcentration of U(VI) ions on few-layered graphene oxide nanosheets from aqueous solutions. *Dalton Trans.*, 41, 6182, 2012.
56. Zhang, Y.R., Shen, S.L., Wang, S.Q., Huang, J., Su, P., Wang, Q.R., Zhao, B.X., A dual function magnetic nanomaterial modified with lysine for removal of organic dyes from water solution. *Chem. Eng. J.*, 239, 250, 2014.
57. Zhu, H., Fu, Y., Jiang, R., Yao, J., Liu, L., Chen, Y., Xiao, L., Zeng, G., Preparation characterization and adsorption properties of chitosan modified magnetic graphitized multi-walled carbon nanotubes for highly effective removal of a carcinogenic dye from aqueous solution. *Appl. Surf. Sci.*, 285, 865, 2013.
58. Seema, H., Kemp, K.C., Chandra, V., Kim, K.S., Graphene–SnO_2 composites for highly efficient photocatalytic degradation of methylene blue under sunlight. *Nanotechnology*, 23, 355705, 2012.
59. Zhu, M., Chen, P., Liu, M., Graphene oxide enwrapped Ag/AgX (X = Br, Cl) nanocomposite as a highly efficient visible-light plasmonic photocatalyst. *ACS Nano*, 5, 4529, 2011.
60. Cong, H.P., Ren, X.C., Wang, P., Yu, S.H., Macroscopic multifunctional graphene-based hydrogels and aerogels by a metal ion induced self-assembly process. *ACS Nano*, 6, 2693, 2012.
61. Sreeprasad, T., Maliyekkal, S.M., Lisha, K., Pradeep, T., Reduced graphene oxide–metal/metal oxide composites facile synthesis and application in water purification. *J. Hazard. Mater.*, 186, 921, 2011.
62. Chandra, V., Kim, K.S., Highly selective adsorption of Hg^{2+} by a polypyrrole–reduced graphene oxide composite. *Chem. Comm.*, 47, 3942, 2011.
63. Li, S., Lu, X., Xue, Y., Lei, J., Zheng, T., Wang, C., Fabrication of polypyrrole/graphene oxide composite nanosheets and their applications for Cr (VI) removal in aqueous solution. *PLoS One*, 7, e43328, 2014.
64. Vasudevan, S., Lakshmi, J., The adsorption of phosphate by graphene from aqueous solution. *Rsc Adv.*, 2, 5234,2012.
65. Li, Y., Zhang, P., Du, Q., Peng, X., Liu, T., Wang, Z., Xia, Y., Zhang, W., Wang, K., Zhu, H., Adsorption of fluoride from aqueous solution by graphene. *J. Colloid Interface Sci.*, 363, 348, 2011.
66. Zhang, L., Luo, H., Liu, P., Fang, W., Geng, J., A novel modified graphene oxide/chitosan composite used as an adsorbent for Cr(VI) in aqueous solutions. *Int. J. Biol. Macromol.*, 87, 586, 2016.
67. Liu, Q., Yang, B., Zhang, L., Huang, R., Adsorptive removal of Cr (VI) from aqueous solutions by cross-linked chitosan/bentonite composite. *Korean J. Chem. Eng.*, 32, 1314, 2015.

68. Wang, F., Ge, M., Fibrous mat of chitosan/polyvinyl alcohol/containing cerium (III) for the removal of chromium (VI) from aqueous solution. *Tex. Res. J.*, 83, 628, 2013.
69. Adamczuk, A., Kołodyńska, D., Equilibrium thermodynamic and kinetic studies on removal of chromium copper zinc and arsenic from aqueous solutions onto fly ash coated by chitosan. *Chem. Eng. J.*, 274, 200, 2015.
70. Chen, D., Li, W., Wu, Y., Zhu, Q., Lu, Z., Du, G., Preparation and characterization of chitosan/montmorillonite magnetic microspheres and its application for the removal of Cr (VI),. *Chem. Eng. J.*, 221, 8, 2013.
71. Hu, X.-j., Wang, J.-s., Liu, Y.-g., Li, X., Zeng, G.m., Bao, Z.-1., Zeng, X-x., Chen, A.-w., Long, F., Adsorption of chromium (VI) by ethylenediamine-modified cross-linked magnetic chitosan resin: isotherms kinetics and thermodynamics. *J. Hazard. Mater.*, 185, 306, 2011.
72. Xiao, Y., Liang, H., Wang, Z., $MnFe_2O_4$/chitosan nanocomposites as a recyclable adsorbent for the removal of hexavalent chromium. *Mater. Res. Bull.*, 48 , 3910, 2013.
73. Gomes, E.G., de Sousa, A.F., Vasconcelos, P.H., Melo, D.Q., Diógenes, I.C., de Sousa, E.H., do Nascimento, R.F., San Gil, R.A., Longhinotti, E., Synthesis of bifunctional mesoporous silica spheres as potential adsorbent for ions in solution. *Chem. Eng. J.*, 214, 27, 2013.
74. Sun, X., Peng, B., Ji, Y., Chen, J., Li, D., Chitosan (chitin)/cellulose composite biosorbents prepared using ionic liquid for heavy metal ions adsorption. *AICHE J.*, 55, 2062, 2009.
75. Zhang, Y., Yan, T., Yan, L., Guo, X., Cui, L., Wei, Q., Du, B., Preparation of novel cobalt ferrite/chitosan grafted with graphene composite as effective adsorbents for mercury ions. *J. Mol. Liq.*, 198, 381, 2014.
76. Li, Y.-H., Di, Z., Ding, J., Wu, D., Luan, Z., Zhu, Y., Adsorption thermodynamic kinetic and desorption studies of Pb2+ on carbon nanotubes. *Water Res.*, 39, 605, 2005.
77. Guo, Z., Luo, X.-K., Li, Y.-h., Zhao, Q.N., Li, M.-m., Zhao, Y.T., Sun, T.S., Ma, C., Simultaneous determination of trace Cd (II), Pb (II) and Cu (II) by differential pulse anodic stripping voltammetry using a reduced graphene oxide-chitosan/poly-l-lysine nanocomposite modified glassy carbon electrode. *J. Colloid Interface Sci.*, 490, 11,2017.
78. Ge, H., Ma, Z., Microwave preparation of triethylenetetramine modified graphene oxide/chitosan composite for adsorption of Cr (VI). *Carbohydr. Polym.*, 131, 280, 2015.
79. Pan, D., Fan, Q., Fan, F., Tang, Y., Zhang, Y., Wu, W., Removal of uranium contaminant from aqueous solution by chitosan@ attapulgite. *Sep. Purif. Technol.*, 177, 86, 2017.
80. Chandra, V., Park, J., Chun, Y., Lee, J.W., Hwang, I.-C., Kim, K.S., Water-dispersible magnetite-reduced graphene oxide composites for arsenic removal. *ACS Nano*, 4, 3979, 2010.

81. Ziaei, E., Mehdinia, A., Jabbari, A., A novel hierarchical nanobiocomposite of graphene oxide–magnetic chitosan grafted with mercapto as a solid phase extraction sorbent for the determination of mercury ions in environmental water samples. *Anal. Chim. Acta*, 850, 49, 2014.
82. Zhang, N., Qiu, H., Si, Y., Wang, W., Gao, J., Fabrication of highly porous biodegradable monoliths strengthened by graphene oxide and their adsorption of metal ions. *Carbon* 49 , 827, 2011.
83. Li, L., Fan, L., Sun, M., Qiu, H., Li, X., Duan, H., Luo, C., Adsorbent for chromium removal based on graphene oxide functionalized with magnetic cyclodextrin–chitosan. *Colloids Surf.*, B 107, 76, 2013.
84. Deng, J.H., Zhang, X.-R., Zeng, G.-M., Gong, J.-L., Niu, Q.-Y., Liang, J., Simultaneous removal of Cd (II) and ionic dyes from aqueous solution using magnetic graphene oxide nanocomposite as an adsorbent. *Chem. Eng. J.*, 226, 189, 2013.
85. Huang, B., Liu, Y., Li, B., Liu, S., Zeng, G., Zeng, Z., Wang, X., Ning, Q., Zheng, B., Yang, C., Effect of Cu (II) ions on the enhancement of tetracycline adsorption by Fe 3 O 4@ SiO 2-Chitosan/graphene oxide nanocomposite. *Carbohydr. Polym.*, 157, 576, 2017.
86. Sheshmani, S., Nematzadeh, M.A., Shokrollahzadeh, S., Ashori, A., Preparation of graphene oxide/chitosan/FeOOH nanocomposite for the removal of Pb(II) from aqueous solution. *Int. J. Biol. Macromol.*, 80, 475, 2015.
87. Fan, L., Luo, C., Sun, M., Qiu, H., Synthesis of graphene oxide decorated with magnetic cyclodextrin for fast chromium removal. *J. Mater. Chem.*, 22, 24577, 2012.
88. Wang, C., Feng, C., Gao, Y., Ma, X., Wu, Q., Wang, Z., Preparation of a graphene-based magnetic nanocomposite for the removal of an organic dye from aqueous solution. *Chem. Eng. J.*, 173, 92, 2011.
89. Nguyen-Phan, T.-D., Shin, E.W., Pham, V.H., Kweon, H., Kim, S., Kim, E.J., Chung, J.S., Mesoporous titanosilicate/reduced graphene oxide composites: layered structure high surface-to-volume ratio doping effect and application in dye removal from water. *J. Mater. Chem.*, 22, 20504, 2012.
90. Li, B., Cao, H., Yin, G., Mg (OH) 2@ reduced graphene oxide composite for removal of dyes from water. *J. Mater. Chem.*, 21, 13765, 2011.
91. Li, B., Cao, H., Yin, J., Wu, Y.A., Warner, J.H., Synthesis and separation of dyes via Ni@ reduced graphene oxide nanostructures. *J. Mater. Chem.*, 22, 1876, 2012.
92. Zhang, M., Gao, B., Yao, Y., Xue, Y., Inyang, M., Synthesis characterization and environmental implications of graphene-coated biochar. *Sci. Total Environ.*, 435, 567, 2012.
93. Jiao, T., Zhao, H., Zhou, J., Zhang, Q., Luo, X., Hu, J., Peng, Q., Yan, X., Self-assembly reduced graphene oxide nanosheet hydrogel fabrication by anchorage of chitosan/silver and its potential efficient application toward dye

degradation for wastewater treatments. *ACS Sustain. Chem. Eng.*, 3, 3130, 2015.

94. Kyzas, G.Z., Bikiaris, D.N., Recent modifications of chitosan for adsorption applications: A critical and systematic review. *Marine Drugs*, 13, 312, 2015.
95. Travlou, N.A., Kyzas, G.Z., Lazaridis, N.K., Deliyanni, E.A., Functionalization of graphite oxide with magnetic chitosan for the preparation of a nanocomposite dye adsorbent. *Langmuir*, 29, 1657, 2013.
96. Sheshmani, S., Ashori, A., Hasanzadeh, S., Removal of Acid Orange 7 from aqueous solution using magnetic graphene/chitosan: a promising nano-adsorbent. *Int. J. Biol. Macromol.*, 68, 218, 2014.
97. Debnath, S., Maity, A., Pillay, K., Magnetic chitosan–GO nanocomposite: synthesis characterization and batch adsorber design for Cr (VI) removal. *J. Environ. Chem. Eng.*, 2, 963, 2014.

6

Decolorization of Textile Wastewater Using Composite Materials

Sharf Ilahi Siddiqui[1], Rangnath Ravi[1], Geetanjali Rathi[1], Nusrat Tara[1], Shahid-ul-Islam[2] and Saif Ali Chaudhry[1,*]

[1]*Department of Chemistry, Jamia Millia Islamia, New Delhi, India*
[2]*Department of Textile Technology, IIT Delhi, New Delhi, India*

Abstract
Over the past few decades, large discharge of dyes from textiles industries in water bodies has been posing serious threat to environmental safety and society. Textile industries are the largest and fastest growing industries, therefore consuming a large amount of water and active ingredients such as dyes and coloration for the finishing of different fibers. These active ingredients enhance the chemical oxygen demand (COD), biological oxygen demand (BOD), and total organic compounds (TOC) of water and show toxic effects toward the human body. Therefore, reprocessing the polluted water is the only way to minimize the hazardous effects of these textile effluents. In this chapter, the issue of various dye discharges in water, their effect on human health, and advanced water treatment process using composite materials has been covered.

Keywords: Color water, dyes, remediation, adsorption, composites

6.1 Introduction

The major environmental problem currently is the generation of effluents (pollutants) from the textile industry, leather industry, food industry, pulp and paper industry, and paint industry. These industries use large quantities of water for their processing and produce wastewaters with different chemical compositions [1]. Color water may be the most problematic.

**Corresponding author*: saifchaudhary09@gmail.com

Shahid-ul-Islam and B.S. Butola (eds.) Nanomaterials in the Wet Processing of Textiles, (187–218) © 2018 Scrivener Publishing LLC

In general, textile industries largely discharge the color-containing water during the dyeing process. Asian countries, particularly India, have large textile industries that play a major role in their economics. More than 100,000 commercial dyes with over 7×10^5 tons of color stuff are produced and consumed every year. China is the largest exporter of color stuffs, whereas India is the second one. Worldwide, textile industries alone discharge more than approximately 10% of total discharged dyes [2, 3].

Color stuffs discharged in water enhance chemical oxygen demands (COD), biological oxygen demand (BOD), and total organic compounds (TOC) of water. These enhancements in BOD, COD, and TOC affect the aesthetic properties of water and are environmentally unacceptable. Moreover, these dyes may cover up the water surface, and reduce the amount of dissolved oxygen, and also reduce the penetration of light, which affect the photosynthetic activities of living creatures present in the water, thereby severely affecting the food source of aquatic organisms [3]. Dyes are non-biodegradable organic pollutants, and therefore, their degradation has become a major concern, which is not easy. Despite this, the removal

Table 6.1 Some selective dyes and their structure.

Dyes	Chromophore	Auxo-chrome	Structure
Methylene Blue	Thiazine 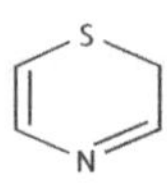	Amine $((CH_3)_2N–)$	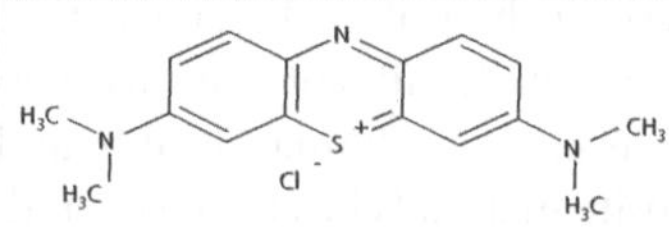
Congo Red	Azo (–N=N–)	Amine $(–NH_2)$ Sulfonate $(–SO_3^-)$	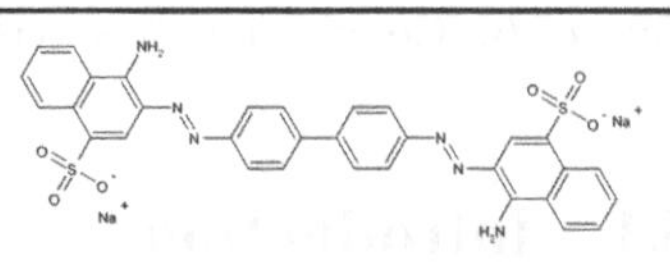
Rhodamine B	Pyrylium 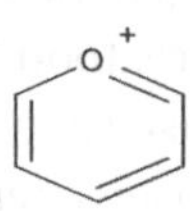	Amine $((CH_3)_2N–)$ Carboxyl (–COOH)	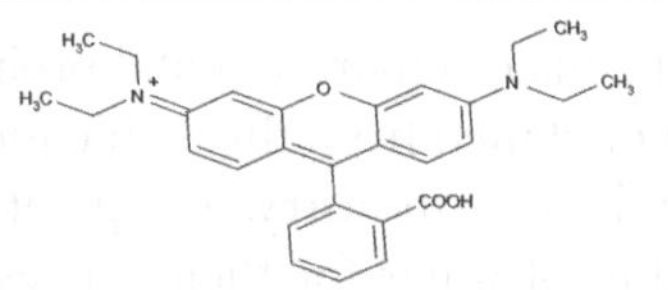
Methyl Violet	Aryl $(C_6H_4)^{2+}$	Carbonium center $(–{}^+C=)$	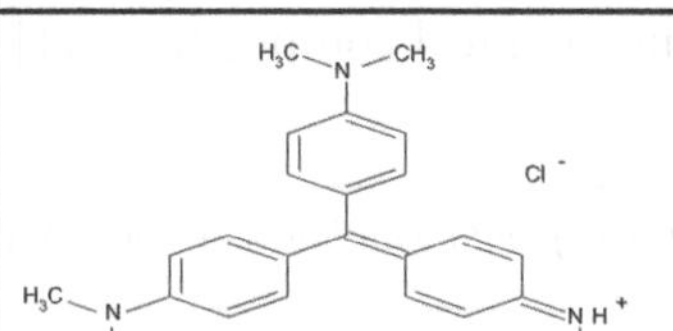

of color stuffs from water or effluents is of great concern. To achieve the momentum in dye removal, it is necessary to develop more suitable treatment technologies with respect to cost, feasibility, practicability, reliability, stability, environmental impact, sludge production, operational difficulty, pre-treatment requirements, the extent of the removal, and the potential toxic by-products. Complete removal of color stuffs from water is sometime associated with the degradation of dye molecules, but it may not be applicable to all types of color stuffs [4–6]. It is a challenge for researchers to develop the economically and environmentally effective treatment technology.

6.2 Classification of Dyes and Their Toxicity

Color stuffs, referred to as dyes, can absorb light in the visible range of the spectrum at a certain wavelength and has high affinity towards substrates on which applied. A large number of natural and synthetic dyes are being used in various industries [7, 8]. The food industry frequently uses natural dyes, while the textile industries use synthetic dyes. The chromophore groups such as –N=N–, =C=O, =C=C=, –C=NH, –CH=N–, NO, N–OH, $-NO_2$ or –NO–OH, and –C=S are responsible for the color of dyes [9, 10] (Table 6.1).

Dyes are classified as acidic, basic, direct, disperse, mordant, pigment, reactive, solvent, sulfur, and vat dyes (Figure 6.1). Acidic, direct, mordant, and reactive dyes are anionic; basic dyes are cationic; whereas pigments, disperse, and solvent dyes are non-ionic [11, 12] (Table 6.2).

Anionic dyes, having amine (NH_2), carboxy (COO^-), hydroxy (OH^-), and sulfonate (SO_3^-) groups, acquire net negative charge, while cationic dyes has net positive charge due to protonation of amine or sulfur-containing groups. These dyes are highly visible, even at very low concentration, and absorb and reflect the sunlight entering water, thus hindering photosynthesis [13].

Moreover, they show acute and chronic effects toward living organisms depending on their concentration and length of exposure. Anionic and cationic dyes are used for the coloring of natural and synthetics fibers. Dyes containing **–N=N–** groups are called azo dyes which have a wide application in textile, leather tanning, paper production, food, color photography, pharmaceuticals and medicine, cosmetic, hair coloring, wood staining, agricultural, biological, and chemical research, light-harvesting arrays, and photo electrochemical cells [14].

Consequently, a large number of synthetic dyes having azo groups are being released into the environment. Various synthetic azo dyes having low toxicity are being applied to various substances. These dyes usually

Table 6.2 Various types of dyes.

Dyes	Examples	Application of dyes
Acid dyes	Congo red Methyl (orange and red) Orange (I,II) Acid (blue, black, violet, yellow)	Wool, silk, nylon, polyamide, polyurethane, fibers
Direct dyes	Martius yellow Direct black Direct orange Direct blue Direct violet Direct red	Cotton, wool, flax silk, leather
Reactive dyes	Reactive red Reactive blue Reactive yellow Reactive black Remazol (blue, yellow, red, etc.)	Cellulosic fibers, wool, polyamide
Disperse dyes	Disperse blue Disperse red Disperse orange Disperse yellow Disperse brown	Polyamide, fibers, polyesters, nylon, polyacrylonitriles
Vat dyes	Indigo, Benzanthrone Vat blue Vat green	Wool, flax, wool rayon fibers
Basic dyes	Methylene blue Basic red Basic brown Basic blue Crystal violet Aniline yellow Brilliant green	Polyester ,wool, silk, mod-acrylic nylon

produce aromatic amines on anaerobic reduction of **–N=N–** groups which can be toxic and harmful [15].

Large exposure of dyes shows allergic, mutagenic, teratogenic, and carcinogenic effects toward the human body; moreover, it causes bladder cancer, splenic sarcomas, and hepato carcinomas [16–18]. Generally, complex

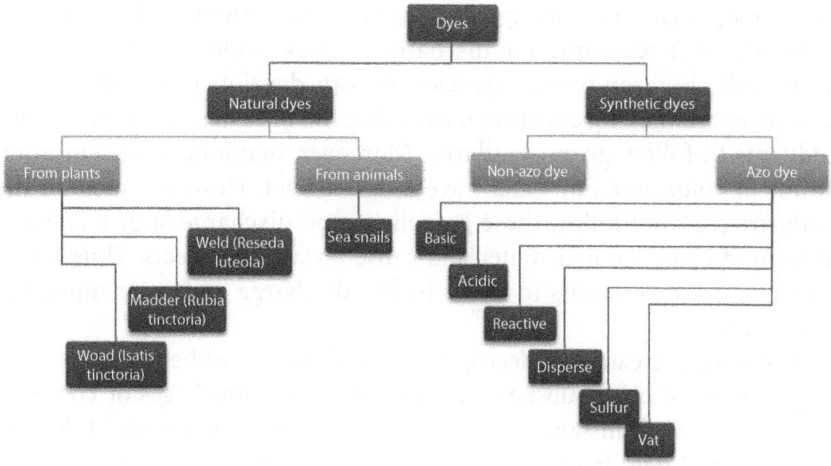

Figure 6.1 Classification of dyes.

dyes and basic dyes are carcinogenic due to the presence of metals and aromatic rings.

These dyes have a high intensity of color which decrease the sunlight penetration in water and affect the photosynthetic activity of aquatic life. Azo dyes have toxic amines groups; anthraquinone-based dyes do not degrade in water; and reactive dyes are highly soluble in water, which produce serious troubles in the environment. These colorants can also produce numerous dysfunctions in the kidneys, reproductive system, liver, brain, and central nervous system in the human body. Moreover, the release of dyes also alter the pH values to either extreme acidic or alkaline, and high temperatures, high BOD, COD, solid, oil, and possibly toxic organics that include phenols into water which may change the water quality and cause large physiological and biochemical changes in water. These changes in water quality show hazards toward aquatic plant and fishes [19, 20]. Therefore, focus on specific methods and technologies to remove dyes from different kinds of wastewater streams is desired.

6.3 Decolorization of Colored Water

The treatment of colored water, before discharge in water bodies, and reuse of dye-contaminated water after the treatment process have become an emerging issue for cleaning the environment. The removal of color from dye-contained wastewater is a subject of prime concern for scientist. Government agencies from various countries have become very stringent regarding the discharge of dyes as effluents from industrial units. Environmental agencies in both developed and developing countries are making efforts to ensure that the industries treat their color effluents to follow given standards. Moreover, maximum concentration limits of color stuffs in water have also been set. However, most of the industries do not follow these guidelines and discharge large a concentration of color stuffs in water and cause hazardous effects. Thus, treatment of these colorants in water before discharge and consumption is necessary.

Color water treatment (decolorization) is based on either photocatalytic degradation of color substance through the bond breakages or complete removal of color substance. Decolorization of water is the result of shifts of absorption of light from the visible to the ultraviolet or infrared region of the electromagnetic spectrum.

Various technologies such as nano-filtration, reverse osmosis, electrodialysis, coagulation, flocculation, electro-flotation, electro-kinetic coagulation, oxidation, irradiation, or electrochemical processes, and aerobic and anaerobic microbial degradation have been utilized for the decolorization of water [21–25] (Table 6.3). However, various disadvantages, such as high cost, low efficiency, limited versatility, need for specialized equipment, interference by other wastewater constituents, large production of sludge, and poor handling of the generated waste, are associated with these methods. Various technologies used in decolorization of colored water have been discussed in several reports [26]. The development of efficient, economic, and eco-friendly technologies for decolorization of the colored water is very important. Being a surface phenomenon, adsorption is being utilized as an eco-friendly and economical treatment process to reduce dye contents in the water system.

Table 6.3 Methods used for decolorization of colored water.

Physical	Chemical	Biological
Sedimentation	Neutralization	Stabilization
Filtration	Reduction	Aerated lagoons
Floatation	Oxidation	Trickling filters
Coagulation	Catalysis	Activated sludge
Reverse osmosis	Ion exchange	Fungal treatment
Solvent extraction	Electrolysis	Flocculation
Adsorption		Anaerobic digestion

6.4 Sorption Technology

The adsorption process is a surface phenomenon, and involves the interaction between the adsorbate (solute) and the adsorbent (solid surface). Small-sized particle, high surface area, large number of active sites, and magnetic character of adsorbent respond to the higher removal capacity of dyes [27].

This phenomenon is known to be simple, inexpensive, and efficient for the dye removal from water or wastewater. Moreover, it does not require the pre-treatment step before its application. Besides, sorption needs less working area, has lower sensitivity to diurnal variation, does not produce toxic chemicals, and is flexible in design and operation.

Adsorbent and dye interact either physically or chemically during the sorption process. Physical interactions involve weak bonding such as van der Waals, hydrogen, and dipole–dipole, while chemical interactions are characterized by strong inter-particle bonds, such as covalent, and ionic bonds, due to an exchange of electrons. Adsorption mechanism and their kinetics can be devised from the application of both linear and nonlinear isotherms to equilibrium data, and various kinetic relationships, respectively [28, 29]. Langmuir and Freundlich isotherms are most commonly used to quantify the amount of adsorbate adsorbed by an adsorbent.

The enhanced remediation of dye contents has been observed by using various solid supports. As already stated, various factors such as dye/sorbent interaction, sorbent surface area, particle size, temperature, pH, and contact time largely affect the sorption capacity of the solid surface [30].

The molecular size of dye and the number of groups attached to the dye are also responsible for adsorption fluctuation. Adsorbents having a large surface area, large active sites, and porous in nature may significantly remove dye molecules; however, the smaller size of dye particles may effectively be adsorbed than bigger molecules. This is due to diffusion limitations [31].

6.5 Recent Development in Adsorption Technology

Numerous adsorbents, such as activated carbon [32], biomasses [33], clay minerals [34], zeolites [35], waste materials [36], and metal-based adsorbents [37], have been used for the adsorptive removal of dyes. Metal-based adsorbents, particularly metal oxides, have expertise in effective removal of dyes [38].

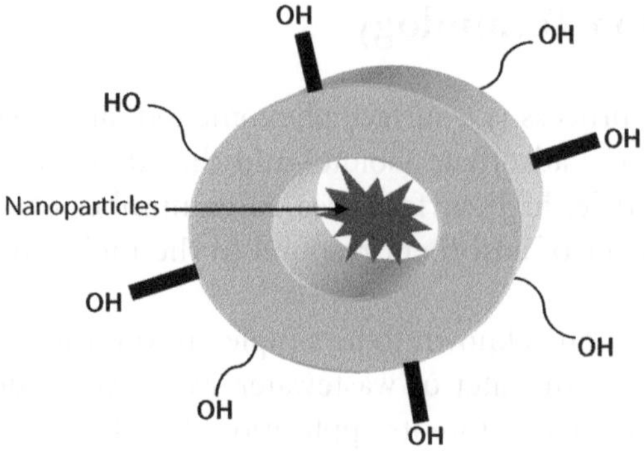

Figure 6.2 Unmodified adsorbent having hydroxyl groups.

The repeating cycle of the used adsorbent enhances its adsorption efficiency. Attempts have been made for the regeneration of exhausted adsorbents via desorption of dyes using an acid or a base [39, 40]. However, this step requires the separation of exhausted adsorbents from water. Filtration and centrifugation are adopted for the separation of exhausted adsorbents but now is avoided due to high cost processing. Surprisingly, newly developed magnetic adsorbents are having an advantage over filtration and centrifugation due to low cost, easy handling, and response to a low magnetic field.

In addition, it has been estimated that functionalized and stable magnetic nano-sized adsorbents can be most attractive and an advanced technique for water purification [41–43]. Functionalized adsorbents, having large hydroxyl or other function groups on their surface, can strongly trap the dye contents at various pH [44] (Figures 6.2 and 6.3). Therefore, a large number of eco-friendly, functionalized nano adsorbents have been developed with remarkable potential for dye remediation from water, waste water, and aqueous solution. Stabilized and functionalized adsorbents could be achieved by functionalization of nanoparticles through organic materials such as polymer and higher acids.

However, the drawback associated with functionalized adsorbents is the detachment of NPs from organic parts during the treatment process. This may cause toxicity on long-term accumulation of NPs in water. Therefore, various environmentally sound and stable adsorbents were developed and preferred for the dye remediation [45].

Recent development in advance water treatment technology is the production of flexible, thermally, and mechanically stable and magnetic multi-functional units, having a large number of special features for

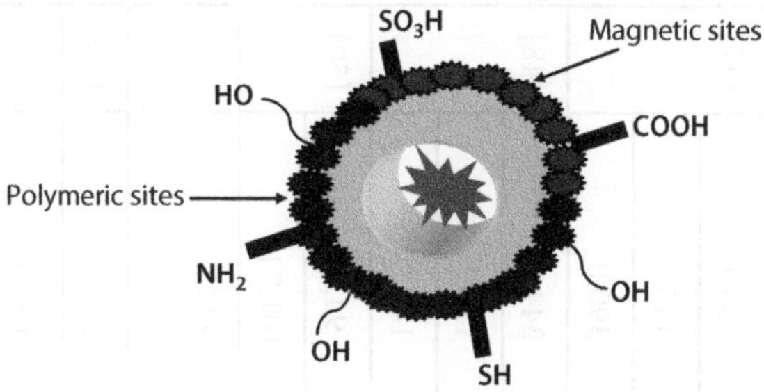

Figure 6.3 Modified adsorbent having other groups.

decolorization of colored water. These developed multifunctional units are referred to as composites, which could be achieved by certain size scale mixing of various materials such as inorganic particles and biopolymer, polymer, oxidizing, or reducing agents, and other carbonaceous materials through methods such as co-precipitation, sol–gel, etc. [46].

Recently, the mixing of inorganic and organic components at the nanometric scale provided the advancement in water purification techniques. These types of materials are referred to as organic–inorganic hybrid nanocomposite. Therefore, application of this sorption technique would be the nourishing step, under above-mentioned factors, to meet discharge color standards. For easing the way of young scientists for the search and development of more adsorbents, herein, we have listed (Table 6.4) the most latest available literature on recently used composites for dyes removal. This chapter will help the scientific community to improve the quality of the water treatment process.

6.6 Removal of Dyes Using Composites

Recently, Zn-Al double-layered hydroxides (ZnAl–LDHs-3) have been developed *via* hydrothermal method for the removal of three different types of dyes such as Congo red, Methyl orange, and Active red (X-3B) [47]. The maximum absorption capacity of ZnAl–LDO–3 was observed as 1540, 1153, and 390 mg/g for Congo red, Methyl Orange, and Active Red (X-3B), respectively. These adsorbents have been proved the most potential adsorbent for dye containing wastewater treatment. Long *et al.* [48] developed the magnetic composite, Fe_3O_4@catechol/PEI, by the co-polymerization of catechol and branched polyethylenimine (PEI) onto Fe_3O_4 nanoparticles under the alkaline condition. The prepared hybrid material was used for

Table 6.4 Various recently used adsorbents for dye remediation.

S.No.	Year	Adsorbent name	Specific dye	Removal efficiency (mg/g)	Ref.
1.	2017	ZnAl–LDHs	Congo red (CR)	1540	[47]
			Methyl orange (MO)	1153	
			Acid red (AR)	390.0	
2.	2017	Fe_3O_4@catechol/PEI	Methylene blue (MB)	344.8	[48]
			MO	192.3	
			Amaranth	146.2	
3.	2017	Modified gum tragacanth/graphene oxide	Crystal violate (CV)	94.0	[50]
			CR	101.74	
4.	2017	Cross-linked polyhedral oligomeric silsesquioxane (POSS) nanohybrid	MO	237.5	[52]
5.	2017	Peganum harmala-L Seeds	BG	18.65	[63]
6.	2017	ZnO/AC	OG	153.8	[59]
			Rhodamine-B (RhB)	128.2	
7.	2017	Hydrogen titanate nanosheets	MB	81.5	[64]
			RhB	52.9	

8.	2016	Carboxymethyl cellulose-grafted polymethacrylic acid CMC-g-PMAA	MB	671.0	[55]
9.	2017	MnFe2O4/diatomite nanocomposite	MB	104.06	[56]
			Basic Fuchsine (BF)	284.09	
10.	2017	Chitosan/SiO_2/carbon nanotubes nano-composite	Direct blue (DB)	61.35	[57]
			Reactive blue (RB19)	97.08	
11.	2017	Zinc (II) oxide nanorods loaded activated carbon ZnO-NRs-AC	MB	83.9	[60]
			CV	81.6	
12.	2016	MSA	MB	65.74	[62]
		HSA		47.21	
		MSA	RhB	34.25	
		HSA		185.61	
13.	2016	Modified iron stick yam skin with polyethyleneimine	Sunset yellow (SY)	476.31(SS)	[65]
				303.31(TS)	
			Lemon yellow (LY)	138.92(SS)	
				36.63(TS)	
			Carmine (CM)	500.13(SS)	
				294.12(TS)	

(Continued)

Table 6.4 Cont.

S.No.	Year	Adsorbent name	Specific dye	Removal efficiency (mg/g)	Ref.
14.	2016	ZnO nanorod-loaded activated carbon	Bromocresol Green (BG)	57.80	[66]
			Eosin Y (EY)	61.73	
15.	2015	Microgel based on nano-cellulose@PVA	CR	869.1	[67]
			AR	1469.7	
16.	2015	Chitin/graphene oxide hybrid composite	Remazol Black (RB)	9.3×10^{-2} mmol/g	[68]
			Neutral Red (NR)	57×10^{-2} mmol/g	
17.	2015	Sorel's cement nanoparticles	MO	23.21	[69]
			Bromophenol blue (BPB)	4.88	
18.	2017	Malachite@clay nanocomposite	MB	277.77	[70]
			CR	238.09	
19.	2016	Alkaline-treated timber sawdust	MB	694.44	[71]
			Methyl Green (MG)	1928.31	
20.	2015	Mixed silica–alumina oxide	Acid Orange, (A)7	41.4	[72]
			Reactive Black (RB)	47.1	
			DB	49.2	

21.	2015	Magnetic nickel zinc ferrite nanocomposite	Alizarin	250.0	[42]
			Reactive Blue/Janus Green B	333.3	
			CR/CV and Methyl Violet	500	
22.	2015	Mesoporous carbon nanospheres with empty cavity	Direct Red 23 (DR23)	769.2	[73]
			Direct Green 97 (DG97)	312.5	
			Basic Yellow 28 (BY28)	909.1	
23.	2014	Chitosan/PAA multilayer/magnetic microspheres	MB	305.8	[74]
			CV	243.6	
24.	2014	Tungstate oxide nanourchins	MB	150.3	[75]
			RhB	81.78	
25.	2017	Ultrathin MoSe2 Nanosheets	RhB	133	[76]
26.	2017	PDA/PEI@PVA/PEI NFMs	Ponceau s	1180	[77]
			MB	1290	
27.	2017	Mn-Fe_3O_4-NP-AC	BG	101.215	[78]
			MG	87.566	

(Continued)

Table 6.4 Cont.

S.No.	Year	Adsorbent name	Specific dye	Removal efficiency (mg/g)	Ref.
28.	2017	ZnS:Ni-NPs-ACATW	MB	21.79	[79]
			Janus Green B	28.01	
29.	2016	amine/Fe_3O_4-resin	Reactive Brilliant Red K-2BP (RBR)	101.0	[80]
			MO	222.2	
			AR	99.4	
30.	2017	$Cu(OH)_2$-NP-AC	EY	26.4	[81]
			MB	32.9	
			Phenol Red (PR),	38.5	
31.	2017	Mesoporous silica (meso-silica MCM-41) and that inorganically modified with CuO (CuO/MCM-41 nano composite)	CV	52.9	[82]
			MB	87.8	
		Mesoporous silica	CV	46.2	
			MB	65.7	
32.	2016	Magnesium hydroxide-coated bentonite, B–$Mg(OH)_2$composite	Procion blue HP (PB)	40.22	[83]
			Remazol brilliant blue (RB)	66.90	

33.	2017	Cellulose nanocrystal-reinforced keratin	Reactive Black 5	1201	[84]
			DR 80	1070	
34.	2017	Pectin beads	MB	1550.3	[85]
35.	2017	Pectin/cellulose microfibers beads	MB	2307.9	
36.	2015	Porous magnetic polyacrylamide microspheres	MB	1990	[86]
			Neutral Red (NR)	1937	
			Gentian Violet (GV)	1850	
37.	2015	Magnetic Fe/Ni nanoparticles doped bimodal meso-porous carbon (MBMC)	MB	959.5	[87]
			MO	849.3	
38.	2017	Chitosan/polyamide nanofibers (CP nanofibers)	Reactive Black 5 (RB5)	502.4	[88]
			Ponceau 4R (P4R)	456.9	
39.	2017	Fe_3O_4@MgAl–LDH	CR	813	[89]
40.	2016	γ-Fe_2O_3-NPs-AC	MB	195.55	[90]
			MG	207.04	
41.	2017	Fe_3O_4-MNPs-AC	SY	76.37	[91]
			MB	78.76	
			EB	102.00	

(Continued)

Table 6.4 Cont.

S.No.	Year	Adsorbent name	Specific dye	Removal efficiency (mg/g)	Ref.
42.	2016	Chitosan/bentonite hybrid composite	Amaranth Red (AR)	362.1	[92]
			MB	496.5	
43.	2016	Magnetite-Mg/Al hydrotalcite (M-Mg/Al HT)	Navy Blue (NB)/ Yellow F3G (YF3G)	63.75	[93]
		Mg/Al hydrotalcite (Mg/Al HT		15.67	
44.	2017	g-Fe_3O_4/C	MB	352.96	[94]

the removal of three different kinds of anionic dyes, Methyl Blue, Orange G, and Amaranth, from aqueous solutions. Adsorption data well fitted to the pseudo-second-order kinetic model and Langmuir isotherm, and the maximum monolayer adsorption capacities for Methylene Blue, Orange G, and Amaranth were 344.8, 192.3, and 146.2 mg/g, respectively. An advance feature of adsorbents is their separation on application of a low magnetic field at equilibrium.

Similarly, magnetic Fe@graphite nanocomposite (Fe@G-N) was also developed for the adsorptive removal of anionic dyes from aqueous solutions [49]. Two anionic dyes, Acid Red 88 (AR88) and Direct Orange 26 (DO26), were removed from aqueous solutions at different parameters, such as initial dye concentration (5–40 mg/L), pH of solution (4–10), and temperature (20–60 °C). The adsorption results obtained from batch experiments suggested that the kinetic of the reaction well fitted to the pseudo-second-order model, and the isotherm well followed both Langmuir and Freundlich adsorption models. Magnetic hydrogel beads, a modified gum tragacanth/graphene oxide, were used for the adsorptive removal crystal violet (CV) and Congo red (CR) from their solutions [50]. The observed result showed that adsorption kinetics was well addressed by the pseudo-second-order kinetic model and isotherm by the Langmuir isotherm model where the maximum monolayer adsorption capacity for CV and CR was 94.0, and 101.74 mg/g, respectively. In addition, the prepared adsorbent could be reutilized for three adsorption–desorption cycles without any significant loss of adsorption capacity.

Another magnetic composite, MCM-41, has been utilized for decolorization of colored water [51]. Composite materials were developed by the surface-initiated atom transfer radical polymerization (SI-ATRP) of (3-aminopropyl) triethoxysilane (APTES) and poly (glycidyl methacrylate) p(GMA), followed by the functionalization with tris(2-aminoethyl)amine (TAEA). The modified material was utilized as an adsorbent for cleaning of Direct Blue 6 (DB-6) and Direct Black-38 (DB-38) dyes from an aqueous medium. The magnetic character of the composite material was investigated by the VSM technique which showed the saturation magnetization at 19.6 emu/g. The adsorbent was found to be highly effective for the removal of both dyes at a wide pH range of 4.0–7.0, and the experimental data well fitted to the Langmuir isotherm and pseudo-second-order kinetic model. The magnetic composite was very stable, and could easily be separated and regenerated.

A cross-linked polyhedral oligosilsesquioxane (POSS)-based hybrid inorganic–organic nanohybrid having a specific surface area of 42.8 m^2/g was also fabricated for adsorptive removal of acidic dyes, Methyl orange (MO) [52]. The cross-linked adsorbent was fabricated through the Schiff base chemistry using octa-amino POSS as building blocks and glutaraldehyde as a dual functional reagent. The hybrid inorganic–organic nanohybrid was confirmed by FTIR spectroscopy. MO could effectively be removed from the aqueous solution with a maximum monolayer adsorption capacity of 237.5 mg/g. The reaction rate was well explained by the pseudo-second-order kinetic model and the reaction was spontaneous and exothermal. The specialty of the adsorbent was that it could effectively be regenerated using the methanol/NaOH (9:1) mixture.

Another cross-linked composite, cross-linked activated organo-bentonite (AO Bent)/sodium alginate (SA), has been developed for cationic and anionic dyes cleanup from aqueous solutions [53]. The AO Bent/sodium alginate (SA) composite was prepared by the intercalation of sodium alginate in activated organo-bentonite, and the cross-linking between sodium alginate and chlorohydric acid was confirmed from the X-ray diffraction (XRD), thermal gravimetric analyses (TGA), and Fourier transform infrared spectroscopy (FTIR). Furthermore, batch experiments were performed to remove cationic dye, Methylene blue, and anionic dye, Methyl orange, from their aqueous solutions. Adsorption data obtained from the batch experiment were well fitted to the pseudo-second-order kinetic model and Langmuir model. The prepared adsorbent was considered as cost effective and ecofriendly.

Similarly, a magnetically recoverable porous cross-linked polyethylenimine (MPEI) composite was developed for enhanced adsorptive removal of six anionic dyes [54]. Sorption mechanism and rates were well explained by the Langmuir monolayer adsorption and the pseudo-second-order kinetic model. The reaction was endothermic and spontaneous. Anionic dyes having the $-SO_3^-$ group were electrostatically adsorbed on to the positively charged MPEI, which was confirmed by the FT-IR and EDX spectra of dye-loaded MPEI. In addition, composite materials could easily be regenerated in an alkaline medium and utilized for a number of cycles. Salama *et al.* [55] fabricated carboxymethyl cellulose-grafted polymethacrylic acid (CMC-g-PMAA) having extraordinary adsorption behavior for Methylene blue, which exhibited extremely high removal capacity. The maximum monolayer adsorption capacity of the hybrid material for Methylene blue was found to be 671 mg/g. Reaction kinetics followed the pseudo-second-order model.

In the same year, Sun *et al.* [56] synthesized magnetic $MnFe_2O_4$/diatomite composite via the precipitation/hydrothermal method for cationic dyes such as methylene blue and basic fuchsine (rosaniline hydrochloride) dyes remediation. The adsorption process was fast and the maximal adsorption capacity for methylene blue and basic fuchsine obtained was 104.06 and 284.09 mg/g, respectively. The sorption process was recorded as an endothermic and spontaneous reaction.

Similarly, magnetic chitosan/SiO_2/CNTs (MNCSC) nano-composite was utilized for direct blue 71 (DB71), and Reactive blue 19 (RB19) cleaning from aqueous solutions under various parameters such as initial dye concentration, contact time, adsorbent dosage, initial pH, etc. [57]. The monolayer maximum adsorption capacity of MNCSC was achieved at 61.35 and 97.08 mg/g for DB71 and RB19, respectively. Three different kinds of kinetic models were applied but the pseudo-second-order equation well fitted the adsorption data. The sorptions of anionic dyes have also been investigated on two prepared adsorbents, isoreticular nanoporous Zn(II)-MOFs (TMU-16) and TMU-16-NH_2 [58]. Results demonstrated that TMU-16-NH_2 showed a higher removal capacity than TMU-16 due to hydrogen bonding between dyes and amine groups.

Low-cost and ecofriendly zinc oxide-loaded activated carbon (ZnO-AC) was also applied for the treatment of organic dyes such as Orange G (OG) and Rhodamine B (Rh-B) [59]. The optimized adsorption capacity of ZnO-AC for OG and Rh-B was 153.8 and 128.2 mg/g, respectively. The sorption reaction was endothermic up to a temperature of 40 °C; however, the removal of OG decreased with increased temperature. Similarly, Dil *et al.* [60] utilized zinc (II) oxide nanorod-loaded activated carbon (ZnO-NRs-AC) for the removal of azo dye from the aqueous solution and optimized their removal capacity under different parameters including initial concentration, adsorbent dosage, pH, temperature, ultrasonic time, etc. At an initial concentration of 18 and 14 mg/L of Methylene blue and Crystal violets, respectively, the maximum adsorption capacity was recorded as 83.9 and 81.6 mg/g for MB and CV dyes, respectively.

Recently, nano-magnetite Fe_3O_4 nanoparticles were impregnated onto the walnut shell (Fe_3O_4-WNS), and used for the removal of methyl violet and rRhodamine 6G dyes from their aqueous solutions [61]. Various models such as multiple linear regression, random forest, and artificial neural network were applied to predict the removal efficiency of magnetic adsorbent against these dyes. Han *et al.* [62] synthesized the hydrophobic surface-modified silica aerogel (MSA) and hydrophilic (hydroxyl-group) silica aerogel (HSA) which could remove Rhodamine B (RhB) and Methylene blue (MB) from aqueous solutions. The surface area of prepared MSA and HSA was 880.47 and 628.52 m^2/g, respectively, which was much higher than silica aerogel. Due to a large surface area, MSA and HSA removed 65.74 and 47.21 mg/g of Methylene blue, and 134.25 and 185.61 mg/g of

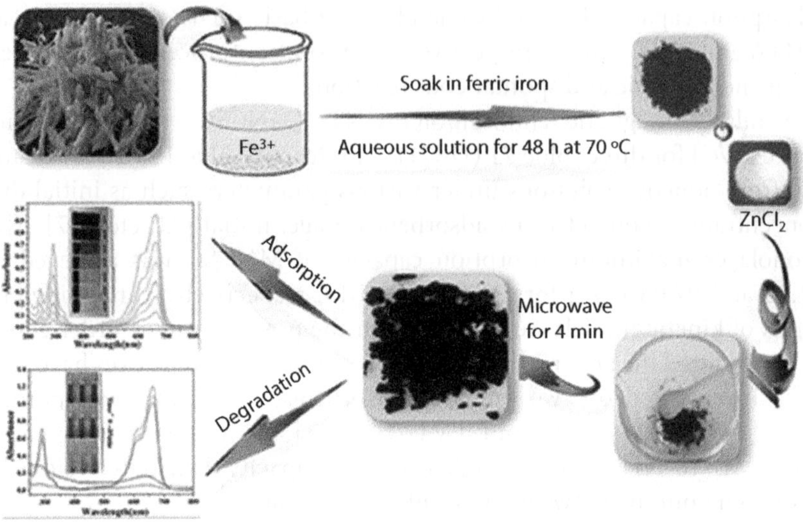

Figure 6.4 Synthesis of g-Fe_2O_3/C and their activity for dye removal and degradation (Adapted from Chen *et al.* [94] Copyright (2017), with permission from the Royal Society of Chemistry).

RhB, respectively. The reaction mechanism of monolayer and kinetics followed the pseudo-second order.

Recently, Fe^{3+} ions ($FeCl_3.6H_2O$) were introduced into bagasse biomass to achieve a novel magnetic porous adsorbent, g-Fe_2O_3/C, for the catalytic degradation and removal of methylene blue (MB) from the aqueous solution (Figure 6.4) [94]. The g-Fe_2O_3/C exhibited handsome adsorption and catalytic degradation abilities for MB. Similarly, quince seed mucilage (QSM)-based magnetic nanocomposite adsorbent (MNCA) has been developed by the *in situ* formation of magnetic iron oxide nanoparticles into the QSM solution for biosorption of MB from the aqueous solution (Figure 6.5a–c) [95]. Batch experiments were performed to optimize the adsorption capacity of MNCA for MB under the influence of various parameters. The process was spontaneous and exothermic following the pseudo-second-order kinetics. A Langmuir adsorption isotherm model was well fitted to the experimental data tested for MB, and the maximum monolayer adsorption capacity was 179 mg/g. These results indicate that the hybrid composite can be employed as efficient low-cost adsorbents with excellent dye adsorption performance in the wastewater treatment process. These were some recent advancements in the dye cleanup technology which made the process significant for dye remediation from water.

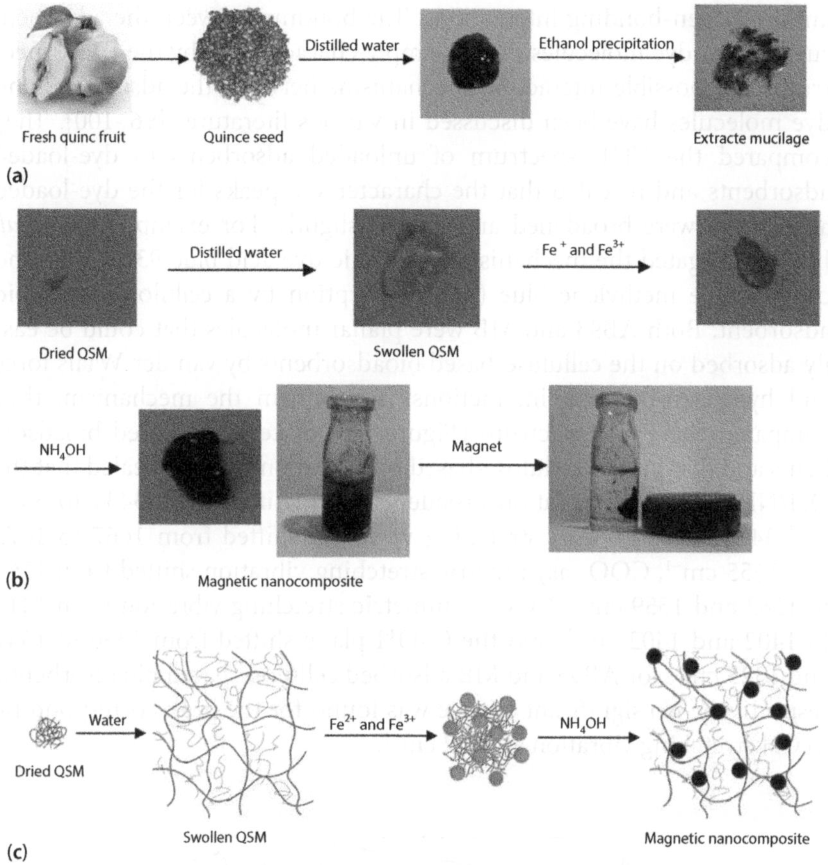

Figure 6.5 (a): Schematic illustration of the extraction of QSM (Adapted from Hosseinzadeh and Mohammadi [95] Copyright (2015), with permission from Elsevier). (b) Schematic illustration of the QSM-MIONs formation and magnetic separation of the nano-composites (Adapted from Hosseinzadeh and Mohammadi [95] copyright (2015), with permission from Elsevier). (c) Schematic illustration of the formation of QSM-based magnetic nanocomposites (Adapted from Hosseinzadeh and Mohammadi [95] copyright (2015), with permission from Elsevier).

6.7 Adsorption Mechanism

The structure and functional behavior of the dye molecule, surface characteristics of the adsorbents, mass transport process, and so on decide the adsorption capacity of adsorbents. Dyes are planar as well as complex molecules, which can be easily adsorbed on to the hybrid composite having a large number of oxygenous functional groups by electrostatic attraction and hydrogen-bonding interactions. The bonding between the adsorbent surface and dye molecules can be simply characterized by the FTIR spectrum. The possible interaction mechanisms between the adsorbents and dye molecules have been discussed in various literatures [96–100]. They compared the FTIR spectrum of unloaded adsorbents to dye-loaded adsorbents and revealed that the characteristic peaks for the dye-loaded adsorbents were broadened and shifted slightly. For example, Liu *et al.* [96] investigated the mechanism of anionic dye acid blue 93 (AB93) and cationic dye methylene blue (MB) adsorption by a cellulose-based bio adsorbent. Both AB93 and MB were planar molecules that could be easily adsorbed on the cellulose-based bioadsorbents by van der Waals force and hydrogen-bonding interactions. To confirm the mechanism, they compared the FTIR spectrum (Figure 6.6) of cellulose-based bioadsorbents and dye-loaded cellulose-based bioadsorbents, and revealed that the OH/NH stretching vibration broadened and shifted from 3434 to 3429 and 3424 cm^{-1}, CO-NH stretching vibration shifted from 1667 to 1672 and 1655 cm^{-1}, COO^- asymmetric stretching vibration shifted from 1564 to 1582 and 1559 cm^{-1}, COO^- symmetric stretching vibration from 1411 to 1402 and 1402 cm^{-1}, and the C–OH plane shifted from 1330 to 1347 and 1325 cm^{-1} for AB93 and MB adsorbed cellulose-based bioadsorbents, respectively. No significant change was found for the peak corresponding to CH stretching vibration at 2942 cm^{-1}.

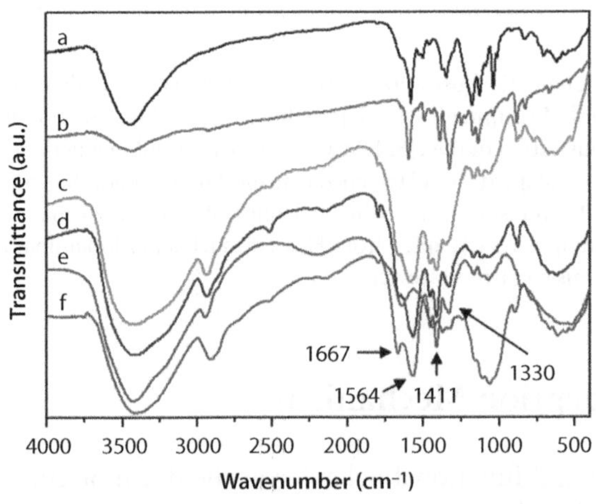

Figure 6.6 FTIR spectra of (a) AB93, (b) MB, (c) AB93 loaded cellulose-based bioadsorbent, (d) MB-loaded cellulose based bioadsorbent, (e) cellulose-based bioadsorbent, and (f) cellulose (Adapted from Liu *et al.* [96] copyright (2015), with permission from the American Chemical Society).

These shiftings in the characteristics peaks for cellulose-based bioadsorbents suggested the specific electrostatic and hydrogen bonding interactions between the functional groups of dye molecules and the adsorbent (Figure 6.7). A number of literature on the basis of FTIR spectrum studies are now available to understand the dye adsorption mechanism for composites which generally suggested that electrostatic and hydrogen-bonding interactions are responsible for dye adsorption on to the composite surface [97–100]. Besides the electrostatic attraction and hydrogen-bonding interactions, complexation between the metallic sites of the adsorbent and electron-donating sites of dyes or π–π interactions between the π electrons in a carbonaceous adsorbent such as graphene oxide and the π electron in the aromatic ring of a dyes are also responsible for the adsorption of dyes onto the adsorbent surfaces.

To understand the π–π interactions between π electrons in a carbonaceous adsorbent such as graphene oxide and the π electron in the aromatic

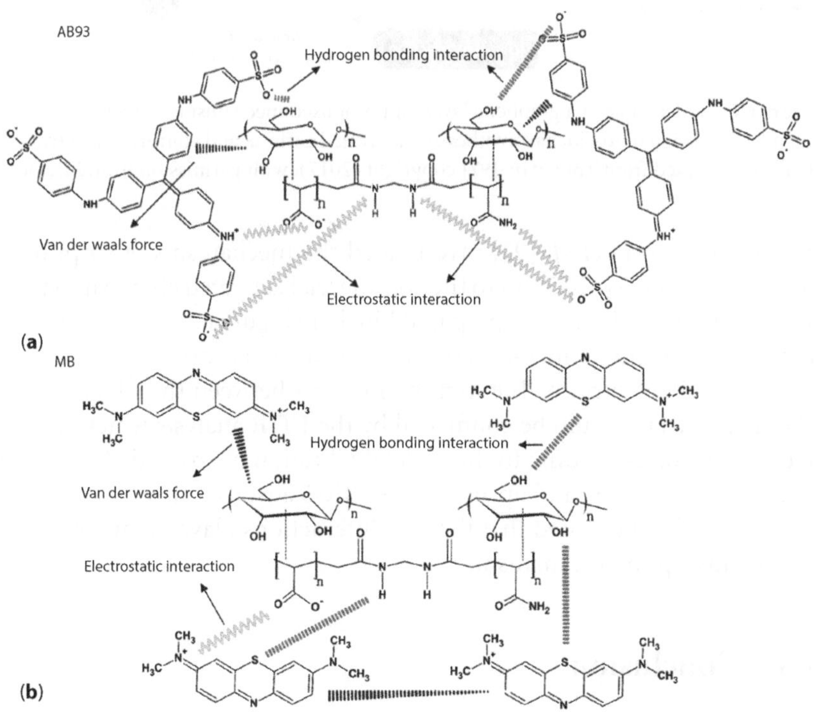

Figure 6.7 Schematic drawing for the possible interactions between the bioadsorbents and (a) AB93 and (b) MB dye molecules (Adapted from Liu et al. [96] Copyright (2015), with permission from the American Chemical Society).

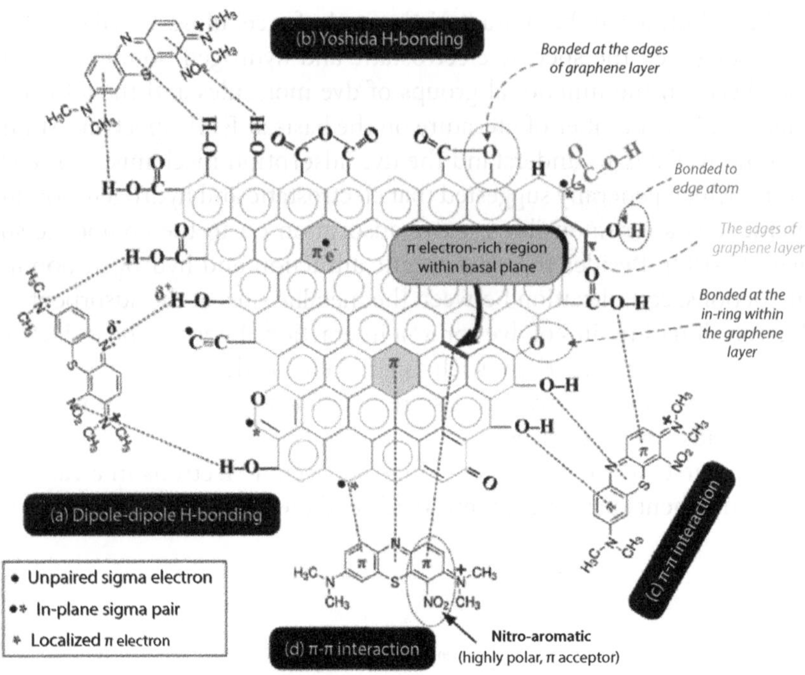

Figure 6.8 An example of a graphene layer and proposed mechanisms of methylene green 5 adsorption onto biochar, synthesized activated carbon, and commercial activated charcoal (Adapted from Tran *et al.* [97] copyright (2017), with permission from Elsevier).

ring of a dye, Tran *et al.* [97] investigated the mechanisms of sorption of methylene green 5 (MG5) on to the commercial activated charcoal, synthesized activated carbons, and prepared biochar (Figure 6.8). They suggested that π–π interactions majorly contributed to the adsorption of MG5 on to the adsorbent surface. The π–π interactions between the MG5 and the adsorbent surface could be confirmed by the FTIR analysis which showed that a peak corresponding to the skeletal vibration of aromatic C-C bonds decreased in intensity and upshifted after MG5 adsorption. Similarly, Chen *et al.* [101] also suggested that the π–π interactions played a major role in the dye adsorption mechanism.

6.8 Conclusion

In summary, we incorporated the study and features of various recently used, novel, inexpensive, magnetic, and some nonmagnetic adsorbents, which efficiently removed and degraded several dyes from aqueous solutions. Adsorbents having large adsorption capacities for several dyes are added herein. The obtained concluding remark is that the porous, high surface sites containing adsorbents such as organic–inorganic nanocomposites show excellent magnetism, superior adsorption capacity, and have excellent catalytic degradation ability for dyes. Thus, on including the list of adsorbents, and the study of inexpensive adsorbents, a more suitable and conventional method can be obtained at a certain interval of time that has great potential for the clean-up of environmental pollutants.

Acknowledgements

The Financial support from the University Grant Commission, UGC, India, and Department of Chemistry, Jamia Millia Islamia, New Delhi, India, is gratefully acknowledged.

References

1. Sarioglu, M., Atay, U.A., Removal of methylene blue by using biosolid. *Global NEST J.*, 8, 113–120, 2006.
2. Attallah, O.A., Al-Ghobashy, M.A., Nebsen, M., Salem, M.Y., Removal of cationic and anionic dyes from aqueous solution with magnetite/pectin and magnetite/silica/pectin hybrid nanocomposites: kinetic, isotherm and mechanism analysis. *RSC Adv.*, 6, 11461–11480, 2016.
3. Ehrampoush, M.H., Ghanizadeh, Gh., Ghaneian, M.T., Namasivayam, C., Kavitha, D., Equilibrium and kinetics study of reactive red 123 dye removal from aqueous solution by adsorption eggshell. *Iran. Environ. Health Sci. Eng.*, 8, 101–108, 2011.
4. Juang, R.S., Tseng, R.L., Wu, F.C., Lin, S.J., Use of chitin and chitosan in lobster shell wastes for color removal from aqueous solution. *J. Environ. Sci. Health, Part A*, 31, 325–338, 1996.
5. Koumanova, B., Peeva, P., Allen, S.J., Gallagher, K.A., Healy, M.G., Biosorption from aqueous solution by egg shell membranes and *Rhizopus oryzae*: Equilibrium and kinetic studies. *J. Chem. Technol. Biotechnol.*, 77, 539–545, 2002.
6. Mckay, G., Porter, J.F., Prasad, G.R., The removal of dye colours from aqueous solution by adsorption on low-cost materials. *Water, Air, Soil Pollut.*, 114, 438–438, 1999.
7. Namasivayam, C., Kanchana, N., Removal of Congo red from aqueous solution by waste banana pith. *J. Sci. Technol.*, 1, 33–42, 1993.
8. Yamjala, K., Nainar, M.S., Ramisetti, N.R., Methods for the analysis of azo dyes employed in food industry–A review. *Food Chem.*, 192, 813–824, 2016.

9. Ibarbia, L.P., Majdanski, T., Schubert, S., Windhab, N., Schubert, U.S., Safety and regulatory review of dyes commonly used as excipients in pharmaceutical and nutraceutical applications. *Eur. J. Pharm. Sci.*, 93, 264–273, 2016.
10. Khatri, A., Peerzada, M.H., Mohsin, M., White, M., A review on developments in dyeing cotton fabrics with reactive dyes for reducing effluent pollution. *J. Clean. Prod.*, 87, 50–57, 2015.
11. Tahir, M.A., Bhatti, H.N., Iqbal, M., Solar Red and Brittle Blue direct dyes adsorption onto Eucalyptus angophoroides bark: Equilibrium, kinetics and thermodynamic studies. *J. Environ. Chem. Eng.*, 4, 2431–2439, 2016.
12. Santhi, T., Manonmani, S., Vasantha, V.S., Chang, Y.T., A new alternative adsorbent for the removal of cationicdyes from aqueous solution. *Arab. J. Chem..*, 9, S466–S474, 2016.
13. Shamraiz, U., Hussain, R.A., Badshah, A., Raza, B., Saba, S., Adsorption of dyes onto palygorskite and its composites: A review. Functional metal sulfides and selenides for the removal of hazardousdyes from Water. *J. Photochem. Photobiol. B: Biol.*, 159, 33–41, 2016.
14. Mu, B., Wang, A., Adsorption of dyes onto palygorskite and its composites: a review. *J. Environ. Chem. Eng.*, 4, 1274–1294, 2016.
15. Khandare, R.V, Govindwar, S.P., Phytoremediation of textile dyes and effluents: Current scenario and future prospects. *Biotechnol. Adv.*, 33, 1697–1714, 2015.
16. Rawat, D., Mishra, V., Sharma, R.S., Detoxification of azo dyes in the context of environmental processes. *Chemosphere,* 155, 591–605, 2016.
17. Abe, F.R., Mendonca, J.N., Moraes, L.A., Oliveira, G.A., Gravato, C., Soares, A.M., Oliveira, D.P., Toxicological and behavioral responses as a tool to assess the effects of natural and synthetic dyes on zebrafish early life. *Chemosphere,* 178, 282–290, 2017.
18. Nesakumar, T., Edison, J.I., Atchudan, R., Sethuraman, M.G., Lee, Y.R., Reductive-degradation of carcinogenic azo dyes using Anacardium occidentale testa derived silver nanoparticles. *J. Photochem. Photobiol. B: Biol.*, 162, 604–610, 2016.
19. Gopinathan, R., Bhowal, A., Garlapati, C., Thermodynamic study of some basic dyes adsorption from aqueous solutions on activated carbon and new correlations. *J. Chem. Thermodyn.*, 107, 182–188, 2017.
20. Nguyen, T.A., Fu, C.C., Juang, R.S., Effective removal of sulfur dyes from water by biosorption and subsequent immobilized laccase degradation on cross-linked chitosan beads. *Chem. Eng. J.*, 304, 313–324, 2016.
21. Liu, M., Chen, Q., Lu, K., Huang, W., Lu, Z., Zhou, C., Yu, S., Gao, C., High efficient removal of dyes from aqueous solution through nanofiltration using diethanolamine-modified polyamide thin-film composite membrane. *Sep. Pur. Technol.*, 173, 135–143, 2017.
22. Velegraki, T., Poulios, I., Charalabaki, M., Kalogerakis, N., Samaras, P., Mantzavinos, D., Photocatalytic and sonolytic oxidation of acid orange 7 in aqueous solution. *Appl. Catal. B: Environ.*, 62, 159–168, 2006.

23. Wang, S., Wu, H., Environmental-benign utilisation of fly ash as low-cost adsorbents. *J. Hazard. Mater.*, 136, 482–501, 2006.
24. Zangeneh, H., Zinatizadeh, A.A.L., Habibi, M., Akia, M., Isa, H., Photocatalytic oxidation of organic dyes and pollutants in wastewater using different modified titanium dioxides: A comparative review. *J. Indus. Eng. Chem.*, 26, 1–36, 2015.
25. Zahrim, A.Y., Tizaoui, C., Hilal, N., Coagulation with polymers for nanofiltration pre-treatment of highly concentrated dyes: A review. *Desalination*, 266, 1–16, 2011.
26. Ahmad, A., Mohd-Setapar, S.H., Chuong, C.S., Khatoon, A., Wani, W.A., Kumar, R., Rafatullah, M., Recent advances in new generation dye removal technologies: novel search for approaches to reprocess wastewater. *RSC Adv.*, 5, 30801–30818, 2015.
27. Siddiqui, S.I., Chaudhry, S.A., Removal of arsenic from water through adsorption onto metal oxide-coated material. *Mater. Res. Found.*, 15, 227–276, 2017.
28. Chaudhry, S.A., Ahmed, M., Siddiqui, S.I., Ahmed, S., Fe (III)–Sn (IV) mixed binary oxide-coated sand preparation and its use for the removal of As (III) and As (V) from water: application of isotherm, kinetic and thermodynamics. *J. Mol. Liq.*, 224, 431–441, 2016.
29. Chaudhry, S.A., Zaidi, Z., Siddiqui, S.I., Isotherm, kinetic and thermodynamics of arsenic adsorption onto iron-zirconium binary oxide-coated sand (IZBOCS): modelling and process optimization. *J. Mol. Liq.*, 229, 230–240, 2017.
30. Shi, K., Zhitomirsky, I., Influence of chemical structure of dyes on capacitive dye removal from solutions. *Electrochim. Acta*, 174, 88–595, 2015.
31. Allen, S.J., Koumanova, B., Decolourisation of water/wastewater using adsorption (review). *J. Univ. Chem. Technol. Metallurgy*, 40, 175–192, 2005.
32. Mezohegyi, G., Van der Zee, F.P., Font, J., Fortuny, A., Fabregat, A., Towards advanced aqueous dye removal processes: A short review on the versatile role of activated carbon. *J. Environ. Manage.*, 102, 148–164, 2012.
33. Kabbout, R., Taha, S., Biodecolorization of textile dye effluent by biosorption on fungal biomass materials. *Phys.. Proc.*, 55, 437–444, 2014.
34. Rahman, A., Urabe, T., Kishimoto, N., Color removal of reactive procion dyes by clay adsorbents. *Proc. Environ. Sci.*, 17, 270–278, 2013.
35. Humelnicu, I., Băiceanu, A., Ignat, M.E., Dulman, V., The removal of Basic Blue 41 textile dye from aqueous solution by adsorption onto natural zeolitic tuff: Kinetics and thermodynamics. *Process. Saf. Environ. Prot.*, 105, 274–287, 2017.
36. Adegoke, K.A., Bello, O.S., Dye sequestration using agricultural wastes as adsorbents. *Water Resour. Ind.*, 12, 8–24, 2015.
37. Raman, C.D., Kanmani, S., Textile dye degradation using nano zero valent iron: A review. *J. Environ. Manage.*, 177, 341–355, 2016.

38. Chowdhury, A.N., Rahim, A., Ferdosi, Y.J., Azam, M.S., Hossain, M.M., Cobalt–nickel mixed oxide surface: A promising adsorbent for the removal of PR dye from water. *Appl. Surf. Sci.*, 256, 3718–3724, 2010.
39. Gupta, V.K., Kumar, R., Nayak, A., Saleh, T.A., Barakat, M.A., Adsorptive removal of dyes from aqueous solution onto carbon nano-tubes: A review. *Adv. Colloid Interface Sci.*, 193–194, 24–34, 2013.
40. Gupta, V.K, Suhas, Application of low-cost adsorbents for dye removal – A review. *J. Environ. Manage.* 90, 2313–2342, 2009.
41. Mahmoodi, N.M., Synthesis of core–shell magnetic adsorbent nanoparticle and selectivity analysis for binary system dye removal. *J. Indus. Eng. Chem.*, 20, 2050–2058, 2014.
42. Afkhami, A., Sayari, S., Moosavi, R., Madrakian, T., Magnetic nickel zinc ferrite nanocomposite as an efficient adsorbent for the removal of organic dyes from aqueous solutions. *J. Indus. Eng. Chem.*, 21, 920–924, 2015.
43. Mehta, D., Mazumdar, S., Singh, S.K., Magnetic adsorbents for the treatment of water/wastewater–A review. *J. Water Process. Eng.*, 7, 244–265, 2015.
44. Shamraiz, U., Hussain, R.A., Badshah, A., Raza, B., Saba, S., Functional metal sulfides and selenides for the removal of hazardous dyes from Water. *J. Photochem. Photobiol. B: Biol.*, 159, 33–41, 2016.
45. Siddiqui, S.I., Chaudhry, S.A., Arsenic Removal from Water Using Nanocomposites: A Review. *Curr. Environ. Eng.*, 4, 81–102, 2017.
46. Siddiqui, S.I., Chaudhry, S.A., Iron oxide and its modified forms as an adsorbent for arsenic removal: A comprehensive recent advancement. *Process Saf. Environ. Protect.*, 111, 592–626, 2017.
47. Huang, G., Sun, Y., Zhao, C., Zhao, Y., Song, Z., Chen, J., Ma, S., Du, J., Yin, Z., Water–n-BuOH solvothermal synthesis of ZnAl–LDHs with different morphologies and its calcined product in efficient dyes removal. *J. Colloid Interface Sci.*, 494, 215–222, 2017.
48. Long, Y., Xiao, L., Cao, Q., Co-polymerization of catechol and polyethylenimine on magnetic nanoparticles for efficient selective removal of anionic dyes from water. *Powder Technol.*, 310, 24–34, 2017.
49. Konicki, W., Hełminiak, A., Arabczyk, W., Mijowska, E., Removal of anionic dyes using magnetic Fe@graphite core-shell nanocomposite as an adsorbent from aqueous solutions. *J. Colloid Interface Sci.*, 497, 155–164, 2017.
50. Sahraei, R., Pour, Z.S., Ghaemy, M., Novel magnetic bio-sorbent hydrogel beads based on modified gum tragacanth/graphene oxide: Removal of heavy metals and dyes from water. *J. Clean. Prod. Part 4,* 142, 2973–2984, 2017.
51. Arica, T.A., Ayas, E., Arica, M.Y., Magnetic MCM-41 silica particles grafted with poly(glycidylmethacrylate) brush: Modification and application for removal of direct dyes. *Micropor. Mesopor. Mat.*, 243, 164–175, 2017.
52. Liu, J., Yu, H., Liang, Q., Liu, Y., Shen, J., Bai, Q., Preparation of polyhedral oligomeric silsesquioxane based cross-linked inorganic-organic nanohybrid as adsorbent for selective removal of acidic dyes from aqueous solution. *J. Colloid Interface Sci.*, 497, 402–412, 2017.

53. Belhouchat, N., Boudiaf, H.Z., Viseras, C., removal of anionic and cationic dyes from aqueous solution with activated organo-bentonite/sodium alginate encapsulated beads. *Appl. Clay Sci.*, 135, 9–15, 2017.
54. Chen, Bo., Liu, Y., Chen, S., Zhao, X., Meng, X., Pan, X., Magnetically recoverable cross-linked polyethylenimine as a novel adsorbent for removal of anionic dyes with different structures from aqueous solution. *J. Taiwan Inst. Chem. Eng.*, 67, 191–201, 2016.
55. Salama, A., Functionalized hybrid materials assisted organic dyes removal from aqueous solutions. *Environ. Nanotech. Monitor. Manage.*, 6, 159–163, 2016.
56. Sun, Z., Yao, G., Liu, M., Zheng, S., In situ synthesis of magnetic $MnFe_2O_4$/diatomite nano-composite adsorbent and its efficient removal of cationic dyes. *J. Taiwan Inst. Chem. Eng.*, 71, 501–509, 2017.
57. Abbasi, M., Synthesis and characterization of magnetic nanocomposite of chitosan/SiO_2/carbon nanotubes and its application for dyes removal. *J. Clean. Prod.*, 145, 105–113, 2017.
58. Roushani, M., Saedi, Z., Beygi, T.M., Anionic dyes removal from aqueous solution using TMU-16 and TMU-16-NH2 as isoreticular nanoporous metal organic frameworks. *J. Taiwan Inst. Chem. Eng.*, 66, 164–171, 2016.
59. Saini, J., Garg, V.K., Gupta, R.K., Kataria, N., Removal of Orange G and Rhodamine B dyes from aqueous system using hydrothermally synthesized zinc oxide loaded activated carbon (ZnO-AC). *J. Environ. Chem. Eng.*, 5, 884–892, 2017.
60. Dil E.A., Ghaedi, M., Asfaram, A., The performance of nano-rods material as adsorbent for removal of azo dyes and heavy metal ions: Application of ultrasound wave, optimization and modeling. *Ultrason. Sonochem.*, 34, 792–802, 2017.
61. Ashrafi, M., Chamjangali, M.A., Bagherian, G., Goudarzi, N., Application of linear and non-linear methods for modeling removal efficiency of textile dyes from aqueous solutions using magnetic Fe3O4 impregnated onto walnut shell. *Spectrochim. Acta Mol. Biomol. Spectrosc.*, 171, 268–279, 2017.
62. Han, H., Wei, W., Jiang, Z., Lu, J., Zhu, J., Xie, J., Removal of cationic dyes from aqueous solution by adsorption onto hydrophobic/hydrophilic silica aerogel. *Colloids Surf. A Physicochem. Eng. Asp.* 509, 539–549, 2016.
63. Agarwal, S., Gupta, V.K., Ghasemi, M., Amin, J.A., Peganum harmala-L seeds adsorbent for the rapid removal of noxious brilliant green dyes from aqueous phase. *J. Mol. Liq.*, 231, 296–305, 2017.
64. Hua, S., Yu, X., Li, F., Duan, J., Ji, H., Liu, W., Hydrogen titanate nanosheets with both adsorptive and photocatalytic properties used for organic dyes removal. *Colloids Surf. A Physicochem. Eng. Asp.*, 516, 211–218, 2017.
65. Zhang, Y.Z., Li, J., Zhao, J., Bian, W., Li, Y., Wang, X.J., Adsorption behavior of modified Iron stick yam skin with polyethyleneimine as a potential biosorbent for the removal of anionic dyes in single and ternary systems at low temperature. *Bioresource Technol.*, 222, 285–293, 2016.

66. Ansari, F., Ghaedi, M., Taghdiri, M., Asfaram, A., Application of ZnO nanorods loaded on activated carbon for ultrasonic assisted dyes removal: Experimental design and derivative spectrophotometry method. *Ultrason. Sonochem.*, 33, 197–209, 2016.
67. Jin, L., Sun, Q., Xu, Q., Xu, Y., Adsorptive removal of anionic dyes from aqueous solutions using microgel based on nano-cellulose and poly vinyl amine. *Bioresource Technol.*, 197, 348–355, 2015.
68. González, J.A., Villanueva, M.E., Piehl, L.L., Copello, G.J., Development of a chitin/graphene oxide hybrid composite for the removal of pollutant dyes: Adsorption and desorption study. *Chem. Eng. J.*, 280, 41–48, 2015.
69. El-Gamal, S.M.A., Amin, M.S., Ahmed, M.A., Removal of methyl orange and bromophenol blue dyes from aqueous solution using Sorel's cement nanoparticles. J. *Environ. Chem. Eng.*, 3, 1702–1712, 2015.
70. Srivastava, V., Sillanpaa, M., Synthesis of malachite@clay nanocomposite for rapid scavenging of cationic and anionic dyes from synthetic wastewater. *J. Environ. Sci.*, 51, 97–110, 2017.
71. Djilali, Y., Elandaloussi, H., Aziz, A., De Ménorval, L.C., Alkaline treatment of timber sawdust: A straightforward route toward effective low-cost adsorbent for the enhanced removal of basic dyes from aqueous solutions. *J. Saudi Chem. Soc.*, 20, S241–S249, 2016.
72. Wawrzkiewicz, M., Wiśniewska, M., Gun'ko, V.M., I. Zarko, V.I., Adsorptive removal of acid, reactive and direct dyes from aqueous solutions and wastewater using mixed silica–alumina oxide. *Powder Technol.*, 278, 306–315, 2015.
73. Konicki, W., Cendrowski, K., Bazarko, G., Mijowska., E., Study on efficient removal of anionic, cationic and nonionic dyes from aqueous solutions by means of mesoporous carbon nanospheres with empty cavity. *Chem. Eng. Res. Des.*, 94, 242–253, 2015.
74. Chen, Y., He, F., Ren, Y., Peng, H., Huang, K., Fabrication of chitosan/PAA multilayer onto magnetic microspheres by LbL method for removal of dyes. *Chem. Eng. J.*, 249, 79–92, 2014.
75. Xiao, L., Zhang, S., Huang, J., Effective removal of organic dyes by tungstate oxide nanourchins. *Powder Technol.*, 258, 297–303, 2014.
76. Jiang, Q., Lu, Y., Huang, Z., Hu, J., Facile solvent-thermal synthesis of ultrathin MoSe2 nanosheets for hydrogen evolution and organic dyes adsorption. *Appl. Surf. Sci.*, 402, 277–285, 2017.
77. Zhu, Z., Wu, P., Liu, G., He, X., Qi, B., Zeng, G., Wang, W., Sun, Y., Cui, F., Ultra high adsorption capacity of anionic dyes with sharp selectivity through the cationic charged hybrid nanofibrous membranes. *Chem. Eng. J.*, 313, 957–966, 2017.
78. Asfaram, A., Ghaedi, M., Hajati, S., Goudarzi, A., Dil, E.A., Screening and optimization of highly effective ultrasound-assisted simultaneous adsorption of cationic dyes onto Mn-doped Fe_3O_4-nanoparticle-loaded activated carbon. *Ultrason. Sonochem.*, 34, 1–12, 2017.

79. Khafri, H.Z., Ghaedi, M., Asfaram, A., Safarpoor, M., Synthesis and characterization of ZnS:Ni-NPs loaded on AC derived from apple tree wood and their applicability for the ultrasound assisted comparative adsorption of cationic dyes based on the experimental design. *Ultrason. Sonochem.*, 38, 371–380, 2017.
80. Song, W., Gao, B., Xu, X., Xing, L., Han, S., Duan, P., Song, W., Jia, R., Adsorption–desorption behavior of magnetic amine/Fe3O4functionalized biopolymer resin towards anionic dyes from wastewater. *Bioresource Technol.*, 210, 123–130, 2016.
81. Dashamiri, S., Ghaedi, M., Asfaram, A., Zare, F., Wang, S., Multi-response optimization of ultrasound assisted competitive adsorption of dyes onto $Cu(OH)_2$-nanoparticle loaded activated carbon: Central composite design. *Ultrason. Sonochem.*, 34, 343–353, 2017.
82. Liang, Z., Zhao, Z., Sun, T., Shi, W., Cui, F., Enhanced adsorption of the cationic dyes in the spherical CuO/meso-silica nano composite and impact of solution chemistry. *J. Colloid Interface Sci.*, 485, 192–200, 2017.
83. Chinoune, K., Bentaleb, K., Bouberka, Z., Nadim, A., Maschke, U., Adsorption of reactive dyes from aqueous solution by dirty bentonite, *Appl. Clay Sci.*, 123, 64–75, 2016.
84. Song, K., Xu, H., Xu, L., Xie, K., Yang, Y., Cellulose nanocrystal-reinforced keratin bioadsorbent for effective removal of dyes from aqueous solution. *Bioresource Technol.*, 232, 254–262, 2017.
85. Lessa, E.F., Gularte, M.S., Garcia, E.S., Fajardo, A.R., Orange waste: A valuable carbohydrate source for the development of beads with enhanced adsorption properties for cationic dyes. *Carbohydr. Polym.*, 157, 660–668, 2017.
86. Yao, T., Guo, S., Zeng, C., Wang, C., Zhang, L., Investigation on efficient adsorption of cationic dyes on porous magnetic polyacrylamide microspheres. *J. Hazard. Mater.*, 292, 90–97, 2015.
87. Liu, Y., Zeng, G., Tang, L., Cai, Y., Pang, Y., Zhang, Y., Yang, G., Zhou, Y., He, X., He, Y., Highly effective adsorption of cationic and anionic dyes on magnetic Fe/Ni nanoparticles doped bimodal mesoporous carbon, *J. Colloid Interface Sci.*, 448, 451–459, 2015.
88. Dotto, G.L., Santos, J.M.N., Tanabe, E.H., Bertuol, D.A., Foletto, E.L., Lima, E.C., Pavan, F.A., Chitosan/polyamide nanofibers prepared by Forcespinning® technology: A new adsorbent to remove anionic dyes from aqueous solutions. *J. Clean. Prod.*, 144, 120–129, 2017.
89. Lu, L., Li, J., Ng, D.H.L., Yang, P., Song, P., Zuo, M., Synthesis of novel hierarchically porous Fe_3O_4@MgAl–LDH magnetic microspheres and its superb adsorption properties of dye from water. *J Indus. Eng. Chem.*, 46, 315–323, 2017.
90. Asfaram, A., Ghaedi, M., Hajati, S., Goudarzi, A., Synthesis of magnetic γ-Fe_2O_3-based nanomaterial for ultrasonic assisted dyes adsorption: Modeling and optimization. *Ultrason. Sonochem.*, 32, 418–431, 2016.

91. Bagheri, A.R., Ghaedi, M., Asfaram, A., Bazrafshan, A.A., Jannesar, R., Comparative study on ultrasonic assisted adsorption of dyes from single system onto Fe_3O_4 magnetite nanoparticles loaded on activated carbon: Experimental design methodology. *Ultrason. Sonochem.*, 34, 294–304, 2017.
92. Dotto, G.L., Rodrigues, F.K., Tanabe, E.H., Fröhlich, R., Bertuol, D.A., Martins, T.R., Foletto, E.L., Development of chitosan/bentonite hybrid composite to remove hazardous anionic and cationic dyes from colored effluents. *J. Environ. Chem. Eng.*, 4, 3230–3239, 2016.
93. Ardhayanti, L.I, Santosa S.J., Synthesis of magnetic Mg/Al hydrotalcite and its application as adsorbent for navy blue and yellow F3G Dyes. *Proc. Eng.*, 148, 1380–1387, 2016.
94. Chen, T., Xiong, Y., Qin, Y., Yang, H., Zhang, P., Fanggui, Y., Facile synthesis of low-cost biomass-based g-Fe_2O_3/C for efficient adsorption and catalytic degradation of methylene blue in aqueous solution. *RSC Adv.*, 7, 336–343, 2017.
95. Hosseinzadeh, H., Mohammadi, S., Quince seed mucilage magnetic nanocomposites as novel bioadsorbents for efficient removal of cationic dyes from aqueous solutions. *Carbohydr. Polym.*, 134, 213–221, 2015.
96. Liu, L., Gao, Z.Y., Su, X.P., Chen, X., Jiang, L., Yao, J.M., Adsorption removal of dyes from single and binary solutions using a cellulose-based bioadsorbent. *ACS Sustain. Chem. Eng.*, 3, 432–442, 2015.
97. Tran, H.N., You, S.J., Bandegharaei, A.H., Chao, H.P., Mistakes and inconsistencies regarding adsorption of contaminants from aqueous solutions: A critical review. Water Res. 120, 88–116, 2017.
98. Ghorai, S., Sarkar, A., Raoufi, M., Panda, A.B., Schonherr, H., Pal, S., Enhanced removal of methylene blue and methyl violet dyes from aqueous solution using a nanocomposite of hydrolyzed polyacrylamide grafted xanthan gum and incorporated nanosilica. *ACS Appl. Mater. Interfaces*, 6, 4766–4777, 2014.
99. Metin, A., Ciftci, H., Alver, E., Efficient removal of acidic dye using low-cost biocomposite beads. *Ind. Eng. Chem. Res.*, 52, 10569–10581, 2013.
100. Zhou, Y.M., Zhang, M., Wang, X.H., Huang, Q., Min, Y.H., Ma, T.S., Niu, J.Y., Removal of crystal violet by a novel cellulose based adsorbent: Comparison with native cellulose. *Ind. Eng. Chem. Res.*, 53, 5498–5506, 2014.
101. Chen, W., Duan, L., Wang, L., Zhu, D., Response to comment on adsorption of hydroxyl- and amino-substituted aromatics to carbon nanotubes. *Environ. Sci. Technol.*, 43, 3400–3401, 2009.

7

Adsorption of Cr (VI) and Textile Dyes on to Mesoporous Silica, Titanate Nanotubes, and Layered Double Hydroxides

Rashmi Acharya*, Brundabana Naik and K. M. Parida*

Centre for Nano Science and Nano Technology, Siksha 'O' Anusandhan University Bhubaneswar, India

Abstract

Wastewater treatment from hazardous pollutants is a promising topic and gaining considerable attention worldwide. The removal of chromium (VI) and noxious textile dyes to get clean drinking water is a concerning issue. Effluents containing Cr (VI) species from various industrial processes such as leather tanning, electroplating, metal finishing, etc. are being discharged into the natural water ecosystems. Contamination of Cr (VI) also occurs from accidental leakage and improper disposal at mining sites. Being cytotoxic, mutagenic, and carcinogenic, it poses threat to both aquatic life and downstream users when its concentration exceeds permissible limits. Dyes are also considered to be toxic or carcinogenic. Due to the complex aromatic structure and synthetic origin of dyes, these possess high chemical stability and low biodegradability in water. It causes major problems in discharge of large quantities of these organic compounds in wastewater and becomes potentially harmful to the environment. Thus, treatment of industrial wastewater containing Cr (VI) and dyes has become an important and urgent environmental issue.

Several techniques such as ion exchange, reverse osmosis, chemical precipitation, electrochemical reduction, solvent extraction, adsorption, photocatalysis, etc. have been developed to remove Cr (VI) and dyes from water. However, high efficiency, simple operation, avoidance of the formation of secondary pollutants, and low energy consumption are some of the advantages that make adsorption an effective and promising technique for the removal of these pollutants from

**Corresponding authors*: kulamaniparida@soauniversity.ac.in; rashmiacharya@soauniversity.ac.in

Shahid-ul-Islam and B.S. Butola (eds.) Nanomaterials in the Wet Processing of Textiles, (219–260) © 2018 Scrivener Publishing LLC

contaminated water bodies. Nanomaterials are being used as potential adsorbents in wastewater treatment due to their large surface area, enough active surface sites, and high adsorption capacity. Silica-based nanomaterials, graphene-based nanomaterials, titanate nanotubes, ferrites, layer double hydroxides, etc. have recently been successfully applied for the removal of Cr (VI) and dyes. In this chapter, we have discussed the adsorption behavior of silica-based nanomaterials, titanate nanotubes, and layer double hydroxides in terms of isotherm models, kinetics, and thermodynamics of adsorption. Maximum adsorption capacities obtained under optimum adsorption conditions such as adsorbent dose, concentration of Cr (VI)/dye solution, pH, and temperature are presented in tables. Mechanisms of adsorption are explained in different schemes.

Keywords: Adsoprtion, Cr (VI), dyes, mesoporous silica, TNT, LDH

7.1 Introduction

The level of environmental pollution has increased alarmingly due to rapid growth of industrialization and urbanization [1–3]. Hexavalent chromium [Cr (VI)] and dyes contribute to a significant extent toward environmental pollution. Wastewater containing approximately 170,000 tons of chromium is being discharged into different water bodies every year all over the world [4]. Similarly, about one million kilograms of dyes released from textile industries are being discharged into rivers and streams each year worldwide [5–7].

Cr (VI) is carcinogenic, mutagenic, and teratogenic. It causes a wide range of clinical/health problems such as damage to the liver and kidneys, lung carcinoma, nausea, dermatitis, nasal membrane inflammation, ulceration, irritation of the gastrointestinal tract, and renal damage as shown in Figure 7.1, when consumed above the permissible limit of 50 $\mu g\ L^{-1}$ in potable water as fixed by the US EPA [8–10]. The compounds of Cr (VI) are mostly used in different industrial processes such as metal plating, leather tanning, and pigment manufacturing. Effluents containing high Cr (VI) concentration from these industries are discharged into the aquatic environments [11–13]. Being highly mobile in an aqueous environment, it contaminates the natural water ecosystems easily.

Dyes are also considered to be toxic or carcinogenic. Due to the complex aromatic structure and synthetic origin of dyes, these possess high chemical stability and low biodegradability in water. It causes a major problem in the discharge of large quantities of organic compounds in wastewater and becomes potentially harmful to the environment. Thus, treatment of industrial wastewater containing Cr (VI) and dyes has become an important and urgent environmental issue. Therefore, various physical, chemical,

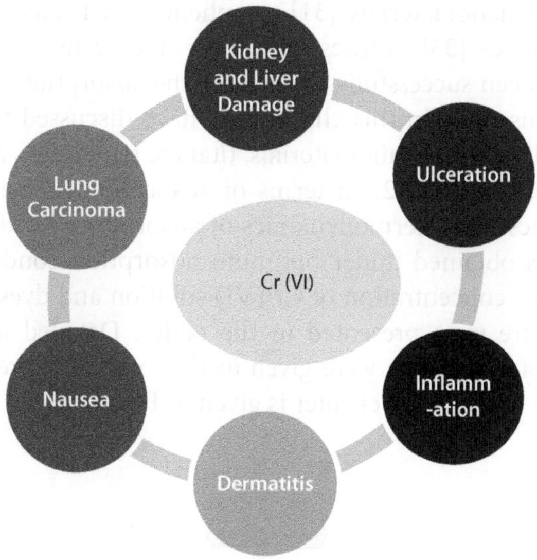

Figure 7.1 Clinical/health problems due to Cr (VI) toxicity.

and biological techniques such as precipitation [14], ion exchange [15], membrane filtration [16], electrochemical approaches [17], reverse osmosis [18], and adsorption [19] have been proposed for the treatment of waste water rich in Cr (VI) and toxic dyes. However, most of these techniques find a number of drawbacks such as high operational cost, low feasibility of scale-up for industrial applications, and ineffectiveness at high concentrations. On the other hand, adsorption has been recognized as one of the most promising techniques for the treatment of Cr (VI) from wastewater due to its several advantages which include high removal efficiency, low energy consumption, less chemical investment, and reusability [20–22]. Hence, different kinds of adsorbents such as activated carbon [23], biomass [24], clay minerals [25], zeolites [26], industrial and agricultural wastes [27, 28] have been used for the removal of Cr (VI) and dyes from the wastewater. Practical applications of these adsorbents are restricted because of low adsorption capacities and selectivities, which is due to less porosity, low surface area, lack of functional groups, and small dye-binding constants [29]. Thus, there was an urgent need for the development of adsorbents having high porosity, large surface area, and high functionalities.

In recent years, nanomaterials have attracted great attention worldwide for various applications. Besides these, they are also being used as potential adsorbents in wastewater treatment due to their large surface area, enough active surface sites, and high adsorption capacity [30].

Silica-based nanomaterials [31], graphene-based nanomaterials [32], titanate nanotubes [33], ferrites [34], layer double hydroxides [35], etc. have recently been successfully applied for the adsorption of Cr (VI) and various noxious dyes. In this chapter, we have discussed the adsorption behavior of silica-based nanomaterials, titanate nanotubes, and layer double hydroxides (Figure 7.2) in terms of adsorption capacities, isotherm models, kinetics, and thermodynamics of adsorption. Maximum adsorption capacities obtained under optimum adsorption conditions such as adsorbent dose, concentration of Cr (VI) solution and dyes solution, pH, and temperature were presented in the tables. Detailed illustrations of mechanisms of adsorption were given in the form of figures. A roadmap describing the scope of the chapter is given in Figure 7.3.

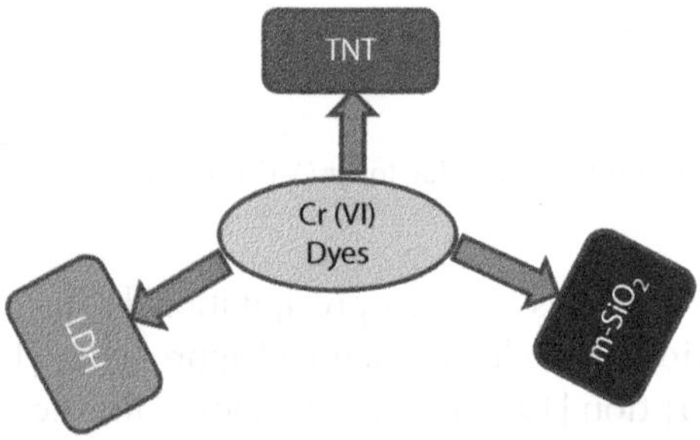

Figure 7.2 Adsorption of Cr (VI) and dyes on to mesoporous silica, TNTS, and LDH.

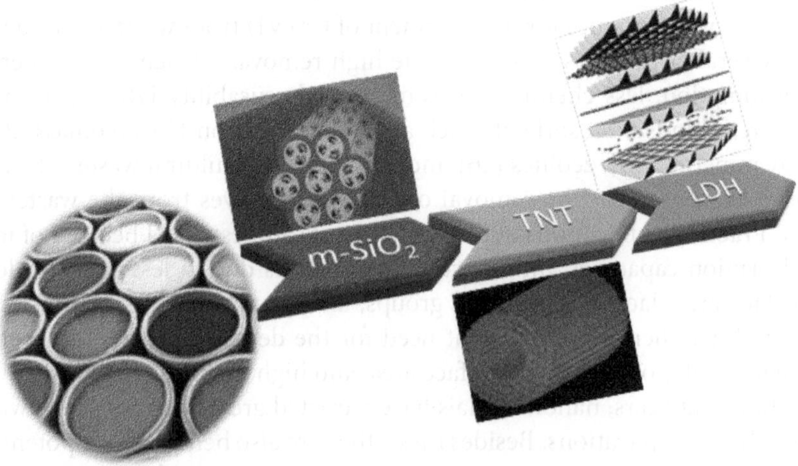

Figure 7.3 Roadmap for the scope of the chapter.

7.2 Mesoporous Silica (m-SiO_2)

Silica exhibits good physical and chemical properties such as water stability (nonswelling), thermal stability (up to 1500 °C), and good mechanical strength. Different kinds of silica materials such as amorphous silica, mesoporous silica, fumed silica, and silica gels have been synthesized. Among these, mesoporous silica is extensively used in various applications due to its high surface area and large number of of silanol (Si–OH) groups. Particularly, these are considered as high-capacity adsorbents as they exhibit enhanced accessibility to active centers [36] and fast mass transfer rates in the regular porous environment [37, 38]. Various ordered mesoporous silica materials such as SBA-15, USG-41, MCM-41, and their modified forms are used as adsorbents for the removal of Cr (VI) and dyes in recent years.

7.2.1 Adsorption of Cr (VI) on to Mesoporous Silica

Barium titanate mesoporous silica ($BaTiO_3$/SBA-15) prepared by an aerosol-assisted solid-state reaction using SBA-15 as a hard template exhibited high adsorption capacity as compared to bulk $BaTiO_3$ and pure SBA-15. It might be attributed to its high porosity and large surface area, which enhanced the oxy-anionic species of chromium to get adsorbed at the internal surface of the pores. Moreover, the development of a positive surface charge on the nanomaterial due to oxygen vacancy adsorbed negatively charged Cr (VI) species effectively by electrostatic interactions [39]. SBA-15 nanoparticles modified with 3-aminopropyltrimethoxy-silane exhibited an improved adsorption capacity of 48.31 mg g^{-1}. These nanoparticles with a large pore size distribution facilitated surface functionalization due to which there was coordination of Cr (VI) species on their surfaces leading to higher adsorption. 3-aminopropyltrimethoxy-silane-functionalized USG-41 nanoparticles with a broad pore size distribution exhibited still higher adsorption capacity of 111.1 mg g^{-1}. MCM-41 nanoparticles moldified with 3-aminopropyltrimethoxy-silane with a narrow pore size distribution exhibited low adsorption capacity of 39.68 mg g^{-1} [40]. Immobilization of TiO_2 onto MCM-41 provides adequate number of active sites on the surface of TiO_2-MCM-41. As a result, the adsorbent showed an enhanced adsorption capacity.

The sample with 20 weight percentage (4.0 mmol g^{-1}) TiO_2 exhibited a maximum adsorption capacity of 120.64 mg g^{-1}. Less uptake of the sample with 25 weight percentage (5.02 mmol g^{-1}) TiO_2 might be attributed to the loss of ordered hexagonal structure of MCM-41that led to irregularity in

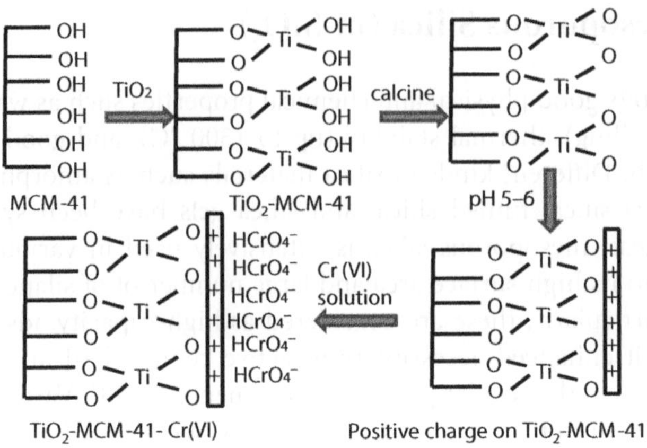

Scheme 7.1 Proposed mechanism for titania loading on MCM-41 and Cr (VI) adsorption on TiO_2-MCM-41 [Reproduced from reference 31].

the surface sites. The mechanism of adsorption was shown in Scheme 7.1, which indicated that positive charges were developed on the surface of the adsorbent below the pH_{ZPC} (i.e., 6.2) due to which negatively charged univalent Cr (VI) species ($HCrO_4^-$) were adsorbed through electrostatic attraction below this pH (i.e., 5.5). At very low and high pH, adsorption is decreased because bivalent Cr (VI) species namely $Cr_2O_7^{2-}$ at lower pH and CrO_4^{2-} at higher pH were the dominant ones which would neutralize twice the number of active sites available on the surface of adsorbents as compared to univalent species [31]. This led to a decrease in adsorption at both lower and higher pH values [41]. Moreover, a decrease in adsorption at higher pH might be attributed to the fact that the typical siliceous hexagonal structure of MCM-41 was destroyed above pH 8.0 [42]. However, maximum adsorption capacity of 904 mg g^{-1} was obtained for MCM-41 samples with a large surface area and uniform diameter [43]. This indicated not only active sites but also large surface area and uniform particle size are responsible for enhanced adsorption capacity. Maximum adsorption capacity and optimum adsorption conditions such as pH, temperature, initial Cr (VI) concentration, adsorbent dose, fitted isotherm, kinetic model, thermodynamic parameters, and mechanism of adsorption for Cr (VI) onto mesoporous silica are given in Table 7.1.

7.2.2 Adsorption of Dyes on to Mesoporous Silica

The three consecutive steps in the adsorption of dye by mesoporous silica adsorbent are: (i) Transport of the dye to the external surface of the

Table 7.1 Maximum adsorption capacity, optimum adsorption condition such as pH, temperature, initial Cr (VI) concentration, adsorbent dose, fitted isotherm, kinetic model, thermodynamic parameters, and mechanism of adsorption for Cr (VI) onto silica-based nanomaterials, titanate nanotubes, and layer double hydroxides.

Name of the adsorbent	Max.Adsn. Capacity (mg g^{-1})	Optimum pH	Temp. (°C)	Initial Cr(VI) conc. (mg L^{-1})	Adsorbent Dose (g L^{-1})	Contact Time (Minutes)	Fitted Isotherm	Kinetic Model	Thermodynamic Parameters	Mechanism of Adsorption	Ref.
MCM-41	904	4.0	35	< 10	0.01	120	Langmuir	Psueudo 2nd order	ΔH° = +ve, ΔG° = -ve, ΔS° = +ve	Electrostatic attraction	43
TiO_2- MCM-41	120.64	5.5	50	400	1.0	240	Langmuir, Freundlich	Psueudo 2nd order	ΔH = +ve, ΔG = -ve, ΔS = +ve,	Electrostatic attraction	31
$BaTiO_3$/SBA-15	-	4.0	50	10	2.5	60	-	Psueudo 2nd order	ΔH = +ve,	Electrostatic attraction	39
AP-USG-41	111.1	3.0	25	10–200	2.0	120	Langmuir	intraparticle diffusion	-	-	40
AP-SBA-15	48.31	3.0	25	10–200	2.0	120	Langmuir	Pseudo 2nd order, diffusion	-	-	40
AP-MCM-41	39.68	3.0	25	10–200	2.0	120	Freundlich	Pseudo 2nd order, diffusion	-	-	40

(*Continued*)

Table 7.1 Cont.

Name of the adsorbent	Max.Adsn. Capacity (mg g^{-1})	Optimum pH	Temp. (°C)	Initial Cr(VI) conc. (mg L^{-1})	Adsorbent Dose (g L^{-1})	Contact Time (Minutes)	Fitted Isotherm	Kinetic Model	Thermodynamic Parameters	Mechanism of Adsorption	Ref.
TNTs	9.8	5.0	25	10	0.6	180	–	–	–	Electrostatic attraction	59
Protonated TNTs	26.60	5.0	30	25–400	1.0	60	Langmuir	Psueudo 2nd order	–	Electrostatic attraction	33
Amino- functionalized TNTs	153.85	5.0	30	25–400	1.0	60	Langmuir	Psueudo 2nd order	–	Electrostatic attraction, reduction of Cr(VI)	33
Amino- modified TNTs	69.1	2-3	25	10–180	1.0	60	Langmuir	Psueudo 2nd order	–	Electrostatic attraction	60
MgAl-CO_3-450C	25	6	30	50	0.5	120	Langmuir	–	–	Reconstruction	76
MgAlZr-CO_3-450C	31	6	30	50	0.5	120	Langmuir	–	–	Reconstruction	76
MgAlCl	74	9	30	15	8.0	1440	Langmuir	–	–	Ion-exchange, reconstruction	77
Uncalcined Mg-Al-CO_3-hydrotalcite	16.3	–	22	11–57	2.0	60	Langmuir	–	–	Electrostatic attraction, ionexchange	78

Calcined Mg-Al-CO_3-hydrotalcite	128	–	22	57–448	2.0	30-1440	Langmuir	–	–	Electrostatic attraction, Ionexchange, recostruction	78
MgAl-CO_3-500C	120	6.0	30	100	0.2	1440	Freundlich	Pseudo1st order	–	Recostruction	80
MgAl-CO_3	17	6	30	10	1	1440	Freundlich	2nd order	–	Electrostatic attraction	81
MgAl-CO_3-150C	29	9	–	10–250	10	1440	Langmuir	–	–	Electrostatic attraction	83
MgAl-CO_3	30	9.5	–	10.6	1	–	–	–	–	Electrostatic attraction	84
MgAl-CO_3-400C	45	6.5	–	200	–	–	–	–	–	Reconstruction	86
Mg/ Fe hydro-talcite – like-compounds	2.53	6.0	30	20.0	2.0	120	Langmuir	1st order	ΔH = +ve, endothermic ΔG = -ve, spontaneous ΔS = +ve,	Electrostatic attraction	35
Al/Mg mixed hydroxide	112	4.0	40	200	2.0	150	Langmuir	Psueudo 2nd order	ΔH = +ve, endothermic ΔG = -ve, spontaneous ΔS = +ve	Anion exchange and elec-trostatic attraction	82

(Continued)

Table 7.1 Cont.

Name of the adsorbent	Max.Adsn. Capacity (mg g^{-1})	Optimum pH	Temp. (°C)	Initial Cr(VI) conc. (mg L^{-1})	Adsorbent Dose (g L^{-1})	Contact Time (Minutes)	Fitted Isotherm	Kinetic Model	Thermodynamic Parameters	Mechanism of Adsorption	Ref.
Calcined Mg/Al hydrotalcite	85.9	3.0	48	20–200	1.0	120	Langmuir	–	ΔH = +ve, endothermic ΔG = -ve, spontaneous ΔS = +ve	Intercalation, Electrostatic attraction and interlayer anionexchange	89
Calcined Mg/Al/Fe hydrotalcite	75.4	3.0	48	20–200	1.0	120	Langmuir	–	ΔH = +ve, endothermic ΔG = -ve, spontaneous ΔS = +ve	Intercalation, adsorption and inter-layer anion exchange	89
Mg-Zn-Al hydro-talcite like materials		6.0	30	10		120	–	Psueudo 1st order	–	–	90
Co-Bi LDH	277.7	7.0	25	100	2.0	120	Freundlich	Psueudo 1st order	ΔH° = -ve, exothermic ΔG° = -ve, spontaneous ΔS° = +ve	–	91

adsorbent (film diffusion). (ii) Transport of the dye within the pores of the adsorbent, with only a small amount of dye being adsorbed on the external surface (particle diffusion). (iii) Adsorption of the dye on an inner surface of the adsorbent.

Huang *et al.* have synthesized ordered SBA-15 using nonionic surfactant as a template in acidic conditions and studied the adsorption behaviors of cationic dyes such as methylene blue (MB) and Janus Green B (JGB), an anionic dye [reactive black 5 (RB 5)], and a neutral dye named dimethyl phthalate (DMP) [7]. The structural formula of (a) MB, (b) JGB (c) RB 5 and (d) DMP are given in Figure 7.4. SBA-15 has shown enhanced adsorption capability toward cationic dyes due to the electrostatic attraction between cationic dyes and the anionic nature of SBA-15 whereas it does not absorb any anionic and neutral dyes. The adsorption capacities for MB and JGB were found to be 49.25 and 66.43 mg g^{-1}, respectively, and adsorption isotherms were well fitted with Langmuir and Freundlich models.

A facile non-polar solvent-assisted Stöber method has been developed by Wang *et al.* to synthesize ordered mesoporous silica materials with tunable pore size, diverse morphologies, and mesostructures by using cetyltrimethyl ammonium bromide (CTAB) as the template and tetraethyl orthosilicate (TEOS) as a silica precursor in a simple aqueous-phase synthesis system [44]. By simply changing the amount of n-hexane and aqueous ammonia in the system, ordered mesoporous silica with a pore size of 2.7–10.5 nm, various morphology (nanocubes, truncated nanocubes, core-shell microspheres, and twisted nanorods), high surface area up to 888 m^2 g^{-1}, and large pore volume of 1.55 cm^3 g^{-1} are synthesized (Figure 7.5). Owing to their highly hydrophilic surface, large and accessible pores, and high surface area, the mesoporous silica materials exhibit an excellent performance in the adsorption of dye molecules of a large dimension (1.6 nm) with a maximum adsorption capacity of 106 mg g^{-1} in 10 min at 200 mg L^{-1} initial Rhodamine B (RhB) concentration.

Extremely ordered mesoporous SBA-15 have been prosperously synthesized by Mirzae *et al.* through a plain sol–gel technique and then functionalized with polyamidoamine (PAMAM) dendrimer for acid blue 62 (AB62) dye adsorption.

The structural formula of (a) RhB and (b) AB 62 are given in Figure 7.6. Appropriate parameters for removing AB62 were at pH 2, adsorbent dosage of 0.03 g L^{-1}, and contact time of 60 min at 25 °C. The experimental results suitably fit the Langmuir model and the adsorption was found to be physisorptive and exothermic [45].

Tsai *et al.* have demonstrated a high density of –COOH groups loading into SBA-16 via a one-pot synthesis route, which involves co-condensation

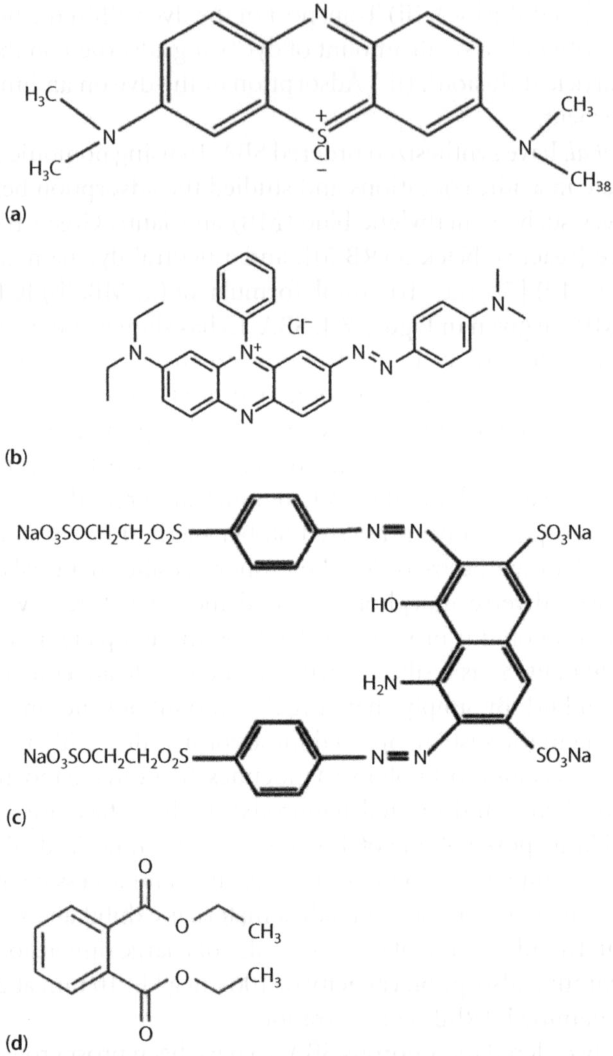

Figure 7.4 Chemical structure of (a) methylene blue (MB), (b) Janus Green B (JGB), (c) reactive black 5 (RB 5), and (d) dimethyl phthalate (DMP).

of carboxyethylsilanetriol sodium salt (CES) and tetraethylorthosilicate (TEOS) templated by Pluronic F127 and P123 in an acidic medium. These functionalized materials are used to effectively remove two cationic dyes methylene blue (MB) and phenosafranine (PF) with the maximum adsorption capacities of -561 and 519 mg g^{-1}, respectively, at pH 9 [46]. The adsorption behavior of MCM-41 on two basic dyes [basic green 5 (BG5) and basic violet 10 (BV10)] was studied by Juang *et al.* They investigated

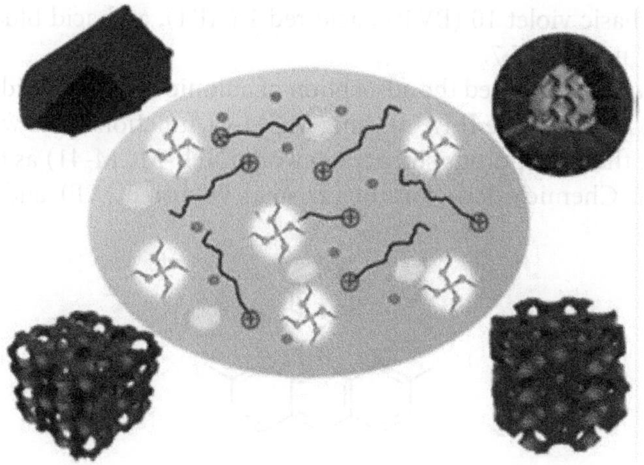

Figure 7.5 Different types of mesoporous silica with varying concentration of surfactants and its monomer precursors (reproduced from reference [44]).

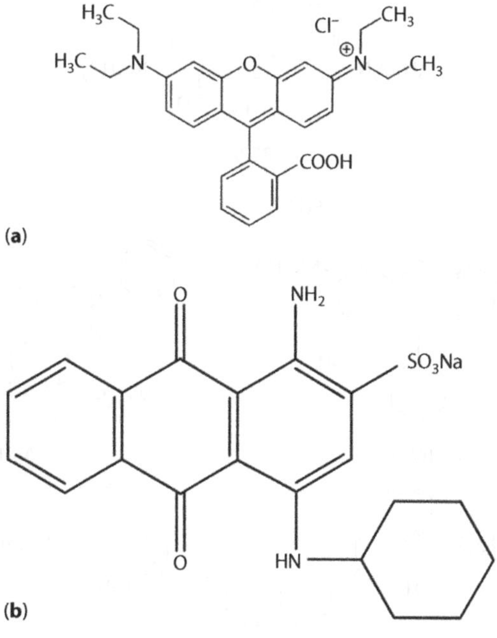

Figure 7.6 Chemical structure of (a) Rhodamine B (RhB) and (b) acid blue 62 (AB62).

that MCM-41 might be a good adsorbent for the removal of basic dye pollutants from effluents, only when the effect of interactions between MCM-41 and large dyes (BV10) on the pore structure stability of MCM-41 is insignificant [47]. Chemical structure of phenosafranine (PF), basic green

5 (BG5), basic violet 10 (BV10), acid red 1 (AR1), and acid blue 9 (AB9) are given in Figure 7.7.

Wu *et al.* have studied the adsorption of anionic dyes like Acid Fuchsine (AF) and Acid Orange II (AO) using MCM-41 functionalized with amine groups in the mesoporous silica framework (NH_2-MCM-41) as the adsorbent [48]. Chemical structure of (a) Acid Fuchsine (AF) and (b) Acid

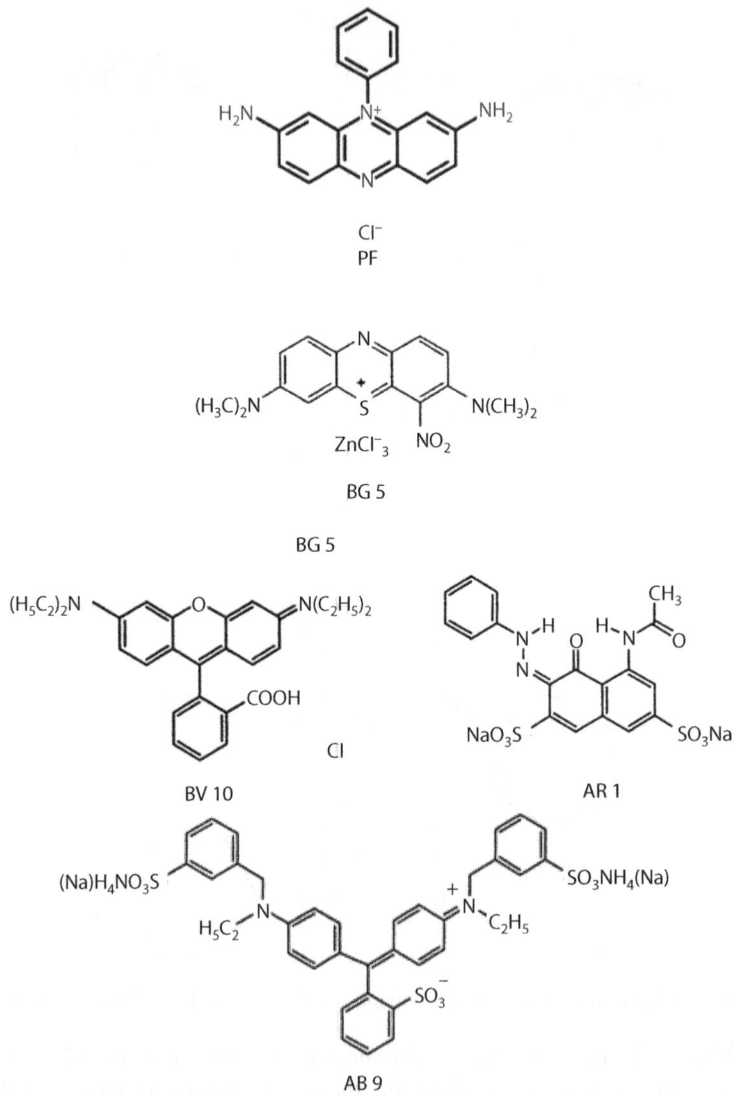

Figure 7.7 Chemical structure of phenosafranine (PF), basic green 5 (BG5), basic violet 10 (BV10), acid red 1 (AR1), and acid blue 9 (AB9).

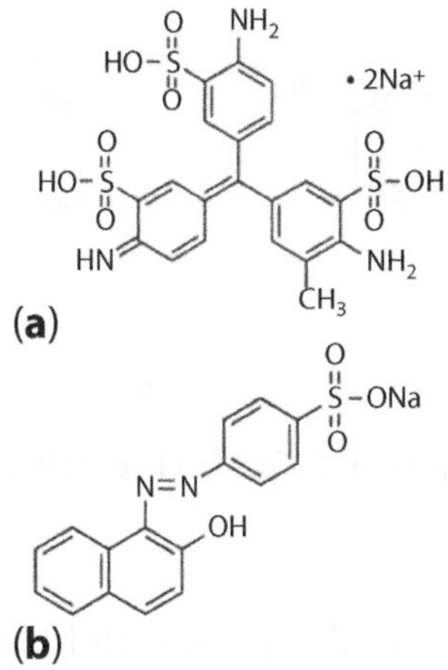

Figure 7.8 Chemical structure of (a) Acid Fuchsine (AF) and (b) Acid Orange II (AO).

Orange II (AO) are given in Figure 7.8. The maximum adsorption capacity for AF was approximately 140.60 mg g^{-1} at pH 2.0 and 25 °C with NH_2-MCM-41 dosage 2.0 g L^{-1}, and 278.38 mg g^{-1} for AO at pH 3.0, respectively. In single-component systems, equilibrium data fitted well the Langmuir and D-R models, suggesting the adsorption to be monolayer and physical in nature. Kinetic studies showed that the adsorption process could be better described by both the Lagergren pseudo-second-order and the Spahn and Schlunder models. Moreover, it was found that the adsorption was governed by film diffusion followed by intraparticle diffusion. Thermodynamic constant values ($\Delta G^o<0$, $\Delta H^o<0$ and $\Delta S^o<0$) demonstrated that the adsorption reactions of AF and AO onto NH_2-MCM-41 were feasible, spontaneous, and exothermic under examined conditions. For binary component systems, AF and AO exhibited competitive adsorption onto NH_2-MCM-41, and adsorption capacity values of AF and AO were reduced compared to those of the corresponding single-component systems. Furthermore, both in single and binary component systems, the experimental data could be better described by the Langmuir isotherm and the pseudo-second-order kinetic models.

The adsorption behavior of the polyamine functionalized mesoporous silica (MS-PEI) and 3-aminopropyltriethoxy silane (APTES) functionalized mesoporous silica was (MS-APTES) studied by Nayab *et al.* toward anionic dyes [49]. The kinetic data support a pseudo-second-order adsorption behavior for both monolayer and polymer functionalized MS. The associated thermodynamic parameters (ΔG^o, ΔH^o, and ΔS^o) reveal that the process of adsorption with MS-PEI is more spontaneous and energetically

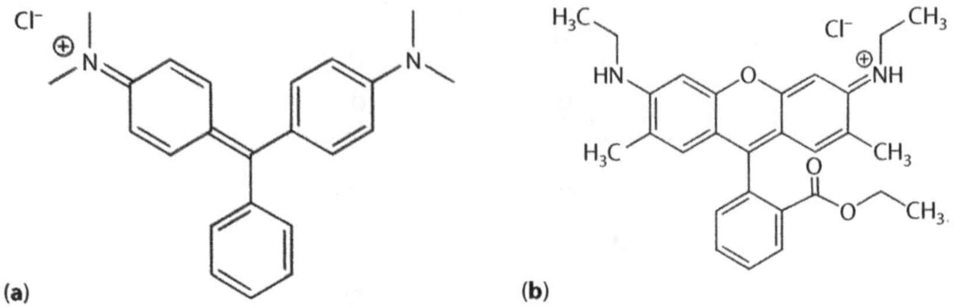

Figure 7.9 Chemical structure of (a) Malachite Green (MG) and (b) Rhodamine 6G (Rd 6G).

favored as compared to the adsorption with MS-APTES. Taken together, the novel adsorbent system derived from a combination of MS and branched polymer (MS-PEI) shows a higher adsorption efficiency and capacity toward the anionic dyes than the monolayer-based adsorbent (MS-APTES).

Nanda *et al.* have synthesized *in situ* preparation of meso-MnO_2-MCM-41 and studied both dye adsorption and photodegradation. The intra-particle mesoporosity and high textural properties have a vital role for an efficient adsorption of dyes such as Rd 6G, MB, MG, and RhB [50]. Chemical structure of (a) Malachite Green (MG) and (b) Rhodamine 6G (Rd 6G) are given in Figure 7.9. Maximum adsorption capacity, optimum adsorption conditions such as pH, temperature, initial dye concentration, adsorbent dose, fitted isotherm, kinetic model, thermodynamic parameters, and mechanism of adsorption for dyes onto mesoporous silica are given in Table 7.2.

7.3 Titanate Nanotubes

After the discovery of carbon nanotubes, tatanium oxide nanotubes gained huge attention owing to its one-dimensional nanostructure and unique physical and chemical properties. In recent years, TNT themselves have attracted much attention due to their potential application in areas of photocatalysis [51, 52], wastewater treatment [53], ion exchange [54], lithium ion storage [55], solar energy conversion [56], and so on. Lower fabrication cost, nanotubular structure, good sedimentation properties, large surface area, and pore volume made titanate nanotubes as promising adsorbents for the removal of organics and heavy metals from aqueous solutions [54, 57, 58]. The adsorption behavior of these materials

Table 7.2 Maximum adsorption capacity, optimum adsorption condition such as pH, temperature, initial dye concentration, adsorbent dose, fitted isotherm, kinetic model, thermodynamic parameters, and mechanism of adsorption for dyes onto silica-based nanomaterials, titanate nanotubes and layer double hydroxides.

Name of the Adsorbent	Name of the dye	Max.Adsn. Capacity (mg g^{-1})	Optimum pH	Temp. (°C)	Initial dye conc. (mg L^{-1})	Adsorbent dose(g L^{-1})	Contact Time (Mins)	Fitted Isotherm	Kinetic model	Thermodynamic parameters	Mechanism of adsorption	Ref.
SBA-15	MB	49.25	5.0–7.0	25	16.0–160.0	2.0	60	Langmuir, Freundlich	Pseudo 2nd order	ΔH^o = -ve, ΔG^o = -ve, ΔS^o =-ve,	Electrostatic attraction	7
SBA-15	JGB	66.43	5.0–7.0	25	25.55–255.5	2.0	60	Langmuir, Freundlich	Pseudo 2nd order	ΔH^o = -ve, ΔG^o = -ve, ΔS^o =-ve,	Electrostatic attraction	7
MS-C	RhB	124	7.0	25	200	1.0	50	–	Pseudo 2nd order	–	Electrostatic attraction	44
C-MS-C	RhB	69.3	7.0	25	200	1.0	20	–	Pseudo 2nd order	–	Electrostatic attraction	44
SBA-15/ PAMAM	AB62	1428.57	2.0	25	40	0.03	60	Langmuir	Pseudo 2nd order	ΔH = -ve, ΔG = -ve ΔS = +ve	Electrostatic attraction	45
S16C-30	MB	561	9.0	25	500	0.25	1440	Langmuir	Pseudo 2nd order	ΔH^o= -ve, ΔG^o = -ve, ΔS^o =-ve	Electrostatic attraction	46
S16C-30	PF	519	9.0	25	500	0.25	1440	Langmuir	Pseudo 2nd order	ΔH^o= -ve, ΔG^o = -ve, ΔS^o =-ve	Electrostatic attraction	46
NH2-MCM-41	AF	140.60	2.0	25	300	2.0	240	Langmuir	Psueudo 2nd order	ΔH^o = -ve, ΔG^o = -ve, ΔS^o =-ve	Electrostatic attraction	48
NH2-MCM-41	AO	278.38	3.0	25	300	2.0	240	Langmuir	Psueudo 2nd order	ΔH^o=-ve, ΔG^o = -ve, ΔS^o =-ve	Electrostatic attraction	48

(*Continued*)

Table 7.2 Cont.

Name of the Adsorbent	Name of the dye	Max.Adsn. Capacity (mg g^{-1})	Optimum pH	Temp. (°C)	Initial dye conc. (mg L^{-1})	Adsorbent dose(g L^{-1})	Contact Time (Mins)	Fitted Isotherm	Kinetic model	Thermodynamic parameters	Mechanism of adsorption	Ref.
MS-PEI	ARS	147.275	6.0	25	15	1.5	15	Langmuir	Psueudo 2nd order	$\Delta H°$= -ve, $\Delta G°$= -ve, $\Delta S°$ = +ve	Electrostatic attraction	49
MS-APTES	ARS	35.71	6.0	25	15	1.5	15	Langmuir	Psueudo 2nd order	$\Delta H°$= -ve, $\Delta G°$= -ve $\Delta S°$= +ve	Electrostatic attraction	49
MS-PEI	XO	186.920	4.0	25	15	2.0	2	Langmuir	Psueudo 2nd order	$\Delta H°$ = -ve, $\Delta G°$= -ve $\Delta S°$ = +ve	Electrostatic attraction	49
MS-APTES	XO	65.23	4.0	25	15	2.0	2	Langmuir	Psueudo 2nd order	$\Delta H°$ = -ve, $\Delta G°$= -ve $\Delta S°$ = +ve	Electrostatic attraction	49
CNTs	MB	133.33	7.0	25	230	0.5	60	Langmuir	Psueudo 2nd order	-	Electrostatic attraction	61
3D-TNT	NR	401.8	-	25	200	0.2	120	Langmuir	Psueudo 2nd order	-	Electrostatic attraction	63
3D-TNT	MB	382.07	-	25	200	0.2	120	Langmuir	Psueudo 2nd order	-	Electrostatic attraction	63
3D-TNT	MG	361.95	-	25	200	0.2	120	Langmuir	Psueudo 2nd order	-	Electrostatic attraction	63
3D-TNT	CV	322.5	-	25	200	0.2	120	Langmuir	Psueudo 2nd order	-	Electrostatic attraction	63
HDTMA-TNT	AR1	400	-	25	2000	1.0	240	-	-	-	Anion exchange	64
TNT	BG5	380	-	25	2000	1.0	240	-	-	-	Cation exchange	64

TNT-160	BG5	211	4.0	25	2000	1.0	180	DKR	Psueudo 2nd order	–	Ion exchange	65
TNT-160	BV10	118	4.0	25	2000	1.0	180	DKR	Psueudo 2nd order	–	Ion exchange	65
ZnAl–CO_3-LDH	AB	261.8	7.0	25	1000	0.5	480	Langmuir	Psueudo 2nd order	ΔH° = -ve, ΔG° = -ve ΔS° = +ve	–	92
Calcined ZnAl–CO_3-LDH	AB	1587	7.0	25	1000	0.5	480	Langmuir	Psueudo 2nd order	ΔH°= +ve, ΔG° = -ve ΔS° = +ve	–	92
MgAl500	azodyeama-ranth	967	6.4	25	48.35–4000	1.0	180	Langmuir	Psueudo 2nd order	–	Anion exchange	93
PDOPA-f-LDH	MB	132	7.0	25	5-50	0.2	30	Freundlich	Psueudo 2nd order	ΔH° = -ve, ΔG°= -ve, ΔS° =-ve	Vander Waals forces, Electrost. attrctn.	94
Ni4Fe1-LDH	MO	205.76	–	–	10	0.2	60	Langmuir	Psueudo 2nd order	–	Anion exchange	95
Mg/Fe-LDH	IC	62.8	9.5					Langmuir		ΔH°=+ve, ΔS°= +ve	–	96
ILs/LDHs (b)	Reactive orange 5	300.9	6.48	25	180	0.5	150	Freundlich	Psueudo 2nd order	ΔH° = +ve ΔG°= -ve ΔS° = +ve	Ion exchange, electrostatic attraction, hydrogen bond interac-tions, van der Waals interactions.	97

(*Continued*)

Table 7.2 Cont.

Name of the Adsorbent	Name of the dye	Max.Adsn. Capacity (mg g^{-1})	Optimum pH	Temp. (°C)	Initial dye conc. (mg L^{-1})	Adsorbent dose(g L^{-1})	Contact Time (Mins)	Fitted Isotherm	Kinetic model	Thermodynamic parameters	Mechanism of adsorption	Ref.
MPL3	MO	624.89	5.6	25	100	0.1	300	Freundlich	Psueudo 2nd order	–	Electrostatic attraction	98
MPL3	CR	584.56	5.6	25	100	0.1	200	Freundlich	Psueudo 2nd order	–	Electrostatic attraction	98
CaAl-LDH-NO_3	Sunset Yellow FCF	398.41	4.0	27	10–1000	1.0	50	Langmuir	–	–	Ion exchange	99
Mg/Fe-CLDH	AB 14	369.71	6.0	25	0.5–50	0.2	1440	Langmuir	Psueudo 2nd order	ΔH^o = +ve, ΔG^o = -ve ΔSo= +ve	Electrostatic attraction, Hydrogen bonding	100
Calcined Ni-Al-LDH	MO	186.39	2.0	50	20–1000	1.0	–	Langmuir, Redlich–Peterson is		ΔH^o = +ve ΔG^o = -ve ΔS^o = +ve	Electrostatic attraction,	101
Mg–Al-LDH	RR	59.49	6.0-9.0	–	–	5.0	60	Langmuir	Psueudo 2nd order	–	Anion exchange	102
Mg–Al-LDH	CR	37.16	6.0-9.0	–	–	5.0	60	Langmuir	Psueudo 2nd order	–	Electrostatic attraction,	102
Mg–Al-LDH	AR1	108.0	6.0-9.0	–	–	5.0	60	Langmuir	Psueudo 2nd order	–	Anion exchange	102

for the removal of Cr (VI) and organic dyes was reviewed in this section [33, 59, 60].

7.3.1 Adsorption of Cr (VI) on to Titanate Nanotubes

The mechanism for adsorption of coexisting Cr (III) and Cr (VI) onto titanate (TiO_6) nanotubes was explained with the help of Scheme 7.2 [59]. According to it, a double-layer adsorption occurred when Cr (III) and Cr (VI) were coexisted. An ionic cluster of Cr (III) and Cr (VI) was formed first and these were bridged with Cr (III), which was adsorbed onto titanate nanotubes mainly by the ion exchange mechanism with the formation of Cr (III)–O–Cr (VI) linkage. The maximum adsorption capacity for Cr (III) and Cr (VI) was found to be 82.1 and 9.8 mg g^{-1}, respectively. It was also observed that the rearrangement of Cr (III) from the second layer to the surface of titanate nanotubes takes place during unsaturation of adsorption sites resulting in the breakage of Cr(III)–O–Cr(VI) linkage which causes decrease in the adsorption of Cr(VI). Adsorption of Cr (VI) from aqueous solution onto amino-functionalized titanate nanotubes was investigated by Wang *et al.* [33]. The adsorbent was prepared by covalently grafting [1-(2-amino-ethyl)-3-aminopropyl] trimethoxysilane onto protonated titanate nanotubes and exhibited much higher adsorption capacity for the removal of Cr (VI) as compared to protonated titanate nanotubes. It is because of high porosity, surface area, and presence of a large number of protonated amino groups which possess good selectivity toward Cr (VI) species due to electrostatic attraction. The authors have suggested that protonated amino groups adsorb negatively charged Cr (VI) species via

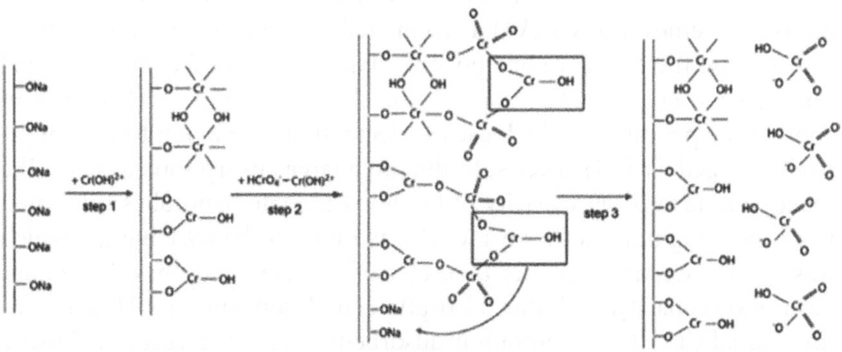

Scheme 7.2 Schematic diagram of the synergetic adsorption of Cr (III) and Cr (VI) in the binary system [Reproduced from reference 59].

electrostatic attraction by exchanging NO_3^- ions present earlier on them according to equations (7.1)–(7.3):

$$-NH_3^+NO_3^- + HCrO_4^- \leftrightarrow -NH_3^+HCrO_4^- + NO_3^- \quad (7.1)$$

$$2(-NH_3^+NO_3^-) + CrO_4^{2-} \leftrightarrow (-NH_3^+)_2CrO_4^{2-} + 2NO_3^- \quad (7.2)$$

$$2(-NH_3^+NO_3^-) + Cr_2O_7^{2-} \leftrightarrow (-NH_3^+)_2Cr_2O_7^{2-} + 2NO_3^- \quad (7.3)$$

The adsorbed Cr (VI) species were reduced partially to Cr (III) with the participation of H^+ ions and electrons from amino groups as per the following chemical equations:

$$Cr_2O_7^{2-} + 14H^+ + 6e^- \rightarrow 2Cr^{3+} + 7H_2O \quad (7.4)$$

$$CrO_4^{2-} + 8H^+ + 3e^- \rightarrow 2Cr^{3+} + 4H_2O \quad (7.5)$$

$$HCrO_4^- + 7H^+ + 3e^- \rightarrow 2Cr^{3+} + 4H_2O \quad (7.6)$$

Cr (III) species so formed were chelated with amino groups and hence were not entered into the solution. The mechanism of adsorption was shown in Scheme 7.3. Experimental results followed the Langmuir isotherm model which suggested the presence of homogeneous active sites and monolayer adsorption of Cr (VI) onto the surface of both the adsorbents. Maximum adsorption capacity computed by the Langmuir isotherm model was found to be 153.85 mg g^{-1} for amino-functionalized titanate nanotubes whereas that for protonated titanate nanotubes was 26.60 mg g^{-1} at initial pH 5.0 and temperature 30 °C. Titanate nanotubes modified with 2-bromo-ethylamine hydrobromide exhibited higher adsorption capacity for the removal of Cr (VI) as compared to titanate nanotubes [60]. It was because the existence of a positive charge on its surface in a wide pH range (pH < 8) as its pH_{zpc} was 8.2 and the presence of $-NH_2$ groups in appropriate pH ranges, which caused electrostatic attraction toward negatively charged Cr (VI) species, leading to a large adsorption capacity. The maximum adsorption capacity of modified titanate nanotubes calculated from the Langmuir isotherm model was found to be 69.1 mg g^{-1}, which was almost five times greater than that of titanate nanotubes. Maximum adsorption capacity, optimum adsorption conditions such as pH, temperature, initial Cr (VI) concentration, adsorbent dose, fitted isotherm, kinetic model, thermodynamic parameters, and mechanism of adsorption for Cr (VI) onto titanate nanotubes are listed in Table 7.1.

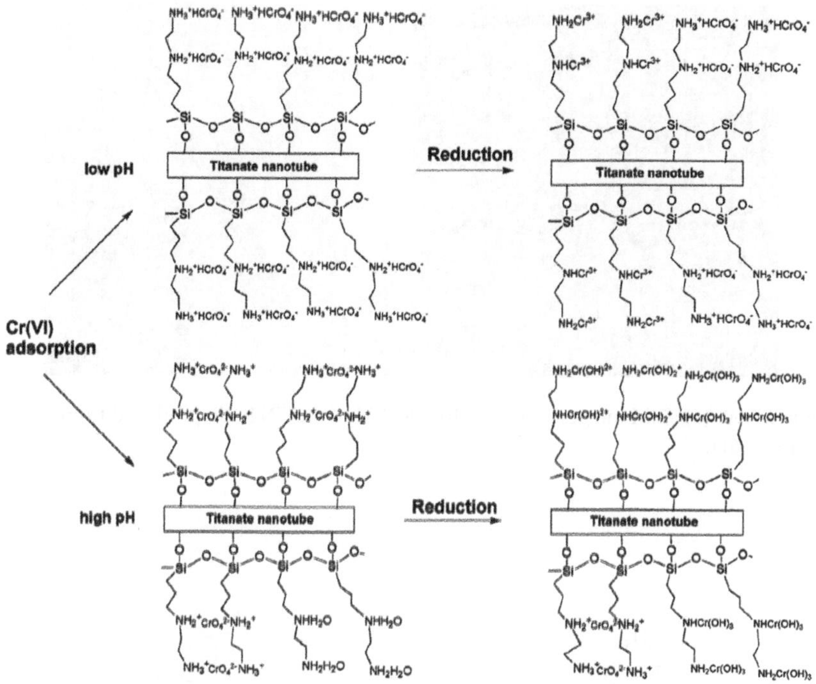

Scheme 7.3 Schematic illustration of Cr (VI) adsorption–reduction mechanism onto amino-functionalized titanate nanotubes (reproduced from reference 33).

7.3.2 Adsorption of Dyes on to Titanate Nanotubes

Calcined titanate nanotubes (TNT) were synthesized by Xiong *et al.* with a hydrothermal treatment of the commercial TiO_2 (Degussa P25) followed by calcination. The adsorption MB onto calcined TNT was studied. The pseudo-second-order model was the best to describe the adsorption kinetics and the intraparticle diffusion was not the rate-limiting step. The equilibrium adsorption data were in good agreement with the Langmuir isotherm model having a correlation coefficient of 0.993 indicating monolayer adsorption. The maximum adsorption capacity of 133.33 mg g^{-1} was obtained.

The mechanism of adsorption can be explained through chemical sorption which involves the formation of methylene blue-calcined titanate nanotube nanocomposite. Electrostatic attraction might be predominant in the initial bulk diffusion [61]. The morphological studies TEM (transmission electron microscopy) and HRTEM (high resolution TEM) of TNTs are shown in Figure 7.10.

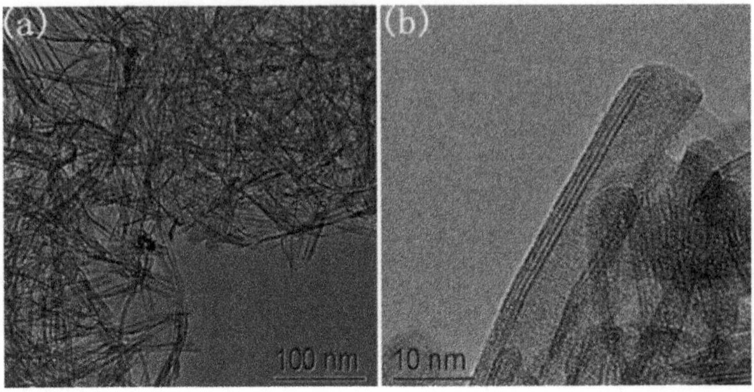

Figure 7.10 (a) TEM images of TNTs. (b) HRTEM of the TNTs [Reproduced from reference 61].

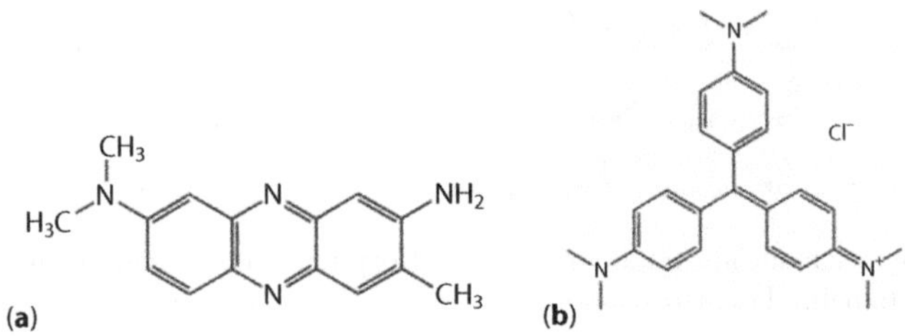

Figure 7.11 Chemical structure of (a) neutral red (NR) and (b) crystal violet (CV).

Hsieh and his coworkers prepared TNTs by the hydrothermal method and investigated the liquid-phase adsorption behavior of MB onto TNTs. Highly porous titania nanotubes with the maximum surface were formed by hydrothermal treatment. The nanotubes are mainly mesoporous and their pore sizes were found to increase with calcination temperature. Adsorption mainly occurs through physisorption process and the maximum adsorption capacity was found to be 290 mg g^{-1} [62]. Hierarchical 3D titanate nanotubes with pompon-like microspheres were synthesized by a simple hydrothermal method in the presence of CH_3COOH and H_2O_2. It indicates that the obtained material is rich in hydroxyl groups. The 3D-TNT microspheres exhibited excellent selective adsorption performance toward cationic dyes such as MB, NR, MG, and CV with very high rate of decolorization process. The saturated adsorption capacities for NR, MB, MG, and CV are found to be 401.8, 382.07, 361.95, and 322.5 mg g^{-1} at 25 °C respectively [63]. Lee *et al.* studied the adsorption behavior of basic dyes [basic green 5 (BG 5) and basic violet 10(BV 10)] on to titanate nanotubes and that of acid dyes (acid red 1 and acid blue 9) onto cetyltrimethylammonium (CTAC) chloride-modified TNTs. The low amount of Na^+ favors good adsorption of TNTs and their CTAC-modified version for the removal of basic and acid dyes from aqueous solution through the cation and anion exchange mechanism [64]. The effect of hydrothermal synthesis temperature of TNTs on adsorption of BG5 and BV10 was investigated by Lee and his coworkers [65]. The sample prepared at 160 °C exhibited a maximum adsorption capacity of 216 mg g^{-1} for BG5, which is comparable to that exhibited by activated carbon. It may be due to the retention of a nanotubular structure with the retention of adequate titanate interlayer spacing, surface area, and pore volume suitable for adsorption at this temperature. Furthermore, the adsorption capacity of TNT for BG5 is greater than that for BV10. It might be attributed to the smaller size of BG5, which can efficiently be packed in TNT [66]. Maximum adsorption capacity, optimum adsorption conditions such as pH, temperature, initial dye concentration, adsorbent dose, fitted isotherm, kinetic model, thermodynamic parameters, and the mechanism of adsorption for dyes onto titanate nanotubes are given in Table 7.2.

7.4 Layered Double Hydroxides

Layered double hydroxides are referred to as hydrotalcite-like materials with a general formula $[M(II)_{1-x}M(III)_x(OH)_2]^{x+}[A^{x-}].nH_2O$, where M(II) is the divalent metal ion, M(III) is the trivalent metal ion, and A^{x-} is the exchangeable interlayer anion [67]. These are a class of layered materials that possess a flat two-dimensional morphology composed of different elementary layers constituted by a hydroxy framework of different metal ions such as Mg^{2+}, Zn^{2+}, Al^{3+}, Fe^{3+}, etc. Isomorphous substitution of bivalent metal ions in the brucite-like hydroxy framework by trivalent metal ions such as Al^{3+}, Cr^{3+}, Fe^{3+}, etc. gives rise to the generation of additional charge over the layers, which is counterbalanced by exchangeable anions such as Cl^-, CO_3^{2-}, SO_4^{2-}, etc. In recent years, these are used extensively for water remediation through the removal of noxious anions (and interestingly sometimes cation) from waste waters [68, 69] because of their large surface area for the adsorption of anion, ability for facile exchange of interlayer anion, and tailorable hydrophobic/hydrophilic nature [70]. HTlcs can remove anions from the solution by three different mechanisms: (a) adsorption on external surfaces, (b) intercalation by anion exchange, and (c) intercalation by reconstruction of the calcined material. The third mechanism follows the calcination of LDH to produce intermediate non-stoichiometric oxides, which undergo rehydration in an aqueous medium and give back the hydroxide structure with different anions in the interlayers (memory effect) [71].

7.4.1 Adsorption of Cr (VI) on to Layered Double Hydroxides

Layered double hydroxides containing nitrates or chlorides in the interlayer, after modification and calcined forms, are generally explored for the removal of Cr (VI) [72, 73]. Houri *et al.* studied the removal of Cr (VI) Mg Al –Cl, Zn Al – Cl and Zn Cr - Cl with chloride as an interlayer anion. The authors reported 55 mg g^{-1} uptake capacity [72]. Bulbulian and co-workers studied the chromate uptake efficiency of MgAl LDHs and reported that the calcined ones are better adsorbents than as-synthesized forms. The authors studied the materials for real systems such as tannery waste and concluded that calcined samples were the best [74]. Goswamee *et al.* carried out the adsorption of Cr (VI) onto MgAl, NiAl, and ZnCr LDHs and found that MgAl was effective than the other ones. Among MgAl samples, those with higher Al content on calcination gave good uptake capacity. The rehydration efficiency of the MgAl sample made it a good adsorbent for Cr (VI) than NiAl and ZnCr [75]. Ternary HTs were used to increase the adsorption capacity. Zr^{4+} was incorporated over ZnAl and MgAl in different amounts and achieved a maximum chromate uptake capacity of 31 mg g^{-1} for calcined MgAlZr-HT (at 450 °C). With the incorporation of Zr^{4+}, the surface charge increased that helped in the removal of anionic species of Cr (VI) [76]. Later in 2007, Carriazo *et al.* explored the anion exchange ability of chloride for chromate uptake and they compared the uptake capacity of chloride-LDHs with its carbonate counterpart. The results reported show that MgAl chloride samples are better adsorbents than even calcined MgAl sample [77]. Calcined MgAl-CO_3 was also explored by Alvarez-Ayuso and Nugteren for the removal of Cr (VI) through reconstruction and reported 128 mg g^{-1} uptake [78]. Similar results were earlier reported by Rhee *et al.* through reconstruction over calcined MgAl-CO_3 [79]. Lazaridis and co-workers investigated the sorption of Cr (VI) over MgAl-CO_3 LDHs. The calcined material gave an uptake capacity of 120 mg g^{-1}. The authors reported that with agitation speed, at low pH, and at high chromate concentration, the uptake capacity enhanced. Detailed kinetic and equilibrium studies were done over as-synthesized MgAl-CO_3 and reported 17 mg g^{-1} uptake capacity [80, 81]. The mechanism of adsorption and factors affecting adsorption of Cr (VI) in an aqueous solution onto aluminum magnesium mixed hydroxide nanoparticles were studied by Li *et al.* [82]. According to the study, adsorption of Cr (VI) species onto the adsorbent is the combination of anion exchange and surface complexation processes. A former mechanism suggested that Cl^- ions associated with the surface or interlayer of the adsorbent exchange with the anionic Cr (VI) species as per Equation (7):

$$\text{Sur-OH}_2^+/\text{Cl}^- + \text{HCrO}_4^- \rightarrow \text{Sur-OH}_2^+/\ \text{HCrO}_4^- + \text{Cl}^- \qquad (7)$$

The later mechanism involved the electrostatic interaction between the protonated surface of the adsorbent and negatively charged Cr (VI) species as mentioned in equation (8).

$$\text{Sur-OH}_2^+ + \text{HCrO}_4^- \rightarrow \text{Sur-OH}_2^+/\text{HCrO}_4^- \qquad (8)$$

The maximum adsorption capacity was found to be 112 mg g^{-1} at pH 4.0 and 40 °C. The nano-sized adsorbents with Mg/Al molar ratio 3 exhibited largest adsorption efficiency due to smallest average particle diameter and highest zeta potential. Ramirez *et al.* studied sol–gel MgAl hydrotalcite for the removal of chromate. MgAl sol–gel sample heated at 350 °C showed high adsorption capacity than 150 °C heated and unheated ones. The authors observed that with surface area the adsorption capacity increased, 350 °C calcined sample exhibits higher surface area and capacity [83]. Recently, Goh *et al.* reported the adsorption of Cr (VI) along with other oxyhalides and anions over MgAl LDHs prepared through fast precipitation. The nitrate and chloride-containing MgAl samples were studied. Among the anions studied, chromate was removed to a greater extent (1.2 meq g^{-1}). The mechanism for uptake of Cr (VI) was also studied by the authors and stated that the occurrence of anion exchange or ligand exchange results in outer-sphere and inner-sphere complexes [84]. The influence of competitive anions such as nitrate, sulfate, carbonate, phosphate over adsorption of chromate, and other oxyanions were also reported by them. The influence of other co-existing anions on the adsorption of chromate had great dependence on the pH of the solution as at a different pH, the co-existing anions as well as chromate moieties will change [85]. Xu *et al.* carried out a column study for chromate uptake over calcined MgAl-CO_3. The results showed that calcined MgAl exhibited around 44 mg g^{-1} uptake capacities over a range of 50–200 mg L^{-1}. The authors claimed that the material could be regenerated and reused effectively [86]. Cocheci *et al.* analyzed the adsorptive capacity of ZnAl-CO_3 and found that as-synthesized samples were effective for chromate removal than its calcined forms. They also stated that in the case of as-synthesized ZnAl-CO_3, the chromate was not exchanged with carbonate in the interlayer, but exchanged at the edges of brucite-like layers [87]. Cocheci and coworkers reported uptake of Cr (VI) onto MgZnAl hydrotalcites with a varying Mg/Zn/Al ratio. The studies over calcined MgZnAl HTs showed that MgAl gave maximum chromate uptake. The results confirmed that incorporation of Zn shows no effect on chromate uptake capacity [88]. The impact of ferric iron in calcined nano-Mg/Al hydrotalcite on the adsorption of Cr (VI) from the aqueous solution was studied with respect to structure characteristics, adsorption properties, and the removal mechanism [89]. Mg/Al and Mg/Al/Fe precursors are nanosized particles and lamellar structure but the latter possessed smaller interlayer space. The mechanism of Cr (VI) removal by both the adsorbents takes place not only through intercalation but also by strong interaction between negative ions and positive external surface and inter layer ion exchange mechanism. After the reconstruction process, calcined Mg/Al/Fe showed a high adsorption rate though its adsorption capacity was less than that of calcined Mg/Al. It was because the replacement of Al^{3+} by Fe^{3+} in layered double hydroxide increased the positive surface charge by strengthening the bonds between double hydroxide layers and the carbonates in the interlayer due to which negatively charged Cr (VI) species were attracted toward it with a faster rate. On the other hand, the anions present in the interlayer were strongly attracted toward Fe^{3+} due to its high polarizing power than Al^{3+} making the ion exchange process more difficult, which decreased the adsorption capacity. Maximum theoretical saturated adsorption capacities calculated from the Langmuir isotherm model for calcined Mg/Al and Mg/Al/Fe were 85.9 and 75.4 mg g^{-1}, respectively, at pH 3.0 and at 48 °C. Though thermodynamic parameter values revealed that the adsorption process was spontaneous and endothermic, yet changes in these values indicated that the addition of Fe^{3+} in calcined Mg/Al will not be favorable for the adsorption process due to the nature of ferric iron, which makes the interlayer anion exchange more difficult, after the reconstruction of the structure. Therefore, the presence of Fe^{3+} in calcined Mg/Al hydrotalcite was unfavorable for the removal of Cr (VI) from the aqueous solution. The efficiency of Mg/Fe hydrotalcite-like compounds for the removal of Cr (VI) from an aqueous medium was studied by Das *et al.* [35]. Adsorption of Cr (VI) increased with the decrease in pH and it is independent of Cr (VI) speciation. Higher adsorption was observed at lower pH due to strong attraction of Cr (VI) species with the positively charged adsorbent surface. However, adsorption decreased significantly at pH $\leq$ 4.0 because of partial dissolution of the adsorbent at this pH. Less diffusion of ions into the pores of the adsorbent, reversible process between the adsorbed species and that in the bulk solution, and increased competition between

Cr (VI) species and OH^- ions are the facts which caused a decrease in the removal of Cr (VI) at higher pH. Kinetic studies revealed that adsorption occurred on the outer surface as well as within the pores of the adsorbent through diffusion. Hydrotalcite-like compounds containing cations such as Mg^{2+}, Zn^{2+}, and Al^{3+} with different proportions in brucite-like sheets and carbonate or benzoate anions in the inter layers were prepared by the coprecipitation method to evaluate their adsorption capacity for the removal of Cr(VI [90]. The most efficient adsorbent was obtained from the calcined Mg-Zn-Al hydrotalcite-like precursor containing interlayer CO_3^{2-} anions. This material exhibited greater performance for the removal of Cr (VI) from the aqueous solution in the presence of competitive anions such as Cl^- and SO_4^{2-}. The adsorption capacity also depended on the metal content. Thus, composition of metal ions and the structural and textural properties influenced the adsorption capacity of these adsorbents. Maximum adsorption occurred at an optimal pH of 6.0. Pseudo-first-order models described well the adsorption kinetics. Mg-Al oxides adsorbed the Cr (VI) species by the reconstruction of a layered structure while Zn-containing materials adsorbed them on their surface. The difference in behavior of Zn-containing materials was due to the separation of oxide phases. Layered double hydroxides with Co/Bi = 3 were synthesized by urea hdrolysis and subjected for adsorption of Cr (VI). The as-prepared Co-Bi LDH samples exhibited a maximum adsorption capacity of 277.7 mg g^{-1} at pH 7.0 [91]. Maximum adsorption capacity, optimum adsorption conditions such as pH, temperature, initial Cr (VI) concentration, adsorbent dose, fitted isotherm, kinetic model, thermodynamic parameters, and mechanism of adsorption for Cr (VI) onto layer double hydroxides are listed in Table 7.1.

7.4.2 Adsorption of Dyes on to Layered Double Hydroxides

Santos *et al.* have studied the thermal decomposition of carbonate-intercalated layered double hydroxide (ZnAl-CO_3-LDH) followed by acid blue 113 adsorption [92]. The kinetics of dye adsorption on calcined-LDH shows a two-step behavior consistent with the dye adsorption on the external and internal surfaces of ongoing recovered LDH lamellar aggregates. The intercalation of anionic dye in the interlayer LDH galleries is only verified after 180 min of contact with the dye solution, contributing to merely 13% of the adsorption capacity. The adsorption isotherms follow the Langmuir model, confirming that the adsorption takes place mainly on the reconstructed LDH external surface. The LDH memory effect allows the efficient recyclability of the used LDH adsorbent.

Magnesium aluminum layered double hydroxides were synthesized by Abdellaoui *et al.* through the co-precipitation method with nitrate and carbonate as an interlayer anion ($MgAlNO_3$ and $MgAlCO_3$) followed by the calcination of a carbonate-containing LDH sample to obtain the mixed oxide MgAl500 [93]. The resulting materials were characterized by several physicochemical techniques. Calcined and uncalcined LDHs were used as adsorbents to remove the amaranth azo dye in an aqueous solution. Adsorption results indicated that MgAl500 had a greater adsorption capacity (1.6 $mmol{\cdot}g^{-1}$) than non-calcined LDHs (0.8 and 0.2 $mmol{\cdot}g^{-1}$ for $MgAlNO_3$ and $MgAlCO_3$, respectively). Isotherms showed that the adsorption of the dye was consistent with the Langmuir model. The kinetic experiments of dye adsorption showed that the system reached the adsorption equilibrium in 3 h for the three adsorbents. The kinetic data fitted well with the pseudo-second-order model. Powder X-ray diffraction patterns of the adsorption products suggest that adsorbed amaranth anion was intercalated in the $MgAlNO_3$ interlayers by anion exchange and in the MgAl500 by the rehydration and reconstruction process, according to the value of the basal space d003 = 17.7 Å. Whereas in the case of $MgAlCO_3$, it was probably adsorbed on the external particle surface.

The influence of common inorganic anions present in the solution was also studied and a decrease in the dye removal was observed in the following order $HPO_4^{2-} > CO_3^{2-} > SO_4^{2-} > Cl^-$. The dye desorption was tested with a carbonate and phosphate solution and it was partial in all cases. The recyclability of the used adsorbents was performed by calcinations, and the percentage of removal decreases after the first cycle for $MgAlNO_3$ and MgAl500. The dye removal improved after the first cycle, and then decreased after the second cycle for $MgAlCO_3$.

A novel strategy for the preparation of poly(levodopa) functionalized MgAl-layered double hydroxide (PDOPA-*f*-LDH) was developed by Zhao *et al.* based on the modified mussel-inspired chemistry [94]. The utilization of PDOPA-*f*-LDH for the removal of methylene blue (MB) from an aqueous solution was also examined. Taken advantage of the self-polymerization of levodopa (DOPA) in an alkaline solution and the strong affinity of catechol groups to the substrate surface, the LDH was covered homogeneously by a layer of polymer coating of DOPA, leading to the functionalization toward LDH. The structure, surface morphology, thermos-stability, and elemental composition of as-prepared PDOPA-*f*-LDH were investigated by the transmission electron microscope, scanning electron microscopy, Fourier transform infrared spectroscopy, thermogravimetric analysis, and X-ray photoelectron spectroscopy. Besides, the surface charge of the PDOPA-*f*-LDH was also investigated using zeta potential. The effects of various parameters, including contact time, initial MB concentration, solution pH, and temperature, on the adsorption of MB onto PDOPA-*f*-LDH were systematically investigated. Results show that the adsorption capacity of functionalized LDH at 25 °C could reach up to 102 mg g^{-1}, which is much higher than that of pure LDH under the same experimental conditions. The adsorption kinetics and isotherm of MB adsorption were studied in batch experiments. The pseudo-second-order model is found to be the best to describe the adsorption kinetics. The isotherm result shows that the Freundlich isotherm is the better-fit-isotherm model to represent the equilibrium data. The values of thermodynamic parameters, including enthalpy change ΔH^0, entropy change ΔS^0, and Gibbs free energy change ΔG^0, were also determined. All the ΔG^0 values are negative; the ΔH^0 and ΔS^0 values of PDOPA-*f*-LDH were −7.824 kJ mol^{-1} and −0.01562 kJ mol^{-1} K^{-1}, respectively. The activation energy of system (E_a) is calculated as 24.69 kJ mol^{-1}. The fact suggests that the MB adsorption on PDOPA-*f*-LDH is a spontaneous and exothermic process. These obtained results indicate that the prepared PDOPA-*f*-LDH could be used as an interesting adsorbent with great potential to adsorb the cationic dyeing pollutants from aqueous media.

The NiFe layered double hydroxides with a different mole ratio of Ni/Fe (4:1, 3:1, 7:3, and 1:1) were prepared through a simple co-precipitation method by Lu *et al.* [95]. The adsorption performance was evaluated by the removal of methyl orange (MO) dye and hexavalent chromium (VI) heavy metal ion as shown in Figure 7.12. It is found that Ni_4Fe_1-LDH can remove

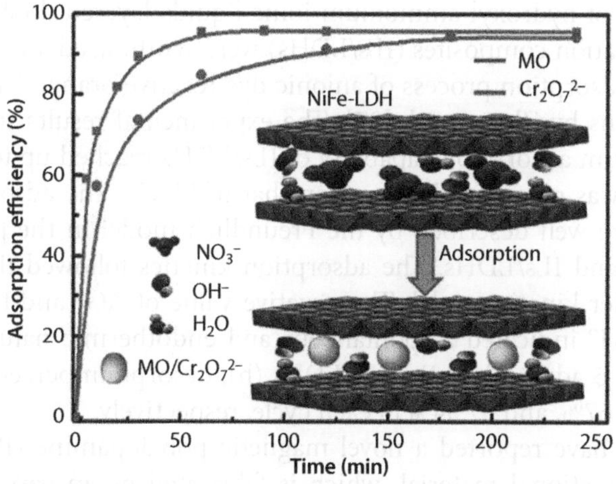

Figure 7.12 NiFe-LDH for Cr (VI) and methgyl orange (MO) dye adsorption [95].

more than 92% of MO in 10 min at the 10 mg L^{-1} MO initial concentration, and 97% of Cr (VI) in 1 h at 4 mg L^{-1} $Cr_2O_7^{2-}$ initial concentration. The saturated adsorption capacity of Ni_4Fe_1-LDH is found to be as large as 205.76 mg g^{-1} for MO and 26.78 mg g^{-1} for Cr (VI). The adsorption behavior of this new adsorbent is fitted well with the Langmuir isotherm and the pseudo-second-order kinetic model, indicative of a monolayer and chemical adsorption that synergistically originates from the exchangeable anion mechanism and layer charge density. Due to the excellent removal capacity of MO and Cr (VI), the NiFe-LDHs could be a promising adsorbent for wastewater treatment.

Ahmed *et al.* adopted a new approach for the removal of indigo carmine blue (IC) dye which is extensively used in jeans manufacture and was successfully performed on novel mesoporous (LDH) nanoparticles prepared by the sol–gel route using CTAB as a shape and pore directing agent [96]. A successful adsorption of IC on LDH nanoparticles of surface area = 85.6 $m^2 g^{-1}$ at various pH with a maximum adsorption capacity of 62.8 mg g^{-1} at pH = 9.5. The Langmuir model is more favorable to describe the adsorption of IC rather than the Freundlich model, which reflects the preferential formation of monolayer on the surface of LDH. Both film diffusion and the intraparticle diffusion affect the dye adsorption. The values of enthalpy change for (ΔH) and (ΔS) are +28.18 and +0.118 kJ mol^{-1}, respectively, indicating that the removal process is endothermic. The results indicated that LDH nanoparticles conserved a good activity even after five consecutive cycles of reuse. The results suggest that mesoporous LDH nanoparticles are considered a potential novel adsorbent for the remediation of wastewater containing IC.

A series of hydroxyl ammonium ionic liquids/layered double hydroxide intercalation composites (ILs/LDHs) were synthesized and adopted to study the adsorption process of anionic dye reactive orange 5 from aqueous solutions by Zhou *et al.* [97]. The experimental results showed that the maximum adsorption capability of ILs/LDHs reached up to 300.9 mg g^{-1}, which was obviously higher than that of LDHs. The adsorption isotherms were well described by the Freundlich model in the presence of the LDHs and ILs/LDHs. The adsorption kinetics followed the pseudo-second-order kinetic model. The negative value of $\Delta G°$ and the positive value of $\Delta H°$ indicated a spontaneous and endothermic nature of reactive orange 5 adsorption. For ILs/LDHs (b), desorption percentages were 57.91%, 46.67%, and 37.34% in each cycle, respectively.

Li *et al.* have reported a novel magnetic polydopamine (PDA)–LDH (MPL) bifunctional material, which is fabricated by an easy and green approach for the simultaneous removal of potentially toxic metals and anionic dyes [98]. In a mono-component system, the removal behaviors showed heterogeneous removal capacities of 75.01, 624.89, and 584.56 mg g^{-1} for Cu (II), methyl orange (MO), and Congo red (CR), respectively. Interestingly, the presence of CR and MO enhanced the removal of Cu (II) significantly in the Cu (II)–dye binary system. However, the presence of Cu (II) had no significant effect on dyes. Such a green and facile synthesis method, efficient removal performance, and superior reusability suggest that the MPL assemblies have a practical application potential for integrative and efficient treatment of coexisting toxic pollutants.

The "batch" method for evaluating the adsorption of Sunset Yellow FCF food dye into a CaAl-LDH-NO_3-layered double hydroxide (LDH) system was investigated by Pereira de Sa *et al.* [99]. The CaAl-LDH-NO_3 was co-precipitated at variable pH and characterized by X-ray diffraction, infrared spectroscopy, and scan electron microscopy. Adsorption experiments were carried out as a function of pH, contact time, and dye concentration. The results showed that the pH affects the adsorbent surface charge and the degree of anionic dye dissociation. The effective pH range for dye removal was between 4.0 and 10.0. A lower pH potentially causes dissolution of CaAl-LDH-NO_3. The Langmuir equation adequately describes the equilibrium data. The best-fits showed a maximum adsorption of 398.41 mg g^{-1}.

The Mg/Fe LDH with carbonate interaction and its calcined product (CLDH) were synthesized by Guo *et al.* [100] and used as adsorbents for the removal of acid brown 14 (AB 14) from aqueous solutions. The synthesized CLDH was efficient and had a much higher adsorption capability of 370.0 mg g^{-1} for AB 14 compared with that of the LDH, 41.7 mg g^{-1}. The X-ray diffraction patterns of the CLDH before and after the adsorption demonstrated that the adsorption was significantly enhanced by reconstruction of its original layered structure in the presence of AB 14 with the memory effect. It was found that the adsorption isotherms were well described by the Langmuir model for both the LDH and the CLDH with a correlation coefficient (R^2) > 0.99, and the adsorption kinetics of AB 14 on these two adsorbents followed the pseudo-second-order kinetic model. The results indicated that the calcined Mg/Fe-layered double hydroxide is an efficient adsorbent for the treatment of waste water with a high concentration of anionic dyes.

Ni-Al-layered double hydroxide (Ni-Al LDH) was synthesized by the co-precipitation method from their nitrate salts and utilized as an adsorbent for the removal of methyl orange (MO) dye from its aqueous solution by Monash *et al.* [101]. The synthesized Ni-Al LDH was characterized using thermogravimetric analysis (TGA), X-ray diffraction (XRD), and

N_2 adsorption–desorption analysis. Batch adsorption isotherm experiments were conducted with methyl orange dye at three different temperatures (30 °C, 40 °C, and 50 °C). Adsorption isotherm data were fitted with Langmuir, Freundlich, and Redlich–Peterson models. It was found that the Langmuir and Redlich–Peterson isotherm models best described the adsorption of MO on calcined Ni-Al LDH. The experimental results revealed that the increase in temperature increases the adsorption capacity of MO on the calcined Ni-Al LDH adsorbent. The maximum adsorption capacity was found to be 5.7×10^{-4} mol g^{-1} at 50 °C. The influence of pH on the adsorption of MO dye indicated that the adsorbent has good structural stability in the studied pH range. Thermodynamic studies authenticated that the adsorption of MO dye on calcined Ni-Al LDH was spontaneous, endothermic, and an entropy-driven process.

The Mg-Al-CO_3-LDH with a Mg^{2+}/Al^{3+} molar ratio of 2 was prepared via the co-precipitation method at a constant pH of 9–10 and used to remove three red dyes by the batch adsorption method [102]. The results showed Mg-Al-LDH were well crystallized and can adsorb the red dyes effectively. The optimal adsorbent dosage and contact time were 1.0 g and 60 min, respectively. Adsorbed amount hardly changed when $pH < 10$. The adsorption kinetics fit the pseudo-second-order kinetic models well and isotherms correspond to the Langmuir model strictly. All the relevant mechanisms were studied and manifested as anion exchange and also electrostatic attraction for CR. Maximum adsorption capacity, optimum adsorption condition such as pH, temperature, initial dye concentration, adsorbent dose, fitted isotherm, kinetic model, thermodynamic parameters, and mechanism of adsorption for dyes onto layer double hydroxides are listed in Table 7.2.

7.5 Conclusion

Nanomaterials emerged as good adsorbents for remediation of Cr (VI) due to their high surface area as well as highly active surface sites and exhibited relatively higher adsorption capacity. Modification of their surfaces was carried out to enhance the surface area and number of active surface sites and, hence uptake capacity. Protonated surfaces or protonated modified surfaces of nanomaterials adsorbed Cr (VI) species through electrostatic attraction. In addition to this, surface functional groups such as -OH, -CO-, etc. reduced Cr (VI) to Cr (III). Methods of preparation were responsible for the formation of nanostructutres with a high surface area and uniform particle size which enhanced the uptake capacity.

Introduction of nanoporous architecture, increasing the surface area, and developing layered functional nanomaterials show promising dye adsorption capacities. Mesoporous silica materials having long range ordered porosity, intraparticle mesoporosity, and high effective surface area favor enhanced adsorption (mostly fitted with the Langmuir model) with second-order kinetics. Titanate nanotubes with their unique arryed nanostructures provide higher surface area and large pore volumes for better adsorption of noxious dyes. Moreover, the layered fuctional structures of LDH favors enhanced adsorption in its interlayer spacings.

Acknowledgment

The authors are thankful to the management of SOA University, Bhubaneswar, for their constant support.

References

1. Ahmaruzzaman, M., Adsorption of phenolic compounds on low-cost adsorbents: A review. *Adv. Colloid Interf. Sci.*, 143, 48, 2008.
2. Rafatullah, M., Sulaiman, O., Hashim, R., Ahmad, A., Adsorption of methylene blue on low-cost adsorbents: A review. *J. Hazard. Mater.* 177, 70, 2010.
3. Sun, Z., Yao, G., Zhang, X., Zheng, S., Frost, R.L., Enhanced visible-light photocatalytic activity of kaolinite/g-C3N4 composite synthesized viamechanochemical treatment. *Appl. Clay Sci.*, X, 7, 2016.
4. Lytras, G., Lytras, C., Argyropoulou, D., Dimopoulos, N., Malavetas, G., Lyberatos, G., A novel two-phase bioreactor for microbial hexavalent chromium removal from wastewater. *J. Hazard. Mater.*, 336, 41, 2017.
5. Órfão, J.J.M., Silva, A.I.M., Pereira, J.C.V., Barata, S.A., Fonseca, I.M., Faria, P.C.C., Pereira, M.F.R., Adsorption of a reactive dye on chemically modified activated carbons – Influence of pH. *J Colloid Interf. Sci.*, 296, 480, 2006.
6. Demirbas, A. Agricultural based activated carbons for the removal of dyes from aqueous solutions: a review. *J. Hazard. Mater.*, 167, 1, 2009.
7. Huang, C.H., Chang. K.P., Ou, H.D., Chiang, Y.C., Wang, C.F., Adsorption of cationic dyes onto mesoporous silica. *Micropor. Mesopor. Mater.*, 141, 102, 2011.
8. Du, Y., Wang, L., Wang, J., Zheng, G., Wu, J., Dai, H., Flower-, wire- and sheet-like MnO_2-deposited diatomites: Highly efficient absorbents for the removal of Cr (VI). *J. Environ. Sci.*, 29, 71, 2015.
9. Huang, J., Zhang, X., Bai, L., Yuan, S., Polyphenylene sulfide based anion exchange fiber: Synthesis, characterization and adsorption of Cr(VI). *J. Environ. Sci.*, 24, 143, 2012.

10. Sun, X., Yang, L., Li, Q., Zhao, J., Li, X., Wang, X., Amino-functionalized magnetic cellulose nanocomposite as adsorbent for removal of Cr (VI): Synthesis and adsorption studies. *Chem. Eng. J.* 241, 175, 2014.
11. Demirbas, A., Heavy metal adsorption onto agro-based waste materials: A review. *J. Hazard. Mater.*, 157, 220, 2008.
12. Ngah, W.S., Hanafiah, M.A., Removal of heavy metal ions from waste water by chemically modified plant wastes as adsorbents: A review. *Bioresour. Technol.*, 99, 3935, 2008.
13. Zhao, D., Gao, X., Wu, C., Xie, R., Feng, S., Chen, C., Facile preparation of amino functionalized graphene oxide decorated with Fe3O4 nanoparticles for the adsorption of Cr (VI). *Appl. Surf. Sci.*, 384, 1, 2016.
14. Chen, Q., Luo, Z., Hills, C., Xue, G., Tyrer, M., Precipitation of heavy metals from wastewater using simulated flue gas: Sequent additions of fly ash, lime and carbon dioxide. *Water Res.*, 43, 2605, 2009.
15. Naeem, A., Saddique, M.T., Mustafa, S., Kim, Y., Dilara, B., Cation exchange removal of Pb from aqueous solution by sorption onto NiO. *J. Hazard. Mater.* 168, 364, 2009.
16. Cheng, X., Li, N., Zhu, M., Zhang, L., Deng, Y., Deng, C., Positively charged microporous ceramic membrane for the removal of Titan Yellow through electrostatic adsorption. *J. Environ. Sci.*, 44, 204, 2016.
17. Liu, Y.X., Yuan, D.X., Yan, J.M., Li, Q.L., Ouyang, T., Electrochemical removal of chromium from aqueous solutions using electrodes of stainless steel nets coated with single wall carbon nanotubes. *J. Hazard. Mater.*, 186, 473, 2011.
18. Owlad, M., Aroua, M.K., Daud, W.A.W., Baroutian, S., Removal of hexavalent chromium-contaminated water and wastewater: a review. *Water Air Soil Pollut.*, 200, 59, 2008.
19. Mallick, S., Dash, S.S., Parida, K.M., Adsorption of hexavalent chromium on manganese nodule leached residue obtained from NH_3- SO_2 leaching. *J. Colloid Interf. Sci.*, 297, 419, 2006.
20. Ahmaruzzaman, M., Gupta, V.K., Rice husk and its ash as low cost adsorbents in water and wastewater treatment. *Ind. Eng. Chem. Res.*, 50, 13589, 2011.
21. Kang, D., Yu, X., Ge, M., Xiao, F., Xu, H., Novel Al-doped carbon nanotubes with adsorption and coagulation promotion for organic pollutant removal. *J. Environ. Sci.*, 39, 1, 2016.
22. Zhang, G., Yi, L., Deng, H., Sun, P., Dyes adsorption using a synthetic carboxymethyl cellulose-acrylic acid adsorbent. *J. Environ. Sci.*, 26, 1203, 2014.
23. Zhang, Y.J., Ou, J.L., Duan, Z.K., Xing, Z.J., Wang, Y., Adsorption of Cr (VI) on bamboo bark-based activated carbon in the absence and presence of humic acid. *Colloids Surf. A*, 481, 108, 2015.
24. Saha, B., Orvig, C., Biosorbents for hexavalent chromium elimination from industrial and municipal effluents. *Coord. Chem. Rev.*, 254, 2959, 2010.
25. Albadarin, A.B., Mangwandi, C., Al-Muhtaseb, A.A.H., Walker, G.M., Allen, S.J., Kinetic and thermodynamics of chromium ions adsorption onto low-cost dolomite adsorbent. *Chem. Eng. J.*, 179, 193, 2012.

26. Yusof, A.M., Malek, N.A.N., Removal of Cr (VI) and As (V) from Aqueous Solutions by HDTMA-modified Zeolite. *J. Hazard. Mater.*, 162, 1019, 2009.
27. Gupta, V.K., Rastogi, A., Nayak, A., Adsorption studies on the removal of hexavalent chromium from aqueous solution using a low cost fertilizer industry waste material. *J. Colloid Interf. Sci.*, 342, 135, 2010.
28. Jain, M., Garg, V.K., Kadirvelu, K., Chromium (VI) removal from aqueous system using *Helianthus annuus* (sunflower) stem waste. *J. Hazard. Mater.*, 162, 365, 2009.
29. Sarkar, B., Xi, Y., Megharaj, M., Krishnamurti, G.S.R., Rajarathnam, D., Naidu, R., Remediation of hexavalent chromium through adsorption by bentonite based Arquad 2HT-75 organoclays. *J. Hazard. Mater.*, 183, 87, 2010.
30. Hua, M., Zhang, S., Pan, B., Zhang, W., Lv, L., Zhang, Q., Heavy metal removal from water/ waste water by nanosized metal oxides: A review. *J. Hazard. Mater.*, 211-212, 317, 2012.
31. Parida, K.M., Mishra, K.G., Dash, S.K., Adsorption of toxic metal ion Cr (VI) from aqueous state by TiO_2MCM-41: Equilibrium and kinetic studies. *J. Hazard. Mater.*,241–242, 395, 2012.
32. Chen, J.H., Xing, H.T., Guo, H.X., Weng, W., Hu, S.R., Li, S.X., Huang, Y.H., Sun, X., Su, Z.B., Investigation on the adsorption properties of Cr (VI) ions on a novel graphene oxide (GO) based composite adsorbent. *J. Mater. Chem. A*, 2, 12561, 2014.
33. Wang, L., Liu, W., Wang, T., Ni, J., Highly efficient adsorption of Cr (VI) from aqueous solutions by amino-functionalized titanate nanotubes. *Chem. Eng. J.*, 225, 153, 2013.
34. Jia, Z., Peng, K., Xu, L., Preparation, characterization and enhanced adsorption performance for Cr (VI) of mesoporous $NiFe_2O_4$ by twice pore-forming method. *Mater. Chem. Phy.*, 136, 512, 2012.
35. Das, J., Das, D., Dash, G.P., Das, D.P., Parida, K., Studies on Mg/Fe Hydrotalcite-like compound (Htlc): Removal of chromium (VI) from aqueous solution. *Int. J. Environ. Studies*, 61, 605, 2004.
36. Brown, J., Richer, R., Mercier, L., One-step synthesis of high capacity mesoporous Hg^{2+} adsorbents by non-ionic surfactant assembly. *Microp. Mesop. Mater.*, 37, 41, 2000.
37. Walcarius, A., Etienne, M., Bessière, J., Rate of access to the binding sites in organically modified silicates, amorphous silica gels grafted with amine or thiol groups. *Chem. Mater.*, 14, 2757, 2002.
38. Walcarius, A., Delacôte, C., Rate of access to the binding sites in organically modified silicates, effect of structure and density of functional groups in mesoporous solids obtained by the co-condensation route. *Chem. Mater.*, 15, 4181, 2003.
39. Kumari, V., Sasidharan, M., Bhaumik, A., Mesoporous $BaTiO_3$ @SBA-15 derived via solid state reaction and its excellent adsorption efficiency for the removal of hexavalent chromium from water. *Dalton Trans.*, 44, 1924, 2015.

40. Idris, S.A., Alotaibi, K.M., Peshkur, T.A., Anderson, P., Morris, M., Gibson, L.T., Adsorption kinetic study: Effect of adsorbent pore size distribution on the rate of Cr (VI) uptake. *Microp. Mesop. Mater.*, 165, 99, 2013.
41. Enshira, D., De Silva, N., Sayari, A., Adsorption of copperon amine functionalized SBA -15 prepared by co-condensation: Kinetic properties. *Chem. Eng. J.*, 166, 454, 2011.
42. Zhao, X.S., Lu, G.Q., Miller, G.J., Advances in mesoporous molecular sieves. *Ind. Eng. Chem. Res.*, 35, 2075, 1996.
43. Tian, L., Xie, G., Li, R., Yu, X., Hou, Y., Removal of Cr (VI) from aqueous solution using MCM-41. *Desalin. Water Treat.*, 36, 334, 2011.
44. Wang, X., Zhang, Y., Luo, W., Elzatahry, A.A., Cheng, X., Alghamdi, A., Abdullah, A.M., Deng, Y., Zhao, D., Synthesis of ordered mesoporous silica with tunable morphologies and pore sizes via a non-polar solvent-assisted Stöber method. *Chem. Mater.*, 28, 2356, 2016.
45. Mirzaie, M., Rashidi, A., Tayebi, H-A., Yazdanshenas, M.E., Removal of anionic dye from aqueous media by adsorption onto SBA-15/Polyamidoamine dendrimer hybrid: Adsorption equilibrium and kinetics. *J. Chem. Eng. Data*, 62, 1365, 2017.
46. Tsai, C.-H., Chang, W-C., Saikia, D., Wu, C-E., Kao, H-M., Functionalization of cubic mesoporous silica SBA-16 with carboxylic acid via one-pot synthesis route for effective removal of cationic dyes. *J. Hazard. Mater.*, 309, 236, 2016.
47. Juang, L.C., Wang, C.C., Lee, C.K., Adsorption of basic dyes onto MCM-41. *Chemosphere*, 64, 1920, 2006.
48. Wu, Y., Zhang, M., Zhao, H., Yang, S., Arkin, A. Functionalized mesoporous silica material and anionic dye adsorption: MCM-41 incorporated with amine groups for competitive adsorption of Acid Fuchsine and Acid Orange II. *RSC Adv.*, 4, 61256, 2014.
49. Nayab, S., Farrukh, A., Oluz, Z., Tuncel, E., Tariq, S.R., Rahman, H., Kirchhoff, K., Duran, H., Yameen, B., Design and Fabrication of branched polyamine functionalized mesoporous silica: An effcient absorbent for water remediation. *Appl. Mater. Interfaces*, 6, 4408, 2014.
50. Nanda, B., Pradhan, A.C., Parida, K., A comparative study on adsorption and photocatalytic dye degradation under visible light irradiation by mesoporous MnO_2 modified MCM-41 nanocomposite. *Microp. Mesop. Mater.*, 226, 229, 2016.
51. Grandcolas, M., Louvet, A., Keller, N., Layer-by-layer deposited titanate-based nanotubes for solar photocatalytic removal of chemical warfare agents from textiles. *Angew Chem Int Edit.*, 48, 161, 2009.
52. Zhai, Q.Q., Bo, T., Hu, G.X., High photoactive and visible-light responsive graphene/titanate nanotubes photocatalysts: Preparation and characterization. *J. Hazard. Mater.*, 198, 78, 2011.
53. Huang, J., Cao, Y., Liu, Z., Efficient removal of heavy metal ions from water system by titanate nanoflowers. *Chem. Eng. J.*, 180, 75, 2012.

53. Li, N., Zhang, L., Chen, Y., Fang, M., Zhang, J.X., Wang, H.M., Highly efficient, irreversible and selective ion exchange property of layered titanate nanostructures. *Adv. Funct. Mater.*, 22, 835, 2012.
54. Liu, J.H., Chen, J.S., Wei, X.F., Sandwich-like, stacked ultrathin titanate nanosheets for ultrafast lithium storage, *Adv Mater.*, 23, 998, 2011.
55. Bavykin, D.V., Friedrich, J.M., Walsh, F.C., Protonated titanates and TiO_2 nanostructured materials: Synthesis, properties, and applications. Adv Mater., 18, 2807, 2006.
56. Chen, Q., Peng, L.M., Structure and applications of titanate and related nanostructures. *Int. J. Nanotechnol.*, 4, 44, 2007.
57. Xiong, L., Chen, C., Chen, Q., Ni, J.R., Adsorption of Pb (II) and Cd (II) from aqueous solutions using titanate nanotubes prepared via hydrothermal method. *J. Hazard. Mater.*, 189, 741, 2011.
58. Liu, W., Chen, H., Borthwick, A. G. L., Han, Y., Ni, J., Mutual promotion mechanism for adsorption of coexisting Cr(III) and Cr(VI) onto titanate nanotubes. *Chem. Eng. J.*, 232, 228, 2013.
59. Niu, G., Liu, W., Wanga, T., Ni, J., Absorption of Cr(VI) onto amino-modified titanate nanotubes using 2-Bromoethylamine hydrobromide through SN2 reaction. *J. Colloid Interf. Sci.*, 401, 133, 2013.
60. Xiong, L., Yang, Y., Mai, J., Sun, W., Zhang, C., Wei, D., Chen, Q., Ni, J., Adsorption behavior of methylene blue onto titanate nanotubes. *Chem. Eng. J.*, 156, 313, 2010.
61. Hsieh, C., Fan, W., Chen, W., Impact of mesoporous pore distribution on adsorption of methylene blue onto titania nanotubes in aqueous solution. *Microp. Mesop. Mater.*, 116, 677, 2008.
62. Jiang, R., Zan, R., Zeng, J., Wen, X., Preparation of 3D Pompon-like titanate nanotube microspheres and their adsorption property on cationic dyes. *RSC Adv.*, 6, 97899, 2016.
63. Lee, C., Liu, S., Juang, L., Wang, C., Lyu, M., Hung, S., Application of titanate nanotubes for dyes adsorptive removal from aqueous solution. *J. Hazard. Mater.*, 148, 756, 2007.
64. Lee, C., Lin, K., Wu, C., Lyu, M., Lo, C., Effects of synthesis temperature on the microstructures and basic dyes adsorption of titanate nanotubes. *J. Hazard. Mater.*, 150, 494, 2008.
65. Wang, C.C., Juang, L.C., Hsu, T.C., Lee, C.K., Lee, J.F., Huang, F.C., Adsorption of basic dyes onto montmorillonite. *J. Colloid Interf. Sci.*, 273, 80, 2004.
66. Reichle, W.T., Synthesis of anionic clay minerals mixed metal hydroxides, hydrotalcite. *Solid State Ionics*, 22, 135, 1986.
67. Goh, K., Lim, T., Dong, Z., Application of layered double hydroxides for removal of oxyanions: A review. *Water Res.*, 42, 1343, 2008.
68. Perez, M.R., Pavlovic, I., Barriga, C., Cornejo, J., Hermosin, M.C., Ulibarri, M.A., Uptake of Cu^{2+}, Cd^{2+} and Pb^{2+} on Zn–Al layered double hydroxide intercalated with edta. *Appl. Clay Sci.*, 32, 245, 2006.

69. Cavani, F., Trifiro, F., Vaccari, A., Hydrotalcite-type anionic clays: Preparation, properties and applications. *Catal. Today*, 11, 173, 1991.
70. Miyata, S., Physico-chemical properties of synthetic hydrotalcites in relation to composition. *Clays Clay Miner.*, 28, 50, 1980.
71. Houri, B., Legrouri, A., Barroug, A., Forano, C., Besse, J.P., Removal of chromate ions from water by anionic clays. *J. Chem. Phys.*, 96, 455, 1999.
72. Prasanna, S.V., Rao, R.A.P., Kamath, P.V., Layered double hydroxides as potential chromate scavengers. *J. Colloid Interf. Sci.*, 304, 292, 2006.
73. Gallegosa, S.M., Martinez, V., Bulbulian, S., Chromium (VI) separation from tannery wastes utilizing hydrotalcites. *Sep. Sci. Technol.*, 39, 667, 2004.
74. Goswamee, R.L., Sengupta, P., Bhattacharyya, K.G., Dutta, D.K., Adsorption of Cr (VI) in layered double hydroxides. *Appl. Clay Sci.*, 13, 21, 1998.
75. Das, N.N., Konar, J., Mohanta, M.K., Srivastava, S.C., Adsorption of Cr (VI) and Se (IV) from their aqueous solutions onto Zr^{4+} substituted ZnAl/MgAl-layered double hydroxides: effect of Zr^{4+} substitution in the layer. *J. Colloid Interf. Sci.*, 270, 1, 2004.
76. Carriazo, D., Arco, M.D., Martin, C., Rives, V., A comparative study between chloride and calcined carbonate hydrotalcites as adsorbents for Cr (VI). *Appl. Clay Sci.*, 37, 231, 2007.
77. Ayuso, E.A., Nugteren, H.W., Purification of chromium(VI) finishing waste waters using calcined and uncalcined Mg-Al-CO_3-hydrotalcite. *Water Res.*, 39, 2535, 2005.
78. Rhee, S.W., Kang, M.J., Kim, H., Moon, C.H., Removal of aquatic chromate ion involving rehydration reaction of calcined layered double hydroxide (Mg-Al-CO_3). *Environ. Technol.*, 18, 231, 1997.
79. Lazaridis, N.K., Asouhidou, D.D., Kinetics of sorptive removal of chromium (VI) from aqueous solutions by calcined Mg–Al–CO_3 hydrotalcite. *Water Res.*, 37, 2875, 2003.
80. Lazaridis, N.K., Pandi, T.A., Matis, K.A., Chromium (VI) removal from aqueous solutions by Mg-Al-CO3 hydrotalcite: sorption-desorption kinetic and equilibrium studies. *Ind. Eng. Chem. Res.*, 43, 2209, 2004.
81. Li, Y., Gao, B., Wu, T., Sun, D., Li, X., Wang, B., Lu, F., Hexavalent chromium removal from aqueous solution by adsorption on aluminum magnesium mixed hydroxide. *Water Res.*, 43, 3067, 2009.
82. Ramirez, E.R., Ortega, N.L.G., Soto, C.A.C., Gutierrez, M.T.O., Adsorption isotherm studies of chromium (VI) from aqueous solutions using sol–gel hydrotalcite-like compounds. *J. Hazard. Mater.*, 172, 1527, 2009.
83. Goh, K.H., Lim, T.T., Banas, A., Dong, Z., Sorption characteristics and mechanisms of oxyanions and oxyhalides having different molecular properties on Mg/Al layered double hydroxide nanoparticles. *J. Hazard. Mater.*, 179, 818, 2010.
84. Goh, K.H., Lim, T.T., Influences of co-existing species on the sorption of toxic oxyanions from aqueous solution by nanocrystalline Mg/Al layered double hydroxide. *J. Hazard. Mater.*, 180, 401, 2010.

85. Xu, Y., Zhang, J., Qian, G., Ren, Z., Xu, Z.P., Wu, Y., Liu, Q., Qiao, S., Effective Cr(VI) removal from simulated groundwater through the hydrotalcite-derived adsorbent. *Ind. Eng. Chem. Res.*, 49, 2752, 2010.
86. Cocheci, L., Barvinschi, P., Pode, R., Seftel, E.M., Popovici, E., Chromium (VI) ion removal from aqueous solutions using a Zn–Al-type layered double hydroxide. *Adsorpt. Sci. Technol.*, 28, 267, 2010.
87. Cocheci, L., Pode, R., Popovici, E., Dvininov, E., Iovi, A., Sorption removal of chromate in single batch systems by uncalcined and calcined Mg/Zn-Al-type hydrotalcites. *Environ. Eng. Manage. J.*, 8, 865, 2009.
88. Xiao, L., Ma, W., Han, M., Cheng, Z., The influence of ferric iron in calcined nano-Mg/Al hydrotalcite on adsorption of Cr (VI) from aqueous solution. *J. Hazard. Mater.*, 186, 690, 2011.
89. Dudek, B., Kustrowski, P., Białas, A., Natkanski, P., Piwowarska, Z., Chmielarz, L., Kozak, M., Michalik, M., Influence of textural and structural properties of Mg Al and Mg Zn Al containing hydrotalcite derived oxides on Cr(VI) adsorption capacity. *Mater. Chem. Phys.*, 132, 929, 2012.
90. Jaiswal, A., Mani, R., Banerjee, S., Gautam, R.K., Chattopadhyaya, M.C., Synthesis of novel nano-layered double hydroxide by urea hydrolysis method and their application in removal of chromium(VI) from aqueous solution: Kinetic, thermodynamic and equilibrium studies. *J. Molec. Liq.*, 202, 52, 2015.
91. Santos, R., Tronto, J., Briois, V., Santilli, C., Thermal decomposition and recovery properties of ZnAl–CO3 layered double hydroxide for anionic dye adsorption: insight into the aggregative nucleation and growth mechanism of the LDH memory effect. *J. Mater. Chem. A*, 5, 9998, 2017.
92. Abdellaoui, K., Pavlovic, I., Bouhent, M., Benhamou, A., Barriga, C., A comparative study of the amaranth azo dye adsorption/desorption from aqueous solutions by layered double hydroxides. *Appl. Clay Sci.*, 143, 142, 2017.
93. Zhao, J., Huang, Q., Liu, M., Dai, Y., Chen, J., Huang, H., Wen, Y., Zhu, X., Zhang, X., Wei, Y., Synthesis of functionalized MgAl-layered double hydroxides via modified mussel inspired chemistry and their application in organic dye adsorption. *J. Colloid Interf. Sci.*, 505, 168, 2017.
94. Lu, Y., Jiang, B., Fang, L., Ling, F., Gao, J., Wu, F., Zhang, X., High performance NiFe layered double hydroxide for methyl orange dye and Cr (VI) adsorption. *Chemosphere*, 152, 415, 2016.
95. Ahmed, M., Mohamed, A., An efficient adsorption of indigo carmine dye from aqueous solution on mesoporous Mg/Fe layered double hydroxide nanoparticles prepared by controlled sol–gel route. *Chemosphere*, 174, 280, 2017.
96. Zhou, Q., Chen, F., Wu, W., Bu, R., Li, W., Yang, F., Reactive orange 5 removal from aqueous solution using hydroxyl ammonium ionic liquids/layered double hydroxides intercalation composites. *Chem. Eng. J.*, 285, 198, 2016.
97. Li, J., Fan, Q., Wu, Y., Wang, X., Chen, C., Tang, Z., Wang, X., Magnetic polydopamine decorated with Mg–Al LDH nanoflakes as a novel bio-based

adsorbent for simultaneous removal of potentially toxic metals and anionic dyes. *J. Mater. Chem. A*, 4, 1737, 2016.
98. de Sá, F.P., Cunha, B.N., Nunes, L.M., Effect of pH on the adsorption of Sunset Yellow FCF food dye into a layered double hydroxide (CaAl-LDH-NO_3). *Chem. Eng. J.*, 215, 122, 2013.
99. Guo, Y., Zhu, Z., Qiu, Y., Zhao, J., Enhanced adsorption of acid brown 14 dye on calcined Mg/Fe layered double hydroxide with memory effect. *Chem. Eng. J.*, 219, 69, 2013.
100. Monash, P., Pugazhenthi, G., Utilization of calcined Ni-Al layered double hydroxide (LDH) as an Adsorbent for removal of methyl orange dye from aqueous solution. *Env. Progress Sustain. Energy*, 33, 154, 2014.
102. Shan, R., Yan, L.-G., Yang, Y.-M., Yang, K., Yu, S., Yu, H., Zhu, B., Du, B., Highly efficient removal of three red dyes by adsorption onto Mg–Al-layered double hydroxide. *J. Indust. Eng. Chem.*, 21, 561, 2015.

8

Ultrasonic Synthesis of Zero Valent Iron Nanoparticles for the Efficient Discoloration of Aqueous Solutions Containing Methylene Blue Dye

Mohammadreza Kamali[1,2], Isabel Capela[1,*] and Maria Elisabete Costa[2]

[1]*Department of Environment and Planning, Center for Environmental and Marine Studies, CESAM, University of Aveiro, Aveiro, Portugal*
[2]*Department of Materials and Ceramics Engineering, Aveiro Institute of Materials, CICECO, University of Aveiro, Aveiro, Portugal*

Abstract

Dyes are complex toxic and non-biodegradable aromatic compounds used as colorants in many industries such as textile, pulp and paper, printing, leather, etc. For environment safety purposes, the separation of dyes from polluted effluents is a major concern. The current study aims at ultrasonic synthesis of nano-sized zero valent iron (nZVI) particles and their application for the removal of methylene blue (MB) from the aqueous solution. The removal of MB dye can be optimized by varying the operating pH and the contact time between the dye molecules and nZVI particles. UV-Vis spectroscopy, employed to study the MB removal efficiency and the underlying mechanisms, indicates high efficiency of the nanomaterials regarding MB removal as compared to literature data. The synthesized nanomaterials possess an excellent removal potential for the treatment of industrial wastewater containing dyes.

Keywords: Nano-sized zero valent iron (nZVI), ultrasonic irradiation, methylene blue, wastewater, treatment

**Corresponding author*: icapela@ua.pt

Shahid-ul-Islam and B.S. Butola (eds.) Nanomaterials in the Wet Processing of Textiles, (261–284) © 2018 Scrivener Publishing LLC

8.1 Introduction

Dyes are major organic compounds of effluents discharged from industries such as textile, pulp and paper, cosmetic and pharmaceuticals, food and packaging, etc. [1, 2]. Methylene blue (MB), that is, methylthioninium chloride (Figure 8.1), is a member of the thiazine class of dyes with many applications [3–5] which was first prepared by Heinrich Caro in 1876 [6]. The toxic effects from thiazine dyes have been previously investigated and documented [7]. Severe toxic effects (1%) on Iris epithelium and corneal endothelium [11], and toxic effects on human luteal cells [12], on fetus [13], and on erythroleukemic cells [14] are only few examples of the harmful impacts on living organisms when exposed to MB. Hence, the toxicity and carcinogenic effects of this type of dyes raise major concerns about human health and also about the safety of aquatic environments where dye-contaminated effluents might end up [8–10].

The removal of MB from industrial effluents is an important environmental issue because most of the conventional biological, physical, and chemical treatment methods are not able to efficiently degrade this compound [15]. In this context, exploring efficient methods and optimizing its operating conditions for MB removal are goals of high significance. The application of engineered nanomaterials (NMs) has proved to be a successful route for the removal of MB dye from contaminated environments when appropriate conditions are met. Recent studies have shown that the performance of NMs is highly conditioned by the experimental conditions, that is, contact time, pH, and temperature among others, the removal efficiency being very dependent on the combination of used conditions. In addition, the application of additional measures such as UV irradiation has shown the ability to improve the final removal efficiency when assisted by NMs. Table 8.1 summarizes literature data supporting these findings. Despite this pool of knowledge, there is still a need to find out the optimum conditions under which maximum removal efficiency can be achieved under reasonable investments of time, materials, and input energy.

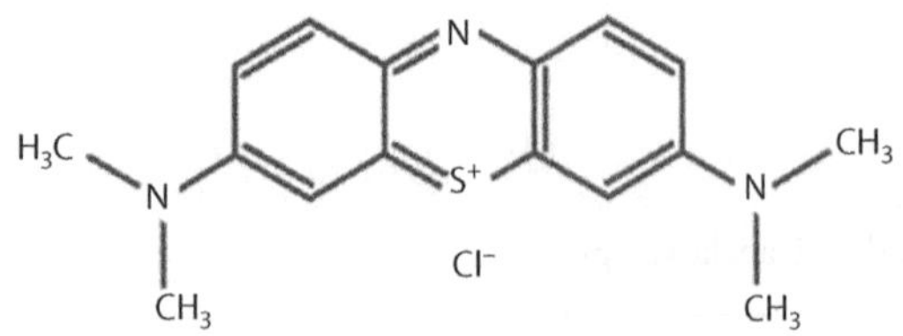

Figure 8.1 Chemical structure of MB.

Table 8.1 Performance of nanomaterials used for the removal of MB from dye solution.

Nanomaterials	Operating conditions							Removal efficiency (%)	Ref.
	Dye conc. (mg/L)	NMs dosage. (g/L)	Time (min)	Temp. (˚C)	pH	Reactor type	Other conditions		
Ti/Sb–SnO_2-TiN[1]	50	NG[2]	60	23	NG	Three-electrode cell	-	73.9[3]	[32]
	50	NG	120	23	NG	Three-electrode cell	-	100	
Ti/Sb-SnO_2	50	NG	60	23	NG	Three-electrode cell	-	34.3	
	50	NG	120	23	NG	Three-electrode cell	-	65.1	
Fe_2O_3/TiO_2	1×10^{-5} mol/L	1	60	NG	7	NG	UV irradiation	95	[8]
Poly Acrylonitrile	5	4	60	30-70	5-9	NG	-	70	[33]
Hydroximated poly acrylonitrile	5	4	60n	30-70	5-9	NG	-	90	
h-MoO_3	10	1	180	Room	NG	NG	Dark	67	[34]
h-MoO_3	10	1	180	Room	NG	NG	UV light[4]	71	
h-MoO_3	10	1	120	Room	NG	NG	Visible light[5]	98	
h-MoO_3	10	1	180	Room	NG	NG	Without catalyst	4	

(Continued)

Table 8.1 Cont.

Nanomaterials	Operating conditions							Removal efficiency (%)	Ref.
	Dye conc. (mg/L)	NMs dosage. (g/L)	Time (min)	Temp. (°C)	pH	Reactor type	Other con-ditions		
β-Cyclodextrin multi-conjugated Magnetic graphene oxide	40	0.7	150	Room	13	Vial reactor	-	93	[35]
$NixZn_{1-x}Fe_2O_4$[6]	200	0.3	90	Room	3	Photocatalytic reactor	UV irradiation		[36]
MnO_2	100	0.35	90	Room	11	Flask reactor	-	90	[37]
CuO/nano TiO_2	10	0.2	300	NG	NG	Total reflection quartz reactor	UV irradiation	99	[38]
SiO_2[7]	100	2.3	180	20	5.5	NG	-	9.4 (mg/g)8	[39]

1 Supporting electrolyte: 0.25 M $Na2S_{O4\ solution.}$
2 NG: not given.
3 The removal process was carried out using electrodes.
4 6 W, λ = 365 nm.
5 350 W, λ >400 nm, 833 × 100 lux.
6 X = 0.0 to 1.0.
7 Silica nano-sheets derived from vermiculite.
8 mg of MB to g of nanomaterials.

As an emerging technology, the application of iron-based NMs for the treatment of polluted waters and wastewaters underwent fast development in recent years. Within this context, iron oxides such as magnetite (Fe_3O_4) [16, 17], hematite (Fe_2O_3) [18–20], maghemite (γ-Fe_2O_3) [21, 22], and iron hydroxides such as lepidocrocite (α-FeOOH) [23] are of particular interest to degrade and remove various types of environmental contaminants. Zero valent iron (ZVI) is another type of iron-based material with a potential to be widely adopted in real-scale applications as an effective, non-toxic, and low-cost alternative for conventional clean-up technologies [24, 25]. The application of micro-sized ZVI particles in permeable reactive barriers [26, 27] is a good example of the utilization of this material as a well-known strategy for the treatment of highly polluted ground waters. However, ZVI particles, if used as nanosized particles (nZVI), may provide a significantly larger surface area-to-volume ratio, and also a higher surface energy which result in significant improvement of their reactivity toward contaminants [28–36]. In addition, various attempts have been made to enhance the properties of these NMs in terms of specific surface area (SSA), crystallinity, etc. [29–31]. However, there is still field for improvements. For instance, the simultaneous enhancement of nanomaterial properties such as crystallinity and surface area and the optimization of nZVI synthesis conditions still remain to be explored.

For optimizing the MB removal efficiency, nZVI synthesized with enhanced surface area and crystalline structure via a combined liquid phase reduction–ultrasonic irradiation technique is proposed here. Also, the MB removal can be maximized by correlating the conditions allowing optimum degradation of MB with the properties of the used NMs. The reactivity of the synthesized nZVI toward MB can be traced with a photometric analysis which allows us to identify the mechanisms involved as well.

8.2 Materials and Methods

8.2.1 Materials

Chemical reagents including iron (III) chloride hexahydrate ($FeCl_3.6H_2O$ from Sigma-Aldrich, St. Louis, MO) (99%) and sodium borohydride ($NaBH_4$ from Sigma-Aldrich, St. Louis, MO) (≥98.0%) without further purification are used as precursors and reductants for the synthesis of nZVI. Ethanol from Fisher scientific (99.0%), and deionized water supplied by an ultrapure water purification system (EASYpure RF-Barnstead/Thermolyne) are also used for experiments. MB (99%) is from Sigma-Aldrich (St. Louis, MO). pH adjustments are performed with H_2So_4 (>95%, Fisher Scientific, UK) and NaOH (JMGC co., Portugal).

8.2.2 Synthesis and Characterization of NMs

The synthesis of nZVI is performed in a 500 mL three-open-neck flask reactor installed inside an ultrasonic cleaning bath (Branson, 1510E-MT) with a theoretical dissipation rate of 70 W and operating at 42 KHZ ± 6%. Mechanical overhead stirring (at 250 rpm) with IKA Eurostar 40 stirrer is utilized to continuously stir the reaction medium. Certain amounts of a sodium borohydride solution (0.91 g of sodium borohydride dissolved in 100 mL of deionized water) are injected in 30 mL of an ethanol/deionized water solution (24 mL of ethanol + 6 mL of deionized water) containing the precursor (0.81 g of $FeCl_3.6H_2O$) at room temperature. A Watson Marlow 120U/R peristaltic pump is utilized to control the injection rate of the reductant. After completing the injection of the reductant solution, the final mixture is stirred for more 10 minutes in the reactor. The black precipitates are then collected, washed, and dried before characterization.

Crystallographic characterization of the synthesized particles is carried out by X-ray diffraction (XRD) analysis (Rigaku, Geigerflex, Japan) in the 2θ range from 30° to 100° with a goniometer speed of 2°/minute. The Scherrer equation (equation (2)) is used to determine the crystallite size of the particles:

$$d = K\lambda / \beta \cos\theta \tag{1}$$

where d is the average crystallite size, K is a dimensionless shape factor, λ is the wavelength of the X-ray used, β is the full-width at half maximum intensity (FWHM) of the sample, and θ is the Bragg angle. Transmission electron microscopy by Hitachi H9000na (Japan) operated at 300 kV acceleration voltage is utilized to study the morphology of the synthesized particles. Particle size distribution of the particles is obtained by counting more than 500 individual particles selected randomly using photoshop and Image J software programs. Particles' specific surface area (SSA) is assessed by gas adsorption with a Micromeritics Gemini V2 (USA) equipment. The SSA is calculated based on the Brunauer–Emmett–Teller (BET) isotherm. Zeta potential measurements are carried out using a zetasizer (Zetasizer Nano ZS, Malvern, UK) for assessing the electrical charge characteristics of the particles surface. For that, dilute aqueous suspensions of the NMs (10 mL) are prepared under ultrasonic irradiation in an ultrasonic cleaning bath (Model Sonoswiss 6lt, Switzerland). The pH of the nanoparticle suspensions is adjusted in the acidic and alkaline ranges using HCl or NaOH, respectively. The magnetic hysteresis measurements are performed using a vibrating sample magnetometer-VSM (CryogenicCryofree). The hysteresis curve is recorded at 5K under a magnetic field up to 10 T.

8.2.3 Discoloration of MB

Reactivity of the ZVI NMs prepared under ultrasonic irradiation ($nZVI^{UI}$) is determined using MB, as a model contaminant [40, 41]. The reactions are performed under visible light at 720 lux assisted by a multi-function environment meter (ST-8820, REED, USA) under ambient temperature. Based on a set of preliminary experiments (data not shown), the optimum dosage of the NMs for the removal of MB was identified as 1 g/L. $nZVI^{UI}$ particles are suspended in a 100 mL aqueous solution containing 25 mg/L of MB. The absorbance behavior of the MB solution under various pHs ranging from 4 to 10 at room temperature is monitored for 30 min with a UV-Vis spectrometer (T80+, PG Instruments, UK) from 200 to 750 nm. When the MB (green color) is dissolved in water, a blue color solution is yielded. The intensity of the blue color measured at wave lengths of maximum absorption, i.e., 665 and 605 nm (related to monomeric (0-0 band) and dimeric (0-1 band) forms of MB, respectively [42]) is used to follow the removal process and its efficiency.

8.3 Results and Discussion

8.3.1 Materials' Characterization

The XRD pattern of the particles prepared under ultrasonic irradiation is presented in Figure 8.2. The sample is composed of ZVI (JCPDS No. 07-71-3763), without any evidence of secondary phases. Based on Sherrer equation (equation (1)), the crystallite size of the powders was determined as 23.9 nm.

A TEM micrograph of the $nZVI^{UI}$ samples prepared under ultrasonic irradiation is presented in Figure 8.3. The sample consists of aggregates of enchained spherical particles. The particles clearly present a core-shell structure consisting of a dark core (nZVI) covered with a thin bright layer. Figure 8.4 shows the particle size distribution of the prepared particles. As observed, the average particle size is around 27 nm. The BET-specific surface area of this sample was determined as 50.6 m^2/g; this large value is

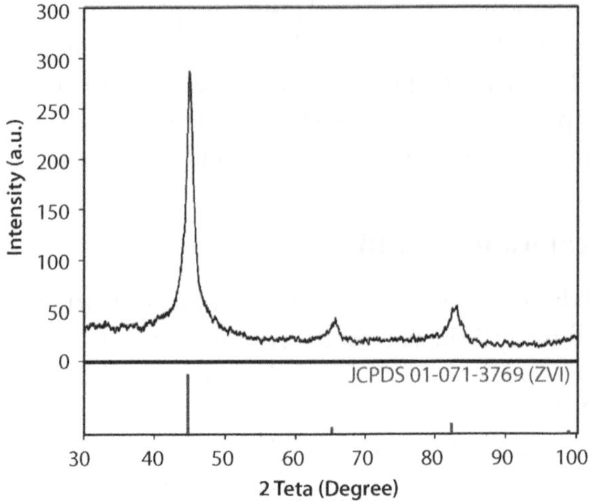

Figure 8.2 XRD pattern of the synthesized nZVI^{UI} particles. A ZVI single-phase can be identified according to the JCPDS database.

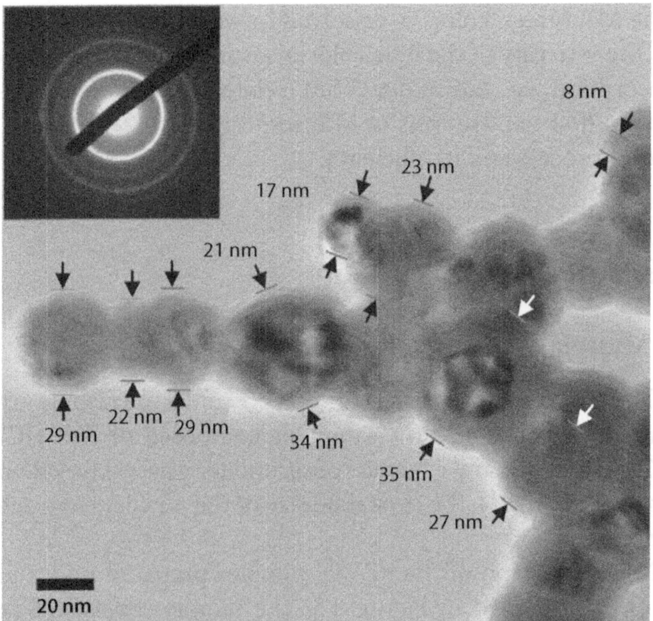

Figure 8.3 TEM and SAED images of the nZVI^{UI} particles.

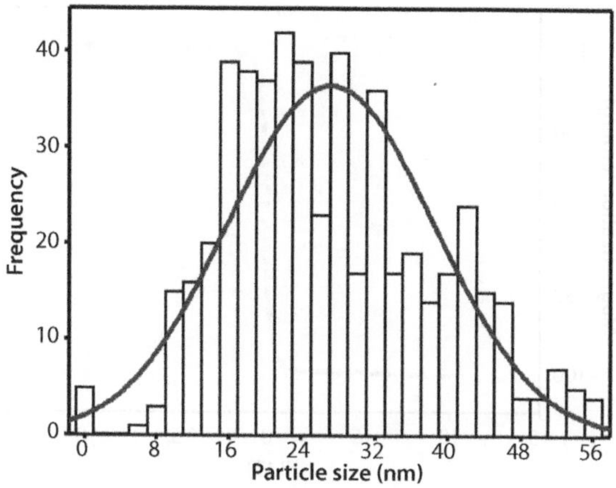

Figure 8.4 Particle size distribution of the synthesized $nZVI^{UI}$. The average particle size is around 27 nm.

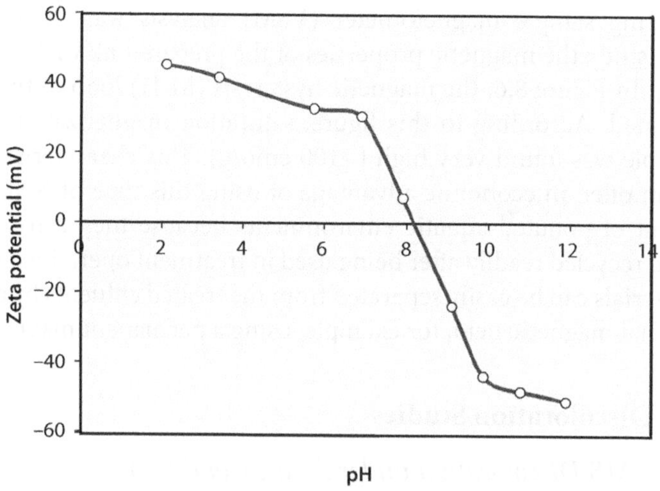

Figure 8.5 pH dependence of the zeta potential of $nZVI^{UI}$. The zero charge point is around 8.

in line with the relatively low average particle size calculated for the sample (Figure 8.4). Also, the SAED image obtained from TEM denotes a relatively high intensity for spots indicating a well-developed crystalline structure.

The pH dependence of $nZVI^{UI}$ zeta potential is shown in Figure 8.5. The results indicate that the zero charge point (zpc) is ~8 and that the surface of the particles is negatively and positively charged at pH above and below 8, respectively.

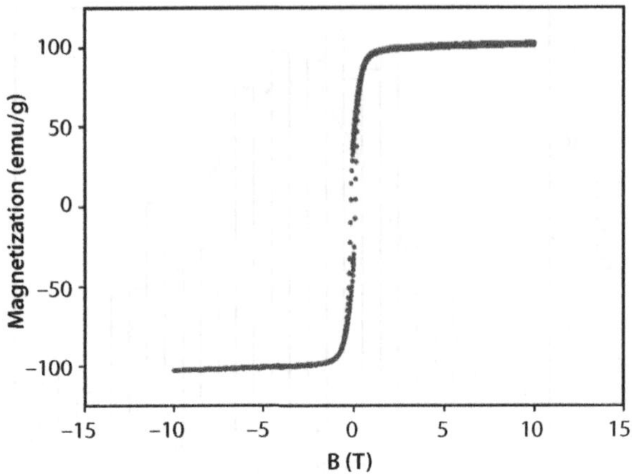

Figure 8.6 Magnetic hysteresis loop of $nZVI^{UI}$ at $T = 5K$.

Vibrating sample magnetometer (VSM) analysis was performed in order to study the magnetic properties of the prepared $nZVI^{UI}$ nanomaterial [35]. In Figure 8.6, the magnetic hysteresis (M-H) loop of the sample is presented. According to this figure, saturation magnetization (Ms) of the sample was found very high (~100 emu/g). This characteristic of the NMs can offer an economic advantage of using this type of NMs for the treatment of polluted aquatic environments because they can be recovered and recycled readily after being used in treatment operations [43–48]. The materials can be easily separated from the treated effluents by applying an external magnetic field, for example, using a permanent magnet.

8.3.2 Discoloration Studies

8.3.2.1 MB Discoloration under Acidic Conditions

A photometric assay is employed for assessing the concentration of MB in synthetic solutions. Adsorption spectra of MB solutions with different concentrations are detailed in Figure 8.7. As observed, two absorption maxima of MB are detected at 610 and 664 nm [49–52], their intensity difference being attenuated as MB concentration increases. In this study, the conversion of MB to any colorless products such as MLB is termed as discoloration whereas the irreversible conversion of MB to products such as CO_2 or H_2O is referred as MB degradation [53].

In order to study the extent of MB discoloration using nZVI particles prepared under ultrasonic irradiation and also the mechanisms involved

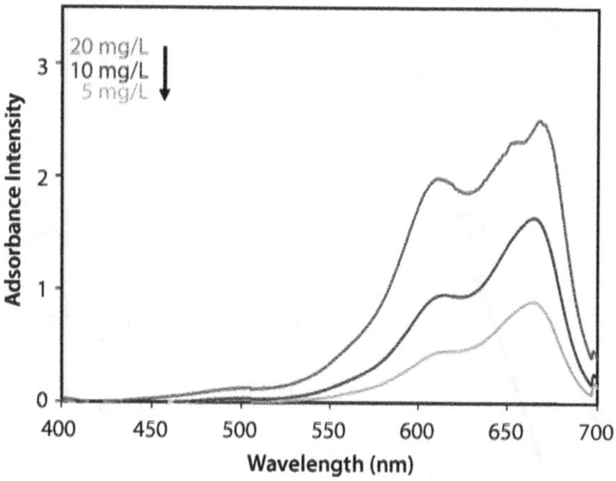

Figure 8.7 The absorption spectra of various concentrations varying from 5 mg/L to 20 mg/L.

in the process, assays at various pHs of the reaction medium, ranging from 4 to 10, are important in order to cover the effects of pH variation.

The assay at pH = 4 is used to assess the effects of acidic conditions on the process of MB discoloration. Under this condition, the percentage of color removal (here identified as the percentage of reduction of the adsorption intensity measured at 664 nm) from the solution reached only 29% after 30 min of reaction and the pH slightly increased to 4.1. Moreover, after re-collecting the nanoparticles from the reaction medium, the solution becomes rapidly re-colored suggesting that there should be a reversible reaction for the removal of MB under acidic conditions where the MB molecules are not attached to the surface of the NMs considering the zeta potential of the synthesized $nZVI^{UI}$.

MB is widely applied for optical oxygen sensing especially in the food industry [54, 55]. It is well known that MB is reduced to leuco methylene blue (LMB) which is a colorless compound in the absence of oxygen. This compound is readily re-oxidized (to MB) by introducing oxygen to the system [56]. When MB is subjected to the treatment with nZVI particles, it is proposed here that the surface layer of ZVI particles undergoes oxidation according to equation (2) [57–60]:

$$Fe^0 \rightarrow Fe^{2+} + 2e^- \tag{2}$$

The released electrons are subsequently used to reduce the MB to LMB, under acidic conditions. A schematic of the involved mechanism is presented in Figure 8.8.

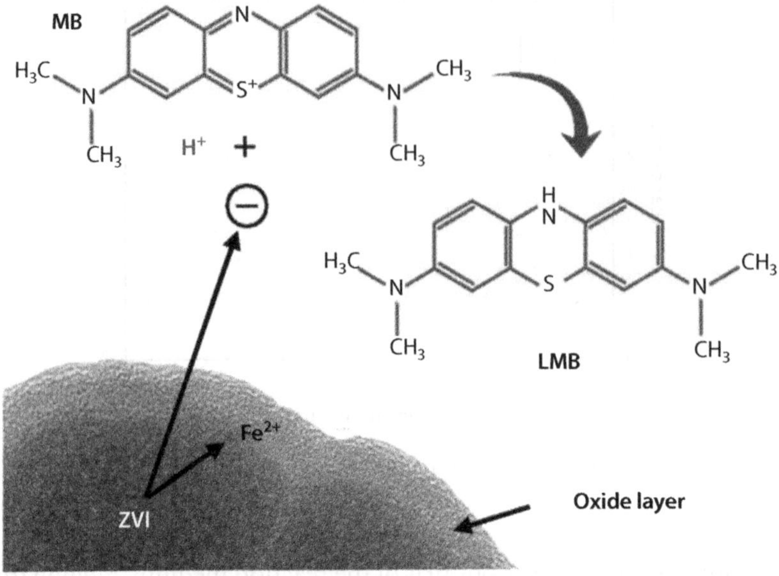

Figure 8.8 Mechanism involved in the discoloration of MB under acidic conditions. The electrons released by the oxidation of the surface layer of the ZVI nanoparticles are used in the reduction of MB to the colorless LMB.

According to Figure 8.5, the point of zero charge (PZC) of the prepared nZVI particles is reached approximately at pH 8. Below this pH, the surface of nZVI[UI] particles is positively charged, while above this pH it is negatively charged. Hence, under acidic conditions, repulsive electrostatic interactions between ZVI particles and the positive ions of MB do not favor the adsorption of MB on the surface of ZVI particles. Therefore, the hypothetic contribution of MB adsorption on ZVI particles for the chemical pathway of MB discoloration is considered negligible within the considered acidic pH window.

As referred, rapid re-coloration of the solution takes place suggesting that under acidic condition LMB does not degrade to simpler intermediate compounds or to final products (CO_2+H_2O) [61]. This suggestion is supported by the photometric results obtained after 30 min of reaction presented in Figure 8.9. The results show that the difference between the two maximum absorbances at 290 and 664nm becomes enhanced with the reaction time, being this considered as an evidence for the decrease of MB in the reaction medium. Also the intensity of the adsorption peak at 243 nm which is attributed to LMB [62] becomes enhanced with the reaction time, in line with the reasoning here proposed.

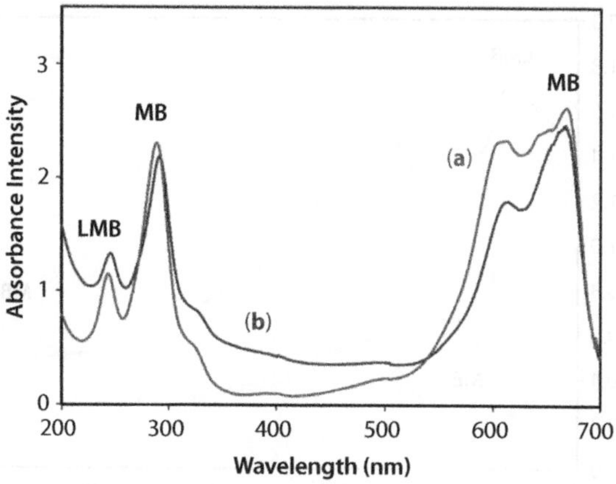

Figure 8.9 The UV-Vis adsorption spectra of the solution containing MB (25 mg/L) at the beginning (a) and after treatment with nZVI^{UI} (1 g/L) for 30 min under acidic conditions, pH = 4 (b).

8.3.2.2 MB Discoloration under Quasi-neutral Conditions

In order to study the effects of increasing the pH till the vicinity of the PZC of the nZVI particles, the pH of the MB solution is adjusted to 7.5 immediately after the addition of the NMs to the suspending medium (water) (1 g/L). The photometric results obtained in a 30 min time interval are presented in Figure 8.10. As observed in Figure 8.9 (a), an apparent effective removal of MB is observed after 5 min of reaction time (blue line), in line with the loss of blue color observed after 5 min. During this short period, the pH of the medium increased slightly to ~8.5. A shoulder at 664 nm (Figure 8.10) indicates that the discoloration of MB was almost achieved but not complete. The shoulder does disappear after 30 min of reaction (Figure 8.10 (b)) (green line), thus indicating that a 100% discoloration of MB was achieved (Figure 8.11). However, a narrow peak observed at 250 nm indicates that part of the organic dye (MB) is converted into LMB. Figure 8.10 (c) presents the UV-Vis adsorption of the solution with regard to 24 h, after collecting the NMs. At this time (red line), a small peak is found at 664 nm, corresponding to a concentration of 3.9 mg/L, which is 15% of the initial amount of the MB. We also observed a decrease in the LMB peak at 250 nm denoting the transformation of the existing LMB to MB in the presence of oxygen in the medium.

When the pH of the medium gets close to the PZC of nZVI particles, MB gets more easily adsorbed on the surface of nZVI particles as the repulsive forces between the surface of the nanoparticles and MB molecules are now much weaker or inexistent. Under these conditions, it is assumed that MB

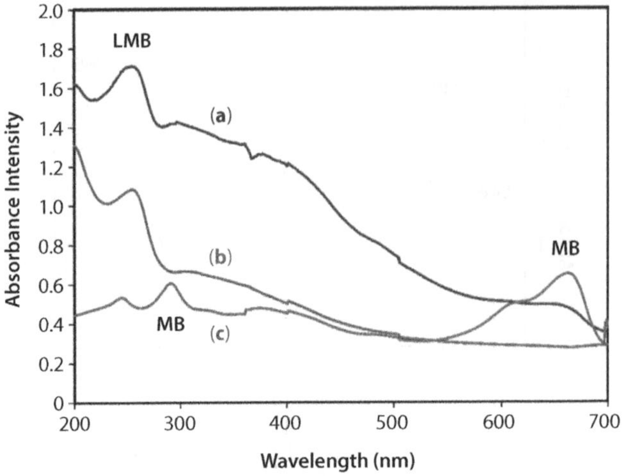

Figure 8.10 UV-Vis spectra of the reaction medium under initial pH = 7.5 after 5 min (a), 30 min (b), and 24 h, being the NMs already separated from the reaction solution (c).

Figure 8.11 Initial MB solution (25 mg/L) (left) and the solution after 30 min of reaction under initial pH=7.5 using nZVI particles (1 g/L) (right).

complexes are formed on nZVI particle surface [61] being very susceptible to chemical attack by species originated at the surface layers of the nZVI particles. According to equations (3) to (7) which describe the reaction possibilities of the nZVI particle surface layers, namely, with the surrounding H_2O molecules, H radicals are generated which may induce the cleavage of the MB molecules [63–66]. This can be suggested as a main reason for the higher efficiency of the MB degradation and cleavage at pH = 7.5

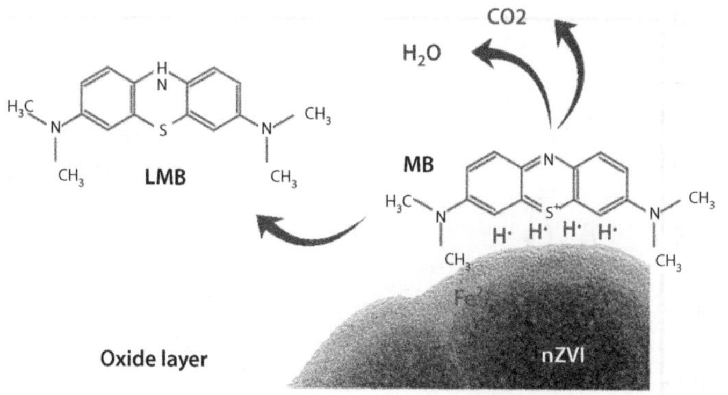

Figure 8.12 Mechanism involved in the discoloration of MB under quasi-neutral conditions.

compared to acidic conditions. A schematic of the involved mechanism under quasi-neutral conditions is presented in Figure 8.12:

$$Fe^0 \rightarrow Fe^{2+} + 2e^- \quad (3)$$

$$Fe^0 + H_2O \rightarrow Fe^{2+} + H^+ + OH^- \quad (4)$$

$$Fe^0 + H^+ \rightarrow Fe^{2+} + H + 2e^- \quad (5)$$

$$Fe^0 + 2H^+ \rightarrow Fe^{2+} + H_2 \quad (6)$$

$$Fe^{2+} \rightarrow Fe^{3+} + e^- \quad (7)$$

Attachment of MB to the surface of the NMs facilitates the degradation of MB molecules.

8.3.2.3 MB Discoloration under Basic Conditions

Figure 8.13 presents the results of the discoloration of MB carried on at an initial pH of 10. As observed, the extent of the discoloration reached only 26% after 5 min (blue line) which is much lower than that recorded at initial pH = 7.5. After 15 min of reaction (red line), the discoloration of MB reached 61% and it increased further to 100% after 30 min of reaction (green line).

Under high pH values, an extensive adsorption of MB on the surface of the used NMs, helped by electrostatic attractive forces, can be expected.

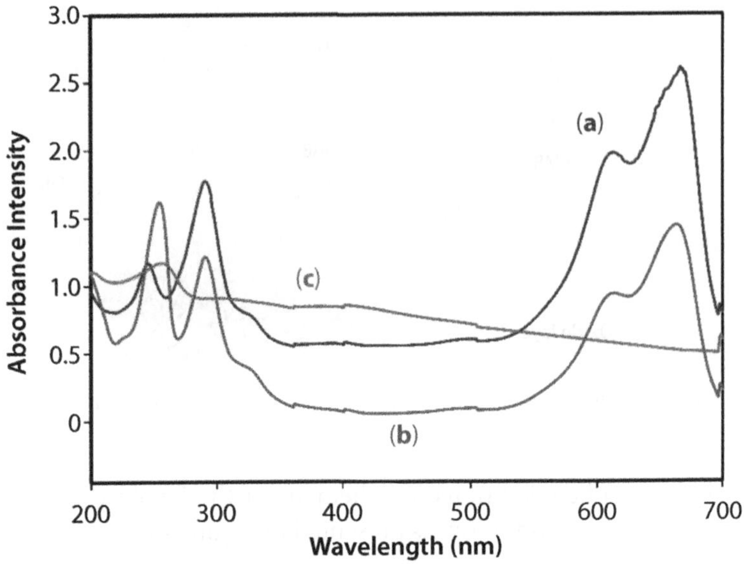

Figure 8.13 UV-Vis spectra of the reaction media under initial pH = 10, after 5 min (a), after 15 min (b), and after 30 min (c) of reaction for the removal of MB (25 mg/L) with ZVI NMs (1 g/L).

Such adsorption provides a favorable condition for the degradation of MB. A lower removal of MB under basic conditions as compared to quasi-neutral conditions can be attributed to the mechanisms involved in the discoloration of MB in an alkaline medium. According to equation (4), the production of H^+ is a key step for the generation of hydrogen radicals which can subsequently degrade and remove MB from the solution. Nonetheless, the presence of abundant OH^- under basic conditions may play a scavenging role leading to the readily consumption of the produced H^+ near the surface of the nZVI particles. Hence, the reactions for the cleavage of MB are restricted.

It has been well documented in the literature that the performance of NMs for the removal of persistent phenolic compounds increases under acidic conditions [67–70]. Most of the phenolic compounds such as chlorophenols are negatively charged which facilitates their attachment to the surface of NMs positively charged as frequently observed in acidic media. Therefore, it can be suggested that the better efficiency of NMs under acidic pH for the removal of such contaminants is probably associated with the formation of pollutants-NMs surface complexes which might be more effective players in the process of pollutants degradation than the release of electrons to the reaction medium by the reduction of Fe nanoparticles with subsequent chemical attack to pollutants.

Table 8.2 Some recent studies for the removal of MB using nZVI materials.

Nanomaterials	Nanomaterials properties			Operating conditions				Discoloration (%)	Ref.
	Size (nm)	Surface area (m^2/g)	Crystallite size (nm)	Dye conc. (mg/L)	pH	NMs dosage (mg/L)	Time (min)		
nZVI[UI]	27	50	23.9	25	4	1	30	29	This study
	27	50	23.9	25	7.5	1	5	100	This study
	27	50	23.9	25	10	1	5	26	This study
	27	50	23.9	25	10	1	15	61	This study
	27	50	23.9	25	10	1	30	100	This study
Iron nanoparticles[9]	40-60	NG[1]0	NG	10	3-4[1]1	1	60	100	[71]
ZVI	NG	NG	NG	20	3	1	30	77	[61]
	NG	NG	NG	20	6	1	40	90	[61]
nZVI	NG	33.7	NG	100	6	0.7	60	96	[71]
nZVI	NG	33.7	NG	100	NG	0.2	60	76	[71]
nZVI-Clinoptilolite	40-60	16.1	NG	100	5	0.1	420	91	[72]

9 A mixture of ZVI, FeOOH, and $Fe3_{O4.}$

10 NG: Not given.

11 Using 5 mL in 50 mL $H2_{O2.}$

The results of this study show that the attachment of MB molecules to the surface of nZVI particles under neutral or basic conditions facilitates the degradation of the dye due to the positive charge of MB species. So, it can be anticipated that the available surface area of the NMs is a key factor, because it can critically determine their efficiency when dealing with complex organic pollutants. The results of this study also indicate that the efficiency of the reactions that occurred in the bulk solution (not on the surface of the NMs) is negligible. So, limiting the release of electrons from the nZVI materials to the reaction medium for the purpose of increasing the lifetime of nZVI particles will probably not affect their efficiency.

Table 8.2 provides the results of some recent studies for the removal of MB using ZVI NMs comparing the efficiency of Fe nanoparticles toward the discoloration of MB, under different operating conditions. In most of the cases, the characteristics of the used NMs have not been fully provided in the respective reports. In the present study, a very high efficiency of the NMs for the degradation of MB can be attributed to the enhanced properties (high crystallinity and large specific surface area) of the used nanoparticles which provide a good condition for the rapid degradation of MB in a short period of time (100% discoloration in 5 min) under appropriate pH conditions.

8.4 Conclusions

nZVI particles with enhanced properties may be successfully synthesized by utilization of ultrasonic irradiation. Such NMs present an efficient performance regarding the degradation of MB and discoloration of MB solutions. The pH of the solution medium is a critical parameter for determining the MB removal efficiency of the synthesized NMs. While acidic pH allows achieving 30% discoloration of MB, quasi-neutral conditions of pH = 7.5 permit the complete removal of the dye color. The properties of the NMs including zeta potential, size, specific surface area, and crystallinity are the main factors determining the efficiency of the NMs for environmental clean-up purposes.

Acknowledgments

Thanks are due, for the financial support, to CESAM (UID/AMB/50017), to CICECO-Aveiro Institute of Materials,

POCI-01-0145-FEDER-007679 (FCT Ref. UID/CTM/50011/2013), to FCT/MEC through national funds, and the co-funding by the FEDER, within the PT2020 Partnership Agreement and Compete 2020. Thanks are also due to FCT for the doctoral scholarship No. SFRH/BD/103695/2014 to the first author (M.K.).

References

1. Brown, M.A., De Vito, S.C., Predicting azo dye toxicity, *Crit. Rev. Environ. Sci. Technol.*, 23, 249–324, 1999.
2. Sala, M., Gutiérrez-Bouzán, M.C., Electrochemical techniques in textile processes and wastewater treatment, *Int. J. Photoenergy*, 2012, 1–12, 2012.
3. Ma, X., Chao, M., Rapid voltammetric determination of maltol in some foods and beverages using a poly(methylene blue)/graphene-modified glassy carbon electrode, *J. Solid State Electrochem.*, 18, no. 3, 621–628, 2014.
4. Lin, S.L., Hu, J.M., Tang, S.S., Wu, X.Y., Chen, Z.Q., Tang, S.Z., Photodynamic inactivation of methylene blue and tungsten-halogen lamp light against food pathogen listeria monocytogenes, *Photochem. Photobiol.*, 88, no. 4, 985–991, 2012.
5. Mahapatra, K., Ramteke, D.S., Paliwal, L.J., Production of activated carbon from sludge of food processing industry under controlled pyrolysis and its application for methylene blue removal, *J. Anal. Appl. Pyrolysis*, 95, 79–86, 2012.
6. Yan, S., Zhang, H., Yang, Z., Tang, M., Zhang, M., Du, C., Cui, H-L., Wei, D., Transformation and dehydration kinetics of methylene blue hydrates detected by terahertz time-domain spectroscopy, *RSC Adv.*, 7, no. 66, 41667–41674, 2017.
7. Stockert, J.C., Juarranz, A., Villanueva, A., Cañete, M. Photodynamic damage to HeLa cell microtubules induced by thiazine dyes, *Cancer Chemother Pharmacol*, 39, 167–169, 1996.
8. Ahmed, M.a., El-Katori, E.E., Gharni, Z.H., Photocatalytic degradation of methylene blue dye using Fe2O3/TiO2 nanoparticles prepared by sol–gel method, *J. Alloys Compd.*, 553, no. MARCH 2013, 19–29, 2013.
9. Akpan, U.G., Hameed, B.H., Parameters affecting the photocatalytic degradation of dyes using TiO2-based photocatalysts: A review, *J. Hazard. Mater.*, 170, no. 2–3, 520–529, 2009.
10. Borker, P., Salker, A.V., Photocatalytic degradation of textile azo dye over Ce1-xSnxO2 series, *Mater. Sci. Eng. B Solid-State Mater. Adv. Technol.*, 133, no. 1–3, 55–60, 2006.
11. Brouzas, D. M., Droutsas, D. M., Charakidas, A. M., Malias, I. M., Georgiadou, E. M., Apostolopoulos, M. M., Moschos, M. M., Severe toxic effect of methylene blue 1% on iris epithelium and corneal endothelium, *Cornea*, 25, 470–471, 2006.

12. Mahadevan M.M., Weitzman G.A., Hogan S., Breckinridge S., Miller M.M., Methylene blue but not indigo carmine is toxic to human luteal cells in vitro, *Reprod. Toxicol.*, 7, 631–633, 1993.
13. Fish, W., Chazen, E., Toxic effects of methylene-blue on the fetus, *Am. J. Dis. Child.*, 146, 1412–1413, 1992.
14. Kirszberg, C., Rumjanek, V.M., Capella, M.A.M., Methylene blue is more toxic to erythroleukemic cells than to normal peripheral blood mononuclear cells: A possible use in chemotherapy, *Cancer Chemother. Pharmacol.*, 56, no. 6, 659–665, 2005.
15. Teixeira, R.S.P., Correa, R.J., Bello Forero, J.S., Silva, M.G.S., Oliveira, R.C.S., Souza, R.S., Comparative study of PEO and PVA hydrogels for removal of methylene blue dye from wastewater, *J. Appl. Polym. Sci.*, 45043, 1–8, 2017.
16. Suppiah, D.D., Abd Hamid, S.B., One step facile synthesis of ferromagnetic magnetite nanoparticles, *J. Magn. Magn. Mater.*, 414, 204–208, 2016.
17. Rad-Moghadam, K., Dehghan, N., Application of cellulose/chitosan grafted nano-magnetites as efficient and recyclable catalysts for selective synthesis of 3-indolylindolin-2-ones, *J. Mol. Catal. A Chem.*, 392, 97–104, 2014.
18. Shipley, H.J., Engates, K.E., Grover, V.A., Removal of Pb(II), Cd(II), Cu(II), and Zn(II) by hematite nanoparticles: Effect of sorbent concentration, pH, temperature, and exhaustion, *Environ. Sci. Pollut. Res.*, 20, no. 3, 1727–1736, 2013.
19. Shuibo, X., Chun, Z., Xinghuo, Z., Jing, Y., Xiaojian, Z., Jingsong, W., Removal of uranium (VI) from aqueous solution by adsorption of hematite, *J. Environ. Radioact.*, 100, no. 2, 162–166, 2009.
20. Shan, C., Ma, Z., Tong, M., Efficient removal of trace antimony(III) through adsorption by hematite modified magnetic nanoparticles, *J. Hazard. Mater.*, 268, 229–236, 2014.
21. Hu, J., Chen, G., Lo, I.M.C., Removal and recovery of Cr(VI) from wastewater by maghemite nanoparticles, *Water Res.*, 39, no. 18, 4528–4536, Nov. 2005.
22. Etale, A., Tutu, H., & Drake, D. C. Application of maghemite nanoparticles as sorbents for the removal of Cu(II), Mn(II) and U(VI) ions from aqueous solution in acid mine drainage conditions. Applied Water Science, 6, 187–197, 2016.
23. Agarwal, A., Joshi, H., Kumar, A., Synthesis, characterization and application of nano lepidocrocite and magnetite in the degradation of carbon tetrachloride, *South African J. Chem.*, 64, 218–224, 2011.
24. Sap, K.A., Demmers, J.A.A., World â€™ s largest Science , Technology & Medicine Open Access book publisher c, *Intech*, 6, 111–133, 2016.
25. Thanh, N.T.K., Robinson, I., Tung, L.D., Dekker Encyclopedia of Nanoscience and Nanotechnology Magnetic Nanoparticles for Biomedical Applications : Synthesis , Characterization and Uses, no. 789671721, 2007.
26. Kong, X., Huang, G., Han, Z., Xu, Y., Zhu, M., Zhang, Z., Evaluation of zeolite-supported microscale zero-valent iron as a potential adsorbent for Cd2+

and Pb2+ removal in permeable reactive barriers, *Environ. Sci. Pollut. Res.*, 24, no. 15, 13837–13844, 2017.
27. Kumar, N., Millot, R., Battaglia-Brunet, F., Omoregie, E., Chaurand, P., Borschneck, D., Bastiaens, L., Rose, J., Microbial and mineral evolution in zero valent iron-based permeable reactive barriers during long-term operations, *Environ. Sci. Pollut. Res.*, 23, no. 6, 5960–5968, 2016.
28. Crane, R.A., Dickinson, M., Popescu, I.C., Scott, T.B., Magnetite and zero-valent iron nanoparticles for the remediation of uranium contaminated environmental water, *Water Res.*, 45, no. 9, 2931–2942, 2011.
29. He, F., Zhao, D., Paul, C., Field assessment of carboxymethyl cellulose stabilized iron nanoparticles for in situ destruction of chlorinated solvents in source zones, *Water Res.*, 44, no. 7, 2360–2370, Apr. 2010.
30. Chen, Z., Jin, X., Chen, Z., Megharaj, M., Naidu, R., Removal of methyl orange from aqueous solution using bentonite-supported nanoscale zero-valent iron., *J. Colloid Interface Sci.*, 363, no. 2, 601–7, Nov. 2011.
31. Fu, F., Dionysiou, D.D., Liu, H., The use of zero-valent iron for groundwater remediation and wastewater treatment: A review, *J. Hazard. Mater.*, 267, 194–205, 2014.
32. Duan, T., Chen, Y., Wen, Q., Duan, Y., Enhanced electrocatalytic activity of nano-TiN composited Ti/Sb–SnO2 electrode fabricated by pulse electrodeposition for methylene blue decolorization, *RSC Adv.*, 4, no. 101, 57463–57475, 2014.
33. Eldin, M.S.M., El-Sakka, S.A., El-Masry, M.M., Abdel-Gawad, I.I., Garybe, S.S., Removal of methylene blue dye from aqueous medium by nano poly Acrylonitrile particles, *Desalin. Water Treat.*, 44, no. 1–3, 151–160, 2012.
34. Chithambararaj, A., Sanjini, N.S., Bose, A.C., Velmathi, S., Flower-like hierarchical h-MoO3: new findings of efficient visible light driven nano photocatalyst for methylene blue degradation, *Catal. Sci. Technol.*, 3, no. 5, 1405, 2013.
35. Cao, X.T., Showkat, A.M., Kang, I., Gal, Y.S., Lim, K.T., β -Cyclodextrin multi-conjugated magnetic graphene oxide as a nano-adsorbent for methylene blue removal, *J. Nanosci. Nanotechnol.*, 16, no. 2, 1521–1525, 2016.
36. Padmapriya, G., Manikandan, A., Krishnasamy, V., Jaganathan, S.K., Antony, S.A., Spinel NixZn1-xFe2O4 ($0.0 \leq x \leq 1.0$) nano-photocatalysts: Synthesis, characterization and photocatalytic degradation of methylene blue dye, *J. Mol. Struct.*, 1119, 39–47, 2016.
37. Peng, L., Liu, B., Zeng, Q., Gu, J.-D., Lei, M., Shao, J., Chai, L., Highly efficient removal of methylene blue from aqueous solution by a novel fishing-net effect of manganese oxide nano-sheets, *Clean Technol. Environ. Policy*, 19, no. 1, 1–9, 2016.
38. Simamora, A.J., Hsiung, T.L., Chang, F.C., Yang, T.C., Liao, C.Y., Wang, H.P., Photocatalytic splitting of seawater and degradation of methylene blue on CuO/nano TiO2, *Int. J. Hydrogen Energy*, 37, no. 18, 13855–13858, 2012.
39. Zhao,M., Tang, M., Liu, P., Removal of methylene blue from aqueous solution with silica nano-sheets derived from vermiculite, *J. Hazard. Mater.*, 158, no. 1, 43–51, 2008.

40. Zuorro, A., Lavecchia, R., Natali, S., Magnetically modified agro-industrial wastes as efficient and easily recoverable adsorbents for water treatment, *Chem. Eng. Trans.*, 38, 349–354, 2014.
41. Kar, P., Sardar, S., Ghosh, S., Parida, M. R., Liu, B., Mohammed, O. F., Lemmens, P., Pal, S. K., Nano surface engineering of Mn_2O_3 for potential light-harvesting application, *J. Mater. Chem. C*, 3, no. 3, 8200–8211, 2015.
42. Cenens, J., Schoonheydt, R.A., Visible spectroscopy of methylene blue on hectorite, laponite b, and barasym in aqueous suspension., *Clays Clay Miner.*, 36, no. 3, 214–224, 1988.
43. Ghulam, M., Hajira, T., Muhammad, S., Nasir, A., Synthesis and characterization of cupric oxide (CuO) nanoparticles and their application for the removal of dyes, *African J. Biotechnol.*, 12, no. 47, 6650–6660, 2013.
44. Bhattacharjee, A., Ahmaruzzaman, M., Green synthesis of 2D CuO nanoleaves (NLs) and its application for the reduction of p-nitrophenol, *Mater. Lett.*, 161, 79–82, 2015.
45. Puddu, V., Choi, H., Dionysiou, D.D., Puma, G.L., TiO2 photocatalyst for indoor air remediation: Influence of crystallinity, crystal phase, and UV radiation intensity on trichloroethylene degradation, *Appl. Catal. B Environ.*, 94, no. 3–4, 211–218, Feb. 2010.
46. Zhao, H., Huang, X., Wang, J., Li, Y., Liao, R., Wang, X., Qiu, X., Xiong, Y., Qin, W., Qiu, G., Comparison of bioleaching and dissolution process of p-type and n-type chalcopyrite, *Miner. Eng.*, 109, 153–161, 2017.
47. Rossetto, E., Petkowicz, D.I., dos Santos, J.H.Z, Pergher, S.B.C., Penha, F.G., Bentonites impregnated with TiO2 for photodegradation of methylene blue, *Appl. Clay Sci.*, 48, no. 4, 602–606, 2010.
48. Zamiri, R., Salehizadeh, S. A., Ahangar, H. A., Shabani, M., Rebelo, A., Suresh Kumar, J., Soares, M. J.,Valente, M. A., Ferreira, J. M. F., Optical and magnetic properties of ZnO/ZnFe2O4 nanocomposite, *Mater. Chem. Phys.*, 192, 330–338, 2017.
49. Kobzi, B. *et al.*, 57Fe-Mössbauer study and methylene blue decomposing effect of nanoparticle mixtures composed of metallic iron and maghemite, *J. Alloys Compd.*, 722, 94–100, 2017.
50. Zhang, S., Li, H., Yang, Z., Controllable synthesis of WO_3 with different crystalline phases and its applications on methylene blue removal from aqueous solution, *J. Alloys Compd.*, 722, 555–563, 2017.
51. Agougui, H., Jabli, M., Majdoub, H., Synthesis, characterization of hydroxyapatite-lambda carrageenan, and evaluation of its performance for the adsorption of methylene blue from aqueous suspension, *J. Appl. Polym. Sci.*, 45385, 1–9, 2017.
52. Rabbani, M., Seghatoleslami, Z.S., Rahimi, R., Selective adsorption of organic dye methylene blue by Cs 4 H 2 PMo 11 FeO 40 $ 6H2O in presence of methyl orange and, *J. Mol. Struct.*, 1146, 113–118, 2017.
53. Rani, M., Shanker, U., Jassal, V., Recent strategies for removal and degradation of persistent & toxic organochlorine pesticides using nanoparticles: A review, *J. Environ. Manage.*, 190, 208–222, 2017.

54. Sandhya, Modified atmosphere packaging of fresh produce: Current status and future needs, *LWT - Food Sci. Technol.*, 43, no. 3, 381–392, 2010.
55. Liu, Y.-N., Zhou, X., Wang, X., Liang, K., Yang, Z.-K., Shen, C.-C., Imran, M., Sahar, S., Xu, A.-W., Hydrogenation/oxidation induced efficient reversible color switching between methylene blue and leuco-methylene blue, *RSC Adv.*, 7, no. 48, 30080–30085, 2017.
56. Impert, O., Katafias, A., Kita, P., Mills, A., Pietkiewicz-Graczyk, A., Wrzeszcz, G., Kinetics and mechanism of a fast leuco-Methylene Blue oxidation by copper(ii)-halide species in acidic aqueous media, *Dalt. Trans.*, 7, no. 3, 348–353, 2003.
57. Choi, K., Lee, W., Enhanced degradation of trichloroethylene in nano-scale zero-valent iron Fenton system with Cu(II), *J. Hazard. Mater.*, 211–212, 146–53, April 2012.
58. Ryu, A., Jeong, S.-W., Jang, A., Choi, H., Reduction of highly concentrated nitrate using nanoscale zero-valent iron: Effects of aggregation and catalyst on reactivity, *Appl. Catal. B Environ.*, 105, no. 1–2, 128–135, Jun. 2011.
59. Crane, R.A., Scott, T.B., Nanoscale zero-valent iron: future prospects for an emerging water treatment technology., *J. Hazard. Mater.*, 211–212, 112–25, Apr. 2012.
60. O'Carroll, D., Sleep, B., Krol, M., Boparai, H., Kocur, C., Nanoscale zero valent iron and bimetallic particles for contaminated site remediation, *Adv. Water Resour.*, 51, 104–122, Jan. 2013.
61. Sun, X., Kurokawa, T., Suzuki, M., Takagi, M., Kawase, Y., Removal of cationic dye methylene blue by zero-valent iron: Effects of pH and dissolved oxygen on removal mechanisms, *J. Environ. Sci. Heal. - Part A Toxic/Hazardous Subst. Environ. Eng.*, 50, no. 10, 1057–1071, 2015.
62. Pegu, R., Majumdar, K. J., Talukdar, D. J., Pratihar, S. Oxalate capped iron nanomaterial: from methylene blue degradation to bis(indolyl)methane synthesis. *RSC Adv.* 4, 33446–33456, 2014.
63. Raman, C.D., Kanmani, S., Textile dye degradation using nano zero valent iron: A review, *J. Environ. Manage.*, 177, 341–355, 2016.
64. Jin, X., Chen, Z., Chen, Z., Zhou, R., Synthesis of kaolin supported nanoscale zero-valent iron and its degradation mechanism of Direct Fast Black G in aqueous solution, *Mater. Res. Bull.*, 61, 433–438, 2015.
65. Luo, F., Yang, D., Chen, Z., Megharaj, M., Naidu, R., The mechanism for degrading Orange II based on adsorption and reduction by ion-based nanoparticles synthesized by grape leaf extract, *J. Hazard. Mater.*, 296, 37–45, 2015.
66. Yin, W., Li, Y., Wu, J., Chen, G., Jiang, G., Li, P., Gu, J., Liang, H., Liu, C., Enhanced Cr(VI) removal from groundwater by Fe0-H2O system with bio-amended iron corrosion, *J. Hazard. Mater.*, 332, 42–50, 2017.
67. Yehia, F. Z., Eshaq, G., Rabie, A. M., Mady, A. H., El Metwally, A. E. Sonocatalytic degradation of phenol catalyzed by nano-sized zero valent Cu and Ni. Desalin. *Water Treat.* 57, 2104–2112, 2016.
68. Han, C., Jing, M., Shen, X., Qiao, G., Fabrication and properties of core–shell structural nano-TiO2@Fe magnetic photocatalyst for removal of phenol

waste water, *Monatshefte für Chemie - Chem. Mon.*, 148, no. 7, 1165–1170, 2017.

69. Wang, P., Zhou, X., Zhang, Y., Yang, L., Zhi, K., Wang, L., Zhang, L., Guo, X., Unveiling the mechanism of electron transfer facilitated regeneration of active Fe^{2+} by nano-dispersed iron/graphene catalyst for phenol removal, *RSC Adv.*, 7, no. 43, 26983–26991, 2017.
70. Asmaly, H. A., Abussaud, B., Saleh, T. A., Bukhari, A. A., Laoui, T., Shemsi, A. M., Gupta, V. K., Atieh, M. A., Evaluation of micro- and nano-carbon-based adsorbents for the removal of phenol from aqueous solutions, *Toxicol. Environ. Chem.*, 97, no. 9, 1164–1179, 2015.
71. Shahwan, T., Abu Sirriah, S., Nairat, M., Boyaci, E., Eroĝlu, A. E., Scott, T. B., Hallam, K. R., Green synthesis of iron nanoparticles and their application as a Fenton-like catalyst for the degradation of aqueous cationic and anionic dyes, *Chem. Eng. J.*, 172, no. 1, 258–266, 2011.
72. Nairat, M., Shahwan, T., Eroğlu, A.E., Fuchs, H., Incorporation of iron nanoparticles into clinoptilolite and its application for the removal of cationic and anionic dyes, *J. Ind. Eng. Chem.*, 21, 1143–1151, 2015.

Index

Also of Interest

Check out these published volumes in the Advanced Materials Series

Advanced Materials for Wastewater Treatment
Edited by Shahid-ul-Islam
Published 2017. ISBN 1-119-40776-8

Advanced Magnetic and Optical Materials
Edited by Ashutosh Tiwari, Parameswar K. Iyer, Vijay Kumar and Hendrik Swart
Published 2016. ISBN 978-1-119-24191-1

Advanced Surfaces for Stem Cell Research
Edited by Ashutosh Tiwari, Bora Garipcan and Lokman Uzun
Published 2016. ISBN 978-1-119-24250-5

Advanced Electrode Materials
Edited by Ashutosh Tiwari, Filiz Kuralay and Lokman Uzun
Published 2016. ISBN 978-1-119-24252-9

Advanced Molecularly Imprinting Materials
Edited by Ashutosh Tiwari and Lokman Uzun
Published 2016. ISBN 978-1-119-33629-7

Intelligent Nanomaterials (2nd edition)
Edited by Tiwari, Yogendra Kumar Mishra, Hisatoshi Kobayashi and Anthony P. F. Turner
Published 2016. ISBN 978-1-119-24253-6

Advanced Composite Materials
Edited by Ashutosh Tiwari, Mohammad Rabia Alenezi and Seong Chan Jun
Published 2016. ISBN 978-1-119-24253-6

Advanced Surface Engineering Materials
Edited by Ashutosh Tiwari, Rui Wang, and Bingqing Wei
Published 2016. ISBN 978-1-119-24244-4

Advanced Ceramic Materials
Edited by Ashutosh Tiwari, Rosario A. Gerhardt and Magdalena Szutkowska
Published 2016. ISBN 978-1-119-24244-4

Advanced Engineering Materials and Modeling
Edited by Ashutosh Tiwari, N. Arul Murugan and Rajeev Ahuja
Published 2016. ISBN 978-1-119-24246-8

Advanced 2D Materials
Ashutosh Tiwari and Mikael Syväjärvi
Published 2016. ISBN 978-1-119-24249-9

Advanced Materials Interfaces
Edited by Ashutosh Tiwari, Hirak K. Patra and Xumei Wang
Published 2016. ISBN 978-1-119-24245-1

Advanced Bioelectronics Materials
Edited by Ashutosh Tiwari, Hirak K. Patra and Anthony P.F. Turner
Published 2015. ISBN 978-1-118-99830-4

Graphene
An Introduction to the Fundamentals and Industrial Applications
By Madhuri Sharon and Maheswar Sharon
Published 2015. ISBN 978-1-118-84256-0

Advanced Theranostic Materials
Edited by Ashutosh Tiwari, Hirak K. Patra and Jeong-Woo Choi
Published 2015. ISBN: 978-1-118-99829-8

Advanced Functional Materials
Edited by Ashutosh Tiwari and Lokman Uzun
Published 2015. ISBN 978-1-118-99827-4

Advanced Catalytic Materials
Edited by Ashutosh Tiwari and Salam Titinchi
Published 2015. ISBN 978-1-118-99828-1

Graphene Materials
Fundamentals and Emerging Applications
Edited by Ashutosh Tiwari and Mikael Syväjärvi
Published 2015. ISBN 978-1-118-99837-3

DNA Engineered Noble Metal Nanoparticles
Fundamentals and State-of-the-art-of Nanobiotechnology
By Ignác Capek
Published 2015. ISBN 978-1-118-07214-1

Advanced Electrical and Electronics Materials
Process and Applications
By K.M. Gupta and Nishu Gupta
Published 2015. ISBN: 978-1-118-99835-9

Advanced Materials for Agriculture, Food and Environmental Safety
Edited by Ashutosh Tiwari and Mikael Syväjärvi
Published 2014. ISBN: 978-1-118-77343-7

Advanced Biomaterials and Biodevices
Edited by Ashutosh Tiwari and Anis N. Nordin
Published 2014. ISBN 978-1-118-77363-5

Biosensors Nanotechnology
Edited by Ashutosh Tiwari and Anthony P. F. Turner
Published 2014. ISBN 978-1-118-77351-2

Advanced Sensor and Detection Materials
Edited by Ashutosh Tiwari and Mustafa M. Demir
Published 2014. ISBN 978-1-118-77348-2

Advanced Healthcare Materials
Edited by Ashutosh Tiwari
Published 2014. ISBN 978-1-118-77359-8

Advanced Energy Materials
Edited by Ashutosh Tiwari and Sergiy Valyukh
Published 2014. ISBN 978-1-118-68629-4

Advanced Carbon Materials and Technology
Edited by Ashutosh Tiwari and S.K. Shukla
Published 2014. ISBN 978-1-118-68623-2

Responsive Materials and Methods
State-of-the-Art Stimuli-Responsive Materials and Their Applications
Edited by Ashutosh Tiwari and Hisatoshi Kobayashi
Published 2013. ISBN 978-1-118-68622-5

Other Scrivener books edited by Ashutosh Tiwari

Nanomaterials in Drug Delivery, Imaging, and Tissue Engineering
Edited by Ashutosh Tiwari and Atul Tiwari
Published 2013. ISBN 978-1-118-29032-3

Biomedical Materials and Diagnostic Devices Devices
Edited by Ashutosh Tiwari, Murugan Ramalingam, Hisatoshi Kobayashi and Anthony P.F. Turner
Published 2012. ISBN 978-1-118-03014-1

Intelligent Nanomaterials (first edition)
Processes, Properties, and Applications
Edited by Ashutosh Tiwari Ajay K. Mishra, Hisatoshi Kobayashi and Anthony P.F. Turner
Published 2012. ISBN 978-0-470-93879-9

Integrated Biomaterials for Biomedical Technology
Edited by Murugan Ramalingam, Ashutosh Tiwari, Seeram Ramakrishna and Hisatoshi Kobayashi
Published 2012. ISBN 978-1-118-42385-1

Printed and bound by CPI Group (UK) Ltd, Croydon, CR0 4YY

17/09/2026

14964245-0002